The Oxford Companion to

Cosmology

ANDREW LIDDLE AND JON LOVEDAY

OXFORD
UNIVERSITY PRESS

OXFORD
UNIVERSITY PRESS

Great Clarendon Street, Oxford OX2 6DP

Oxford University Press is a department of the University of Oxford.
It furthers the University's objective of excellence in research, scholarship,
and education by publishing worldwide in

Oxford New York

Auckland Cape Town Dar es Salaam Hong Kong Karachi
Kuala Lumpur Madrid Melbourne Mexico City Nairobi
New Delhi Shanghai Taipei Toronto

With offices in

Argentina Austria Brazil Chile Czech Republic France Greece
Guatemala Hungary Italy Japan Poland Portugal Singapore
South Korea Switzerland Thailand Turkey Ukraine Vietnam

Oxford is a registered trademark of Oxford University Press
in the UK and in certain other countries

Published in the United States
by Oxford University Press Inc., New York

© Oxford University Press 2008, 2009
The moral rights of the authors have been asserted
Database right Oxford University Press (maker)

First published 2008

Published as an Oxford University Press paperback 2009

British Library Cataloguing in Publication Data
Data available

Library of Congress Cataloging in Publication Data
Data available

Typeset by SPI Publisher Services, Pondicherry, India
Printed in Great Britain by
Clays Ltd, St Ives plc

ISBN 978-0-19-956084-4

3

to Jeanette and Catriona
and
to Alan, Audrey, Simon, and Urmi

Contents

Preface

The Universe is a big place, so it's nice to have a guide to help you find your way around. That's what we aim to provide with *The Oxford Companion to Cosmology*. We hope that this *Companion*, written with the interested lay-reader in mind, might act as a bridge between popular accounts of the subject and the more technical literature.

We are approaching the hundredth anniversary of cosmology in its modern sense, made possible when Albert Einstein unveiled his greatest discovery, the general theory of relativity, in 1915. By incorporating his earlier concept of space-time into a theory of gravity, he made it possible to build mathematical models describing the entire Universe. Now, almost a century later, it seems that a detailed description of the Universe is within our grasp, in the form of the standard cosmological model.

In writing this book, we've taken a broad view of the word 'cosmology', encompassing a fair slew of extragalactic astrophysics (i.e. the study of anything outside our own Milky Way Galaxy) as we go along. For instance, you can find descriptions of the properties and evolution of galaxies, the study of which has proven so central in understanding cosmological models. We have also given brief summaries of some of the key physical ideas underlying cosmology, so you can find, for example, short items highlighting the key ideas of quantum mechanics, and of the special and general theories of relativity.

The bulk of the book is an alphabetical list of topics, of varying lengths, each containing numerous cross references to other entries. Our principal aim in creating a reference book in fairly traditional encyclopedic style has been to make it easy for readers to find information on specific topics, either by direct identification of a topic heading or via the index. The aim is for each topic to be a self-contained account, with cross references in place to unravel any unfamiliar terms. Cross references are indicated thus, *cosmology, with an asterisk. Where appropriate we have indicated, at the end of each topic, useful websites and suggestions for further reading.

Nevertheless, it is possible to find a narrative too, albeit not of a traditional linear form. Cosmology is a tight-knit topic, linking theoretical ideas to observational support, and observational indications back to underlying physical concepts. By following the cross references backwards and forwards, it should be possible to build an appreciation of this tight interplay. In order to help the reader find this narrative, we begin with an overview article, on page 1, highlighting the historical threads that have come together to give the modern view of the hot big bang cosmology. We have not included cross references in the overview item, as there would simply be too many, but all the ideas therein are followed up in more detail in the main body of the book.

A book of this length always leaves behind a trail of contributions without which it wouldn't have been possible, from the tolerance of immediate family to a wide variety of professional help. Amongst the latter, we'd like to thank our patient editors, first Michael Rodgers and then Judith Wilson, for initiating the project and keeping it on track to the best of their abilities. We thank Lorna Mitchell for producing some of the diagrams, and google.com for its unfailing ability to locate many others. Andrew thanks the Institute for Astronomy, University of Hawaii, for its hospitality during the final stages of bringing this book together, and the Leverhulme Trust and the Particle Physics and Astronomy Research Council (PPARC) for providing the research fellowships that made it slightly easier to find the time to write it. Jon thanks his students and research collaborators for their patience and understanding while much of his time was taken up with work on this book.

Honolulu, November 2006   ANDREW LIDDLE

Brighton, November 2006   JON LOVEDAY

Glossary

This glossary contains brief definitions of a few technical terms that do not merit full entries in their own right.

21-cm line emission or absorption line at 1 420 MHz (21.1 cm) in the radio spectrum due to the flip in the spin of the electron in a neutral hydrogen atom.

absolute magnitude *magnitude corresponding to *intrinsic* brightness (luminosity).

apparent magnitude *magnitude corresponding to *observed* brightness (flux).

arcminute (abbrev. arcmin) a measure of angle equal to one-sixtieth of a degree.

arcsecond (abbrev. arcsec) a measure of angle equal to one-sixtieth of an arcminute.

astronomical unit (abbrev. au) the average distance between the Sun and the Earth, approximately 150 million kilometres.

Balmer lines a series of emission or absorption lines due to transitions in the hydrogen atom between the second and higher energy levels.

billion one thousand million, or in mathematical notation, 10^9. The prefix G multiplies the following unit by one billion, so 1 Gyr = 1 billion years = 10^9 years.

bolometric as measured over the entire *electromagnetic spectrum rather than through a particular passband.

brown dwarf essentially a failed star; an object more massive than the planet Jupiter, but with insufficient mass to fuse hydrogen into helium in its core and so emit visible radiation.

Caltech California Institute of Technology

CCD charge-coupled device, a solid-state detector used for astronomical imaging.

CERN Centre Europ'een pour la Recherche Nucleaire: European laboratory for experimental particle physics research, located on the Swiss–French border.

convolution an integral that expresses the amount of overlap of one function g as it is shifted over another function f. Mathematically, the convolution is given by $(f * g)(y) = \int f(x)g(y - x)dx$.

correlation length the scale at which the *correlation function has a value of one.

covariance a mathematical way of describing the dependence of the value of one parameter on another.

CPU central processing unit

declination (abbrev. dec; symbol δ) the celestial coordinate of latitude, with $\delta = \pm 90$ degrees corresponding to the north and south celestial poles.

deuterium a heavy isotope of hydrogen, whose nucleus consists of one proton and one neutron.

Doppler broadening the broadening of features in a *spectrum due to the *Doppler effect caused by large relative velocities within the source.

dynamical equilibrium the state reached by a system when the initial motions of its consituents are no longer recognizable. Such a system is also said to be 'relaxed'. *See* VIRI-ALIZATION.

dynamical friction the deceleration of a galaxy moving through an otherwise uniform medium due to the gravitational pull of material in its wake.

dynamics the study of the effects of forces on the motions of objects. In cosmology, the forces are mostly gravitational and so dynamics provides a way of estimating the total mass of a system.

effective radius the radius of a circular aperture, centred on a galaxy, that contains half of the total flux from the galaxy.

ESA European Space Agency: responsible for European space science.

ESO European Southern Observatory: European organization for astronomical research in the southern hemisphere.

extragalactic outside our own Galaxy.

filament term to describe the apparently one-dimensional distribution of *galaxies between *clusters of galaxies in the *cosmic web.

flat Universe a Universe which has zero *curvature (*not* a two-dimensional Universe).

flux amount of *electromagnetic radiation received per unit time from a source.

Galactic coordinates a coordinate system in which the longitude and latitude (l, b) of an object are described relative to the *Milky Way Galaxy.

Galactic latitude (symbol b) gives the observed angle in the range [−90, 90] degrees of an object above the plane of the Galaxy.

Galactic longitude (symbol l) gives the observed angle in the range [0, 360] degrees of an object within the plane of the Galaxy relative to the Galactic centre.

giant galaxy any galaxy that is not classified as a *dwarf galaxy, that is with a luminosity $L \gtrsim 10^9 L_\odot$.

giga (abbrev. G) as a prefix, multiplies the following unit by a factor of one billion, i.e. 10^9, e.g. giga-watt or GW.

inclination the angle between the rotation axis of a disk galaxy and our line of sight to it.

intensity amount of energy detected per unit time, per unit area, per unit solid angle.

inverse square law the law which states that the flux received from a source decreases with the square of distance to the source.

IR abbreviation for *infrared.

kelvin (symbol K) the unit of absolute temperature, with 273 kelvin corresponding to the freezing point of water (zero degrees centigrade).

kinetic energy the energy associated with the motion of mass. For a particle of mass m moving with velocity v, the kinetic energy T is given by $T = \frac{1}{2}mv^2$.

light year the distance travelled by light (and all electromagnetic radiation) in one year, equal to 9.4605×10^{15} metres.

lightcurve a plot of the brightness of an object such as a *supernova as a function of time.

luminosity the energy radiated per second by a source.

mag abbreviation for *magnitude, an astronomical measure of brightness.

main-sequence star a star that is fusing hydrogen to helium in its core.

micron one millionth of a metre, also written as μm.

multipole a term in a spherical harmonic expansion, used for instance in analysing *cosmic microwave background anisotropies.

nano multiplies the following unit by a factor of 10^{-9}, e.g. a nano-watt is one billionth of a watt.

NASA National Aeronautics and Space Administration: US government agency responsible for space exploration and science.

of order a shortened form of 'within an order of magnitude', i.e. within a factor of 10, and equivalent to the symbol $\sim$.

optical refers to that part of the *electromagnetic spectrum visible to the human eye, namely 400–750 nm.

parsec (abbrev. pc) measure of distance, equal to 3.0857×10^{16} metres or 3.2616 light years.

passband a letter (e.g. r) denoting the sensitivity as a function of wavelength for a particular filter–instrument combination.

photometry the quantitative study of the brightness of astronomical sources.

Poisson statistics describe a random distribution whose variance is equal to its mean.

potential energy the energy associated with a *gravitational potential. The sign of potential energy is negative, as a loss of potential energy corresponds to a gain in kinetic energy.

power law a functional dependence of the form $y = Ax^B$, where A and B are constants. A logarithmic plot of log y versus log x will be a straight line with slope B.

proper motion the apparent change in position of an object with time due to its motion perpendicular to the observer's line of sight.

radial velocity the velocity observed along the line of sight to an object.

radian the natural unit of angle, equal to $180/\pi$ degrees, and with 2π representing a full circle.

recession velocity the apparent radial velocity of an object due to its *redshift.

refractive index property of materials which determines the *speed of light through them.

relativistic refers to velocities that are a significant fraction of the *speed of light c, $v \gtrsim 0.1c$, in which case *relativity theory becomes important.

restframe the *inertial frame in which the object being observed is stationary.

right ascension (abbrev. RA; symbol α) the celestial coordinate of longitude, quoted in either degrees (0–360) or hours (0–24).

shock heating the heating of a gas by a super-sonic wave, as might be produced by the merger of two *galaxies.

Solar luminosity the luminosity of the Sun.

solid angle two-dimensional equivalent of angle; this expresses the apparent size of an object.

spectrophotometric standard star a star whose flux as a function of wavelength is known.

standard star a star of known apparent *magnitude.

steradian (abbrev. sr) unit of solid angle, equal to $(180/\pi)^2$ square degrees, and with 4π steradians being the solid angle of the entire sky.

strong interaction the force responsible for binding atomic nuclei together. *See* STANDARD MODEL OF PARTICLE PHYSICS.

surface brightness flux received per unit area from an extended source.

Taylor series a mathematical technique for representing a more complicated function as a sum of simpler terms.

tera (abbrev. T) as a prefix, multiplies the following unit by a factor of 10^{12}, e.g. tera-watt or TW.

UV abbreviation for *ultra-violet.

violent relaxation the process by which energy is transferred between stars in a *gravitational potential that is varying rapidly with time.

watt (abbrev. W) unit of power equivalent to 1 joule per second.

wavelength distance λ between successive peaks in a wave; light of longer wavelengths is perceived as redder.

wavenumber spatial frequency of a wave, $k \equiv 2\pi/\lambda$.

weak interaction the force responsible for radioactive beta decay. *See* STANDARD MODEL OF PARTICLE PHYSICS.

zenith the point directly overhead at any given time.

zodiacal plane the approximate plane in which the planets, including Earth, orbit the Sun.

Commonly used symbols

By convention, symbols in italic font refer to constants or variables, such as t for time, whereas units are given in Roman font, such as s for seconds.

Å	Angstrom: unit of length equal to 10^{-10} metres	Mpc	megaparsec, one million parsecs, the distance unit most commonly used by cosmologists
au	astronomical unit, the average Sun–Earth distance	n	as a prefix, e.g. ns (nanosecond), divides following unit by one billion
b	Galactic latitude		
c	speed of light $\approx 300\,000$ km/s	p	pressure
C_ℓ	angular *power spectrum	pc	parsec, unit of distance
e	the base of natural logarithms $\approx$ 2.71828	$P(k)$	*power spectrum
		q	*deceleration parameter
eV	electron-volt, the *particle physics unit of energy	r	physical separation between two points
G	Newton's gravitational constant, which sets the scale of the gravitational force between two masses	r_p	component of separation perpendicular to the line of sight
		s	second, the SI unit of time
		SI	International System of Units, based on the metre (length), kilogram (mass) and second (time)
G	as a prefix, e.g. Gyr (gigayear), multiplies unit by one billion (10^9)		
h	dimensionless Hubble constant, $h = H_0/(100$ km/s/Mpc) **or** the Planck constant (*see* PLANCK SCALE)	sr	steradian, the SI unit of solid angle
		t	time
		T	temperature
$\hbar$	pronounced 'h-bar', the reduced Planck constant, $\hbar = h/2\pi$	T	as a prefix, e.g. TeV (tera-electron-volt), multiplies following unit by 10^{12} (a million million)
H_0	Hubble constant		
Hz	Hertz, the SI unit of frequency equivalent to one cycle per second	$ugriz$	*magnitude system used by the *Sloan Digital Sky Survey, with each letter denoting one of five passbands from near-UV to near-IR
J	joule, the SI unit of energy		
JHK	near-infrared magnitudes		
k	*wavenumber	v	velocity
k	as a prefix, e.g. kpc (kiloparsec), multiplies following unit by one thousand	w	*equation of state
		$w(\theta)$	angular *correlation function
		W	Watt, the SI unit of power, equal to one joule per second
K	kelvin, the SI unit of temperature		
k_B	the Boltzmann constant	yr	year
kg	kilogram, the SI unit of mass	z	*redshift
l	Galactic longitude	α	right ascension
L	luminosity	δ	declination **or** as a prefix, e.g. δt, the change in the following quantity
m	apparent *magnitude **or** mass		
m	metre, the SI unit of length	θ	angular separation
M	absolute *magnitude, with any subscript denoting passband (e.g. M_V is visual magnitude)	λ	wavelength
		Λ	*cosmological constant
		μ	as a prefix, e.g. μm (micro-metre, of micron), divides following unit by one million
M	as a prefix, e.g. Myr (megayear), multiplies following unit by one million		
		ξ	spatial *correlation function

π the mathematical constant that is the ratio of the circumference to the diameter of a circle ≈ 3.14159 **or** component of separation parallel to the line of sight

ρ *density

σ standard deviation **or** *velocity dispersion **or** component of separation perpendicular to the line of sight

Ω cosmological *density parameter

Ω_Λ contribution of *dark energy to the *density parameter

Ω_M contribution of matter to the *density parameter

$\sim$ equal to within an order of magnitude, i.e. a factor of ten

$\approx$ equal to within a factor of two or better

$\equiv$ equivalent of (equal by definition, not just numerical value)

$\propto$ proportional to: if the quantity on the left-hand side of this symbol is multiplied by a constant factor, the right-hand side increases by the same factor

$<$ less than

$>$ greater than

$\lesssim$ approximately less than

$\gtrsim$ approximately greater than

$\pm$ plus or minus, shows the range of uncertainty around a given numerical value

$\odot$ used as a subscript, this symbol refers to the Sun, thus $1L_\odot$ refers to one Solar luminosity, the luminosity of the Sun

Overview: the hot big bang cosmology

Cosmology is amongst both the oldest and youngest of sciences. In its primitive form, it can be traced back to prehistoric eras, for instance through the well-established astronomical alignments of pyramids and stone circles. As well as religious motivations, study of the skies was essential to develop timekeeping so that, for instance, crops could be planted at the correct point of the seasonal cycle.

In its modern form, however, cosmology is quite a young science. At the beginning of the 20th century, essentially none of our current understanding was in place. The true extent of the Universe was unknown and unsuspected, the existence of galaxies other than the Milky Way had not been demonstrated, and the essential understanding of gravity that would later come with Albert Einstein's development of his general theory of relativity was not yet available. In that light, it is remarkable that, 100 years later, we may have in place a description of the Universe robust enough to outlast us all.

The development of the known Universe

For almost all of history, humankind's view of the cosmos was quite limited. Objects visible to the naked eye are the planets out to Saturn, the Moon, comets and meteorites, and in good conditions a few thousand stars and the shimmer of the Milky Way. Understandably, those stars were seen as marking out the extent of the Universe, and it was not realized that the Milky Way was the combined light of vast numbers of distant stars. Through naked eye observations, notably those of the 16th-century Danish astronomer Tycho Brahe as interpreted by Johannes Kepler in the early 1600s, the laws of motion within the Solar System were determined empirically, later to be explained by Isaac Newton's law of gravity.

Our view of the Universe widened with the development of the telescope. Following its invention in the Netherlands, it was first exploited for astronomy by the legendary Italian scientist Galileo Galilei, widely recognized as the inventor of the modern scientific method with its emphasis on observational tests of hypotheses, and also the first effective popularizer of science through his writings. Around 1610 he used his telescope to discover the moons of Jupiter, and later discovered that the Milky Way was made up of stars packed too closely to be resolved by the naked eye.

In the following years, more and more was learnt about the stars making up the Milky Way. Astronomer William Herschel (the discoverer of infrared radiation) realized around 1800 that it took the form of a disk, though he believed that the Solar System was located at its centre. It would be around a hundred years before the Solar System would be correctly placed as significantly away from the Galactic centre.

Right into the 20th century, the popular view remained that the Milky Way was all there was in the Universe. Or, rather, that the Milky Way *was* the Universe. By then many so-called nebulae—diffuse extended objects—were known but their nature was unclear. A famous discussion, now referred to as the Great Debate, took place in Washington in 1920 between astronomers Harlow Shapley and Heber Curtis, the former arguing that the Milky Way was all there was, the latter that many of the nebulae lay outside the Galaxy. While that debate is widely regarded as having been inconclusive, a few years later Edwin Hubble used the newly constructed Mount Wilson telescope in California to resolve individual variable stars within the Andromeda galaxy. This allowed him to demonstrate that Andromeda lay well outside the Milky Way, even though in fact he significantly underestimated the true distance. This opened the way to the modern view that the Milky Way is just one of a vast (indeed perhaps infinite) number of galaxies throughout the Universe. In 1953 Walter Baade derived a new distance to Andromeda much more in keeping with the modern value, though work on refining these measurements has continued right through to the present.

The expansion of the Universe

The ability to construct proper models of cosmology arrived with Einstein's publication of his general theory of relativity in 1915. Indeed, Einstein created the first such models in 1917, but his belief that the Universe was static and unchanging led him to introduce a quantity called the cosmological constant into his theory, which he later regretted as '... my biggest blunder'. It was left to Alexander Friedmann to discover in 1922 the class of cosmological models, now known as Friedmann cosmologies, that form the basis of modern cosmology. These were models of an expanding Universe. They were also later discovered independently by Georges Lemaître, who recognized that they predicted an initial time when the Universe was of zero size, which he called the 'primeval atom'. This can be regarded as the birth of the big bang cosmology, though the name followed long after.

Friedmann's models tied together the evolution of the Universe and its geometry. Closed Universe models, which contained a high density of material and were finite in extent, would ultimately recollapse. Low-density open Universe models, whose curvature caused parallel lines to diverge, would expand forever. Balanced between these two was the special case of a flat Universe, in which the usual laws of Euclidean geometry applied. Such a Universe contained a particular density of material, known as critical density. It was not known which, if any, of these models might describe our Universe.

Observational developments were proceeding in parallel. Vesto Slipher had begun to measure the redshift of spectral lines of bright nebulae in order to estimate their velocities, obtaining strikingly large values. In 1913 he found that the Andromeda nebula was moving rapidly towards us, but by 1917 his sample of 25 measurements contained 21 which were moving away, most at great velocity. He concluded that his observations supported the idea that the nebulae were 'stellar systems seen at great distance', though his contemporary audience was apparently unconvinced. Nevertheless, there is a compelling case to attribute the discovery of the expansion of the Universe to Slipher, even though he himself did not make this inference.

The definitive work on the expansion was due to Edwin Hubble and Milton Humason (the latter having started his astronomical career as the mule driver who ferried supplies up the mountain to the telescope!). By measuring the redshifts of many nearby galaxies, they were able to show in 1929 that the vast majority of them were receding from the Milky Way, confirming the indications from Slipher's work and establishing beyond reasonable doubt the expansion of the Universe. Moreover, they found that the rate of recession increased in proportion to the distance of the galaxy.

This expansion law, usually known as Hubble expansion and sometimes as uniform expansion, has been confirmed with ever-increasing accuracy since then, and extended to very distant galaxies. The linear law is the unique form such that the same law is seen from *every* galaxy in the Universe—all see that galaxies recede radially with velocity proportional to their distance. The Hubble law therefore respects the cosmological principle, which states that there are no special observers in the Universe. This law underpins all modern cosmological models (though is partly overturned by the anthropic principle which has lately seen increased use).

The hot big bang

The next phase of development of cosmology was the realization that the big bang must have occurred at high temperatures. George Gamow was the early leader of this work, which began in the mid 1940s. Gamow realized that the young big bang would have been such a hot environment that nuclear reactions would have been important, and sought a cosmological theory of the nucleosynthesis process leading to the observed abundances of different chemical elements in the Universe. A key paper, written in collaboration with Ralph Alpher and Hans Bethe (Bethe did not however contribute to the work—*see* GAMOW), indicated that the lighter elements, particularly hydrogen and helium-4, would indeed be a consequence of the big bang model. The majority of heavier elements would later be shown, by Fred Hoyle and collaborators, to be created within stars. Refined by many researchers in the decades since, cosmic nucleosynthesis remains one of the most powerful pillars supporting the big bang cosmology.

As part of this research programme, Alpher and Robert Herman went on in 1948 to predict the existence of relic radiation left over from the hot early phase, which would suffuse the Universe and which we now know as the cosmic microwave background radiation. With hindsight, this had actually already been discovered through Andrew McKellar's 1941 study of rotational excitation of interstellar cyanogen molecules, but Alpher and Herman did not know of this discovery, nor did they think their predicted radiation was detectable, and their ideas were largely forgotten for over a decade.

The cosmic microwave background was directly discovered, by accident, in 1965 by Arno Penzias and Robert Wilson. Once its cosmological interpretation became clear, and its black-body form demonstrated to ever higher accuracy, the hot big bang model became fully established as the basis for cosmological models, with alternatives such as the steady-state cosmology falling by the wayside. It has now dominated cosmological thinking, without serious challenge, for more than 40 years.

The dark stuff, part I

Throughout this time, an undercurrent to cosmology was the growing suspicion that there was more to the Universe than met the eye. Specifically, that there was an invisible component of matter. This was first proposed by Fritz Zwicky (who is almost invariably referred to as 'maverick astronomer Fritz Zwicky'). He realized that a giant nearby cluster of galaxies, the Coma cluster, would fly apart were it not for the existence of much more gravitational attraction, i.e. mass, than could be inferred from studying the luminous stars and galaxies. Zwicky referred to this as 'missing matter', and strengthened his case with further observations throughout the 1930s.

However, the existence of this extra mass, now known as dark matter, did not enter the mainstream until the 1970s. At this time, careful work studying the rotation of spiral galaxies, especially by Vera Rubin and her collaborators, established the existence of dark matter in individual galaxies. Rotation curves, measuring the rotation rate of a galactic disk as a function of distance from the centre, typically indicated ten or more times as much dark matter as visible matter. This mass was essential to stop the galaxy flying apart under its fast rotation. The dark matter was therefore the dominant constituent of any galaxy, and the concentrations within which galaxies reside became known as dark matter halos.

The introduction of dark matter into cosmological models meant that cosmologists had to worry not just about the total density of material in the Universe, but also about how that density is divided up amongst the different constituents, and how that division might alter with time.

Structures in the Universe

The late 1960s and 1970s saw growing interest in the distribution of galaxies in the Universe, stimulated particularly by the work of the 'Soviet school' of astrophysics centred around Yakov Zel'dovich, and a US effort inspired by James Peebles of Princeton (who had also played a key role in the identification of Penzias and Wilson's observations as cosmological in origin). The existence and clustering properties of galaxies, referred to generically as large-scale structures in the Universe, violates the hypothesis of a smooth distribution of material on which the simplest cosmological models hinge. Instead, it replaces it with the notion of 'statistical homogeneity', whereby the distribution is smooth only in an average sense. As well as understanding the properties of the global Universe, it became imperative to understand how and why the Universe deviated from perfect smoothness.

It was soon appreciated that any irregularities in the Universe would be enhanced as the Universe evolved, as the extra gravitational attraction exerted by the more dense regions would draw more material towards them. This process—gravitational instability—is the standard paradigm for how structures form and evolve.

The development of theories of structure formation also sparked a long battle between two rival paradigms for dark matter, known as hot dark matter and cold dark matter. As dark matter had a greater density than visible matter, it would be primarily responsible for the gravitational attraction causing structure formation. In the cold dark matter model, the dark matter particles would be essentially at rest when structure formation began, while in the hot dark matter scenario they would have significant velocities which would oppose gravitational collapse. These alternatives led to quite different predictions for how structures would form. In the 1980s, evidence began to stack up in favour of the cold dark matter hypothesis, though it was perhaps not until the 1990s that the hot dark matter scenario was declared truly dead. More recently, hot dark matter had a small symbolic victory with the discovery that neutrinos (an abundant type of fundamental particle) have a small mass, meaning that they act as a trace component of hot dark matter. But cold dark matter dominates, and such models are called cold dark matter cosmologies.

The early Universe

In the late 1970s and 1980s, attention began to focus on the types of physical process that might have taken place during the very earliest stages of the Universe's evolution, when the temperature and energy would have been so high that the nature of physical laws is uncertain. This is the regime of the particle cosmologist, who seeks to take ideas from modern particle physics and apply them to the early Universe. Such ideas include the unification of fundamental forces, the possibility of extra dimensions, and superstring theory. In particular, are any clues as to the nature of high-energy physics left in the present Universe for us to observe?

The single most influential idea in particle cosmology is cosmological inflation, invented in 1981 by American physicist Alan Guth. Guth postulated that the very early Universe might have undergone a period of rapid, accelerated, expansion, and showed that this could explain a number of otherwise mysterious features of our Universe. Such an expansion, he suggested, might be associated with a phase transition during which the young Universe was changing its physical state. Inflation was soon shown to give a possible explanation for the origin of structure in the Universe, through amplification of small quantum effects taking place during the inflationary era. Inflation is now routinely considered to be part of the standard cosmological model, explaining the large-scale appearance of the Universe and initiating the development of structures. One of its key predictions is that the Universe should have a flat geometry, and hence contain a critical density of material.

Other early Universe ideas are at a less mature stage. Explanations are sought for the origin of the matter–anti-matter asymmetry in the Universe, and more generally for the complete inventory of particles present in our Universe. Theorists speculate on the possible role of phase transitions, and ask whether gravitational waves, topological defects, or primordial black holes might be consequences of particle physics models. Whether ideas such as superstring theory, or its successor M-theory, might have implications for astronomical observations remain unclear.

Galaxies

The 1980s also saw major advances in understanding the detailed properties of galaxies. In 1986 the Center for Astrophysics (CfA) Redshift Survey team published its famous 'stick man' plot of the galaxy distribution. For the first time, the strongly clustered nature of the three-dimensional galaxy distribution became strikingly apparent in a visual representation, with galaxies apparently lying on the surfaces of bubble-like structures up to 50 Mpc in diameter. The Coma cluster of galaxies formed the 'body' of the stick man. This and the other early large redshift surveys of galaxies allowed the intrinsic properties of galaxies, such as their luminosity and three-dimensional clustering statistics, to be reliably measured for the first time.

Around the same time, the Automated Plate Measurement (APM) machine in Cambridge, UK, was being used to digitize 269 wide-field photographic plates, and so obtain a digital map of 3 million galaxies over more than 6 000 square degrees of sky. The vast size and quality of the APM Galaxy Survey allowed, in 1990, the first reliable measurement of the clustering of galaxies at very large separations. This large-scale clustering was significantly stronger than predicted by the then-popular critical density, or 'standard', cold dark matter model, and provided perhaps the first nail in the coffin of a matter-dominated Universe.

Technological developments in the 1980s in manufacturing more sensitive light detectors and more efficient spectrographs allowed the first surveys of galaxies to significant redshifts. It thus became possible to probe the evolution of galaxies for the first time. The Durham/AAT redshift survey, published in 1988, included 200 galaxies out to redshifts of almost one half. The redshift distribution and spectra of these galaxies indicated that intense bursts of star formation were common in distant galaxies.

Galaxy and redshift surveys have continued to develop apace, with state-of-the-art surveys such as the Sloan Digital Sky Survey, the Deep Extragalactic Evolutionary Probe (DEEP), and VIRMOS-VLT Deep Survey currently leading the field.

Cosmic microwave background anisotropies

A landmark discovery took place in 1992, when the team operating the COsmic Background Explorer (COBE) satellite showed that the cosmic microwave background had small temperature variations, known as anisotropies. These small irregularities were predicted by models of structure formation, as the small seeds from which structures eventually formed. The microwave background, having originated in the young Universe, provided a perfect snapshot of conditions in the Universe at that time.

There were now two different views of the Universe, one from when the microwave background formed, and the other from the distribution of galaxies around us. This meant that cosmologists could study the dynamical evolution of structures in the Universe, and in particular how that evolution might depend on their assumptions as to how the Universe works. Ultimately, observations of structure formation, and particularly of the cosmic microwave background, would become the most powerful tool in building and testing cosmological models.

The dark stuff, part II

One more ingredient was yet to emerge, nowadays believed to be the most dominant component of all, dark energy. Throughout the 1990s there were persistent hints of a further new type of material, one which could supply the density necessary to permit a flat Universe but which did not participate in gravitational instability. An example of such a material was the cosmological constant which Einstein had, so many years before, proposed in a misguided attempt to obtain a static Universe. In the new models, the cosmological constant would instead drive a more rapid expansion of the Universe, perhaps even an accelerated one.

In this epoch, discussions of cosmology often referred to three distinct possibilities for the make-up of the Universe, which made quite different predictions but which were indistinguishable to the observations of the time. These were a Universe where the combination of normal and dark matter add up to the critical density needed to make the Universe flat, a model where they added to only about one-third of the critical density giving an open Universe, and finally a model where a cosmological constant made up the gap so that the low matter density could be compatible with spatial flatness. In particular, the last enabled one to square inflation's prediction of a flat Universe with the persistently low matter density measured by observers.

This three-way balance was finally broken in 1998, with the announcement by two separate teams that observations of distant supernovae favoured an accelerating Universe, and hence the existence of a cosmological constant. The supernovae on their own might not have been compelling, but they were a particularly clean observational probe, and strongly supported trends that were already apparent in other data. Soon afterwards, the open Universe model received a fatal blow when cosmic microwave background observations by the Boomerang experiment gave convincing evidence that the Universe's geometry was close to flat.

Cosmologists discovered that, however hard they tried, they were unable to sustain cosmological models without acceleration. The responsible material became known as dark energy, to allow more general possibilities for its properties than just a cosmological constant. Nowadays, almost all cosmologists accept the strength of the observational evidence for dark energy, though often reluctantly on aesthetic grounds.

The standard cosmological model

At or soon after the turn of the millennium, the various ingredients that had been coming together assembled themselves into a standard model of cosmology. Although many things contributed to this, many people regard the 2003 first release of results from the Wilkinson Microwave Anisotropy Probe (WMAP), a NASA satellite which measured the cosmic microwave background anisotropies, as the dawn of the era of precision cosmology. This data, combined with information from other sources, was of sufficient power that it could be used to determine accurate values of many of the cosmological parameters describing our Universe.

| **1910** | 1915: | Einstein introduces general relativity |
| | 1917: | Slipher's measurements indicate that the Universe is expanding |

1920	1920:	Curtis and Shapley's Great Debate on the nature of nebulae
	1922:	Friedmann devises expanding Universe models
	1924:	Hubble measures distances to nebulae proving many are extragalactic
	1929:	Hubble and Humason measure the expansion rate of the Universe

| **1930** | 1933: | Zwicky proposes the existence of dark matter |

1940	1946:	Lifshitz invents cosmological perturbation theory
	1946:	Gamow and collaborators begin study of hot Universe models
	1948:	Alpher and Herman predict existence of the cosmic microwave background (CMB)

| **1950** | 1950s: | Big bang versus steady state debate |
| | 1958: | Abell creates the first galaxy cluster catalogue |

| **1960** | 1965: | Penzias and Wilson discover the CMB; hot big bang becomes the standard paradigm |
| | 1967: | Lick galaxy catalogue completed |

| **1970** | 1970: | Harrison–Zel'dovich perturbation spectrum proposed |
| | 1970s: | Dark matter enters the mainstream |

1980	1981:	Guth introduces cosmological inflation
	1981:	Inflation shown to predict initial perturbations
	1983:	First CfA galaxy redshift survey released
	1980s:	Cold dark matter model developed

| **1990** | 1992: | COBE satellite discovers CMB anisotropies |
| | 1998: | Dark energy becomes a standard part of cosmological models |

| **2000** | 2000: | Boomerang CMB experiment supports flat Universe |
| | 2003: | WMAP satellite makes precision CMB maps |

Fig. 1. Some landmarks in the development of the standard cosmological model.

Intriguingly, the success of WMAP was really that it failed to find any exciting new phenomena—astrophysicist John Bahcall, commenting on the WMAP press announcement, said 'The biggest surprise is that there are no surprises'. Instead, the observations were compatible with existing models, and the strength of the data could purely be focused on accurately determining the parameters of the model. For instance, the total density of the Universe was shown to be within 1% of the critical density, implying a flat or very nearly flat Universe, and the age of the Universe determined to be 13.7 billion years, with an uncertainty of only a couple of hundred million years.

Where next for cosmology?

The standard cosmological model is a striking success, as a phenomenological description of cosmological data. Having such a clear idea of the global properties of the Universe opens the way to accurate studies of processes happening within it, such as galaxy formation, without worrying that some effects might really be due to using the wrong cosmological model. The model's success in explaining high-precision observations has led a clear majority of the cosmological community to accept it as a good account of how the Universe works.

But that doesn't mean that cosmologists are particularly happy, or that the era of cosmological investigation is over. The model may be a good one empirically in terms of explaining observations, but it is very unsatisfying at a deeper level, as hardly any of it is actually understood in a fundamental way. What is dark matter? No one knows, despite its pervasive gravitational influence. Why do protons and neutrons only make up 4% of the total density, and why is there no anti-matter? No compelling theory exists. And dark energy is really just a name for something not understood at all, even apart from the obvious objection that, having played the 'mysterious unknown dark stuff' card once with dark matter, we really shouldn't be allowed to play it again with dark energy. Yet cosmologists, reluctantly, have been forced to do so.

New insight is clearly needed, which may be observational or theoretical. Many new observational programmes are coming on line, especially with the aim of tackling the mystery of dark energy. At the very least, the projects should tell us whether or not the density of dark energy changes with time. And, in other spheres of cosmology, increasing precision always offers the chance of uncovering new phenomena in the details of how the Universe works. As to theoretical insight, you can't really plan for that. You need a lot of smart people, and faith that with enough smart people, thinking about difficult problems in enough different ways, someone can transform our view about how the world works. It's happened many times before, and it will surely happen again.

Kolb, R. *Blind Watchers of the Sky*, Perseus, 1997. An entertaining and insightful history of cosmology from its early days to the present.

Weinberg, S. *The First Three Minutes*, Basic Books, 1993 (originally published 1977). A classic popularization of the hot big bang cosmology, dated in parts but with much that is still relevant.

Silk, J. *The Big Bang*, Times Books, 2000 (3rd edition). A popular-level, though in places challenging, account of modern cosmology.

Rees, M. *Our Cosmic Habitat*, Phoenix, 2003. A discussion of the underlying rules that might decide why our Universe is as it is.

Abell clusters

George Abell is best remembered for his catalogue of *clusters of galaxies, a result of his PhD work at the California Institute of Technology (Caltech), published in 1958. Abell inspected 879 photographic plates (taken through a red filter) from the Palomar Observatory Sky Survey, covering the entire northern sky down to a declination of −27 degrees. Clusters in the catalogue are identified by the letter A and a sequence number, e.g. A1656 is the *Coma cluster.

Since *galaxies cluster together, their projected number density on photographs of the sky is non-uniform. Abell looked for significant over-densities of galaxies that were likely to comprise physical associations, or clusters of galaxies. Abell's putative clusters had to pass a number of criteria in order to make it into his homogeneous catalogue.

Richness—A cluster must contain at least 50 members not more than two *magnitudes fainter than the third brightest member. This criterion excludes smaller groups of galaxies.

Compactness—The cluster members must lie within a radius of 1.5 h^{-1}Mpc of the cluster centre. Note that this criterion requires an estimate of the distance to the cluster.

Distance—A cluster must be sufficiently distant that its members are confined to a single survey plate or pair of neighbouring plates. This corresponds (using the incorrect *Hubble parameter of 180 km/s/Mpc assumed by Abell at the time) to a *redshift limit cz = 6000 km/s. The Coma cluster (cz = 6925 km/s) is thus amongst the closest clusters in Abell's catalogue, whereas the *Virgo cluster (cz = 1080 km/s) is too close to be included. There is also an upper distance limit set by the requirement that a galaxy 2 magnitudes fainter than the third brightest member be visible. This corresponds to a redshift $cz \approx 60000$ km/s.

Galactic latitude—In regions of low Galactic latitude the high stellar density and presence of obscuring dust prevents complete identification of cluster galaxies. The actual latitude limits used vary with Galactic longitude.

Of 2712 clusters inspected by Abell, 1682 pass the above criteria and are included in the homogeneous catalogue.

Once identified, magnitudes were estimated for the cluster galaxies by comparison with calibrated galaxy images, and distances to the clusters were estimated by assuming that the tenth brightest galaxy in each cluster has the same absolute magnitude.

In 1975, Abell proposed that the cluster catalogue be extended to the southern hemisphere, using photographic plates from the United Kingdom Schmidt Telescope at Siding Spring, Australia. With the help of Harold Corwin and Ronald Olowin, the extended Abell cluster catalogue was published in 1989. This all-sky catalogue contains 4073 rich clusters of galaxies with at least 30 members not more than two magnitudes fainter than the third brightest member.

The Abell cluster catalogues have been tremendously important in studies to trace out the distribution of matter on very large scales, despite the systematic selection effects which inevitably bias a catalogue constructed by eye. The original 1958 catalogue paper has received more than 1000 citations in the astronomical literature, and the 1989 paper more than 500.

http://cfa-www.harvard.edu/~huchra/clusters/images.html

acceleration of the Universe

The expansion rate of the *Universe, measured by the *Hubble parameter, would normally be expected to be slowing down, as the gravitational force between objects opposes the expansion. That is to say, the expansion of the Universe should be decelerating. However, the possibility that the Universe might accelerate during some epochs has been raised in two separate contexts.

During the very early stages of the Universe's evolution, a period of acceleration is known as *cosmological inflation. Accelerating cosmologies date back to work in 1917 by Dutch astronomer Willem de Sitter, and his model is still known as *de Sitter space and exploited today. But the benefits of such a period of expansion were not realized until a seminal paper by Alan Guth in 1981, where he showed that such an expansion explained a number of otherwise unexplained features of our Universe. *See* COSMOLOGICAL INFLATION.

The second context is our present Universe. In the late 1990s, studies of distant type Ia *supernovae indicated that the Universe is presently accelerating, a result which has subsequently been supported by independent observations including *cosmic microwave background anisotropies. The reason for this acceleration is not at all understood, though it is attributed to the presence of a material known as *dark energy (giving something a name does not however amount to understanding it). One of the most important goals of *cosmology is to confirm that the acceleration is indeed taking place, and to discover the physical mechanism responsible.

Such understanding might also address whether the present acceleration is expected to continue for ever, or come to an end as did the early Universe inflation. If the Universe does continue to accelerate, then all but the most nearby objects will move ever more rapidly from us and be removed from our sight (*see* FUTURE OF THE UNIVERSE). The distant future of such a Universe would therefore be a rather boring place, with not much for astronomers to see.

acoustic peaks

The term 'acoustic peaks' refers to characteristic features seen in the structure of *inhomogeneities in the *Universe. In the young Universe, before the epoch of *decoupling when the *cosmic microwave background (CMB) was created, the *density perturbations in the Universe were in the form of sound waves. Oscillations in the density of the primordial fluid (comprised of frequently interacting *baryons and *photons, the latter destined to become the CMB) would occur under the competing forces of gravitational attraction and *pressure repulsion, the pressure preventing gravitational collapse. These standing waves

are sound waves, characterized by the primordial sound speed which would be close to the *speed of light, and the oscillations are known as acoustic oscillations.

On particular length scales the acoustic oscillations might lead to an enhancement of the density perturbations, if the waves are at the point of maximum compression or rarefaction, or to a suppression if the waves are (instantaneously) of zero amplitude. Waves on different scales oscillate with different frequencies, given by the time a sound wave takes to travel from one peak to the next, and so at a given instant waves on some scales are enhanced and on others suppressed. When one analyses the typical size of the waves, using for instance the *power spectrum, it therefore exhibits a series of maxima and minima. The maxima are known as the acoustic peaks.

At the epoch of decoupling, the photons which had been providing the pressure support suddenly cease to interact with the baryons and the oscillations cease. The baryons undergo gravitational collapse, while the photons begin to travel freely. The peaks are most prominent in the spectrum of *CMB anisotropies, where they are the most visible feature across a wide range of scales. They have also been detected in the power spectrum of the large-scale *galaxy distribution, where they are usually referred to as *baryon oscillations, by the *Sloan Digital Sky Survey and the *two degree field galaxy redshift survey.

The acoustic peaks occur on length scales that can be calculated for given *cosmological models, providing a *standard ruler which can be used to study the *angular-diameter distance. In the CMB, it can be used to probe the *geometry of the Universe, leading to the conclusion that the Universe is flat or very close to it. In the future, study of the acoustic peaks in the galaxy power spectrum out to *redshifts of at least one may be used to probe the nature of *dark energy.

active galactic nuclei (AGN)

Some *galaxies possess an **active nucleus**: a compact central region emitting substantial radiation that is *not* due to stars or hot gas. These active galactic nuclei, AGN for short, emit strongly over the whole *electromagnetic spectrum including the radio, *X-ray and gamma-ray parts of the spectrum. The most powerful AGN outshine their entire host galaxy, with luminosities $\sim 10^{12}$ times that of the Sun.

AGN are most commonly found at high *redshift, suggesting that nuclear activity may be characteristic of a galaxy's early life. Optical and *ultra-violet (UV) spectra show strong, broad emission lines characteristic of a moderate density gas. The widths of these lines are consistent with *Doppler shifts due to gas clouds travelling at $\sim 10\,000$ km/s—much faster than typical stellar velocities in galaxies (~ 200 km/s). Many AGN are variable, with luminosity and line strengths changing on a timescale of months, days or even hours. This tells us that the size of the emitting region is no larger than the Solar System, which implies that the most likely power source for the AGN is a central *supermassive black hole of $\sim 10^8$ solar masses.

Historically, AGN have been classified into several categories.

Seyfert galaxies—Broad emission lines from a galactic nucleus were first reported in 1907, but no systematic study was made until 1943, when Carl Seyfert published 12 galaxies with nuclear spectra showing broad emission lines of ions that could only be excited by photons more energetic than those of the hottest young stars in star-forming regions. Seyfert subdivided these into two sub-categories: *Seyfert 1* galaxies have very broad emission lines due to Doppler shifts of more than 1 000 km/s. *Seyfert 2* galaxies have moderately broad emission lines due to Doppler shifts of under 1 000 km/s. Most of Seyfert's galaxies were spirals, but one was the giant elliptical galaxy NGC 1275. We now know that $\sim 10\%$ of Sa and Sb galaxies have Seyfert nuclei. Two possible (but not exhaustive) interpretations of this observation are that all of these galaxies spend $\sim 10\%$ of their lives as Seyferts, *or* one in ten has a long-lasting Seyfert nucleus. Around 25% of Sa and Sb galaxies have **low ionization nuclear emission regions** (LINERs), which are less luminous than a *Seyfert 2*. These could arise from a low-powered active nucleus and/or bursts of star formation.

Radio galaxies—Many radio sources observed in the 1950s were identified with luminous elliptical galaxies, now called radio galaxies. Many of these show twin radio-bright lobes up to three Mpc across centred on the galaxy. The radio emission is produced by energetic particles moving through magnetic fields. Radio lobes are thought to result from jets of energy emitted by the galaxy, sometimes visible in the optical as well as the radio. Galaxies with the largest lobes are giant ellipticals. The size of lobes implies that the nucleus has been active at least 10–50 million years.

Quasars—The first *quasars (quasi-stellar radio sources) were discovered in the 1960s as 'radio galaxies with no galaxy'—they appeared point-like (stellar) in optical images. Only their enormous redshifts indicated they were not Galactic stars but extremely distant and luminous sources. **Radio-quiet quasars** (also known as **quasi-stellar objects** or QSOs) were subsequently found by searching for objects that appeared stellar on optical photographs but emitted too strongly at *infrared or ultra-violet wavelengths. Radio-quiet quasars outnumber radio-loud quasars by a factor 10–30. Both radio-loud and radio-quiet QSOs are now thought to be variants of the same type of object, and the term 'quasar' now includes QSOs. Deep imaging in the 1980s revealed that quasars are bright nuclei of galaxies outshining the surrounding stars, and quasars are now regarded as more powerful versions of Seyfert nuclei.

BL Lac objects—These are quasars with very weak emission lines and are possibly the most extreme form of AGN. Their luminosity can fluctuate widely within a few days (one doubled in brightness within three hours), and their radio and optical emissions are strongly *polarized. Quasars with similar variability but stronger emission lines are known as **optically violently variable** (OVV) quasars. These and BL Lacs are collectively known as **Blazars**. All known Blazars are radio loud. Blazars are the most luminous known objects in the Universe. If their radiation was emitted isotropically then they would have luminosities $\sim 10^{14}$ times that of the Sun. In fact, it is likely that most of their radiation is focused along two narrow beams, and so their total luminosities are likely to be considerably less than this.

It is now thought that all AGN are probably manifestations of the same basic process: gas giving up potential energy as it falls into a black hole (see Fig. 1). Any angular momentum of the infalling gas will result in

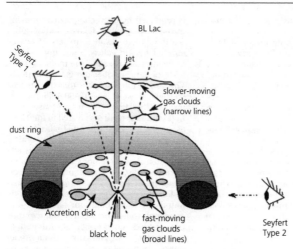

Fig. 1. Schematic diagram illustrating that how we classify an AGN depends on our viewing angle relative to the jet and to the dust ring surrounding the central black hole.

Labels in figure: BL Lac; Seyfert Type 1; jet; slower-moving gas clouds (narrow lines); dust ring; Accretion disk; black hole; fast-moving gas clouds (broad lines); Seyfert Type 2

the formation of an **accretion disk**. Viscosity causes the gas disk to slowly spiral in, heating up, and radiating away potential energy. Magnetic fields are pulled inwards with the flow of hot, ionized gas. Close to the central black hole, the magnetic field may be strong enough to channel twin jets of relativistic plasma moving out along the spin axis at close to the *speed of light. Observed along this jet, we will see a BL Lac. Close to the jet axis, we observe the *Seyfert 1* broad-line region. Closer to the plane of the disk, we observe the Seyfert 2 narrow-line region through a ring of obscuring dust.

It appears that most galaxies have a massive black hole at their centre, and it is conjectured that the AGN phenomenon is closely linked to the process of *galaxy formation and evolution.

http://www.astronomynotes.com

adiabatic perturbations

Adiabatic perturbations are a particular type of perturbation to the *density of material in the *Universe. Such perturbations are proposed to grow under *gravitational instability to form *galaxies and other *cosmic structures. An adiabatic perturbation is a special type whereby the overall perturbation is shared amongst all the different types of material (e.g. *cold dark matter, *baryons and *photons) in the Universe, so that each species has a maximum density at the same

location in space. The simplest models of *cosmological inflation give rise to adiabatic perturbations, and they give the best description of current observational data.

See DENSITY PERTURBATIONS.

age of the Universe

In the hot big bang cosmology (*see* 'Overview'), the *Universe began as a *singularity and has been expanding ever since. In this picture, the age of the Universe is the time from the *big bang to the present.

Before *Hubble demonstrated the expansion of the Universe in 1929, it had often been assumed that the Universe was infinite in age (although if this were the case, one would not expect the night sky to be dark—*see* OLBERS' PARADOX.) Several attempts were made, however, to estimate the age of the Earth, which clearly sets a lower limit on the age of the Universe.

In 1644, the Hebrew scholar and Vice-Chancellor of Cambridge University, Dr John Lightfoot, estimated the age of the Earth from Biblical genealogies. He calculated that the world was created at the equinox in September of 3298 BC. In 1650, Archbishop James Ussher arrived at a revised creation date of Sunday 23 October 4004 BC. Both of the above estimates were regarded by their proposers as exact.

It was not until the 18th century that scientific estimates were made of the age of the Earth. By assuming that the Earth had

Table 1. The estimated age of the Universe as a function of the time the estimate was made.

Proponent	Year of Estimate	Age in Years	Method
Lightfoot	1644	4,942	Biblical genealogy
Ussher	1650	5,654	Biblical genealogy
Buffon	1760	75,000	Geology
Lyell	1831	240 million	Fossil studies
Kelvin	1897	20–400 million	Earth cooling rate
Rutherford and Boltwood	1907	500 million – 1.64 billion	Radioactive dating
Hubble	1929	2 billion	Uniform expansion rate
Gamow	1947	2–3 billion	Uniform expansion rate
Bok	1952	1–10 billion	Globular clusters
Jiminez et al.	1996	13.5 ± 2 billion	Oldest globular clusters
WMAP team	2003	13.84 ± 0.14 billion	CMB fluctuations

cooled from a molten state, the French mathematician Georges Buffon estimated an age of 75 000 years in 1760. In 1831, Charles Lyell arrived at an age of 240 million years based on fossils of marine molluscs. In 1897, William Thomson (Lord Kelvin) used improved knowledge of heat conduction and radiation to improve the calculation of the Earth's cooling rate, concluding the Earth was between 20 and 400 million years old. Between 1905 and 1907, Ernest Rutherford and Bertram Boltwood determined the age of rocks and minerals from measurements of radioactive decay. They found ages of 500 million years to 1.64 billion years. Subsequent work found rock samples as old as 4.3 billion years.

With the discovery of the expansion of the Universe in 1929, attempts were made to estimate the age of the Universe itself. By assuming a uniform expansion rate, Hubble in 1929 arrived at an age of about two billion years, using his estimate of the *Hubble constant of 500 km/s/Mpc. This value for the Hubble constant was later shown to be overestimated by a factor of 5–10, hence Hubble underestimated the age of the Universe by the same factor. In 1947, George *Gamow reanalysed Hubble's data on *Cepheid variables and concluded that the '...expansion must have started about two or three billion years ago. More recent information leads, however, to an estimate of somewhat longer time periods.'

Meanwhile, methods for estimating the ages of astronomical objects were being developed, in particular the ages of star clusters in the *Milky Way Galaxy. The oldest known objects in the Milky Way are the globular clusters. These are compact stellar systems containing $\sim 10^5$ stars. Being formed early, these stars contain few heavy elements (10–100 times fewer than found in the Sun),

were formed at a similar time, and clump together in a compact region of space. Comparison of the observed colours and luminosities of these stars with stellar evolution models yields an estimate for the age of each cluster. In 1952, Bart Jan Bok estimated that galactic clusters must be between one and ten billion years old. The largest source of uncertainty is that in the distance to the cluster. The most recent estimates of the oldest clusters give ages 13.5 ± 2 Gyr.

This history of estimates of the age of the Universe is shown in Fig. 1. On his tongue-in-cheek webpage given below, Donald Simanek has pointed out that as time goes on the age of the Universe not only increases, but does so at an accelerating rate.

Presently, the best estimate of the age of the Universe comes from analysis of data from the *WMAP satellite. The expansion history, and therefore the estimated age, depends on other *cosmological parameters, and so one arrives at slightly different age estimates depending on what, if any, other data are included, and which parameters one allows to vary. Using WMAP data alone, the WMAP team arrive at an age of 13.73 ± 0.16 Gyr. Adding complementary data from other *CMB experiments, galaxy *redshift surveys, *supernova surveys and *Lyman alpha forest observations gives an age 13.84 ± 0.14 Gyr. Note that these ages are consistent within the stated uncertainties and are also consistent with the ages of the oldest known globular clusters (13.5 ± 2 Gyr).

In a *cyclic Universe, which undergoes a sequence of big bangs and *big crunches, the measured age of the Universe would correspond to the time since the most recent big bang.

http://www.lhup.edu/~dsimanek/cutting/ageuniv.htm

AGN
See ACTIVE GALACTIC NUCLEI.

Aitoff projection
See HAMMER–AITOFF PROJECTION.

ALMA (Atacama Large Millimetre Array)
ALMA, the Atacama Large Millimetre Array, is an international project to build a large array of antennae that will study the *Universe in unprecedented detail in the millimetre and sub-millimetre parts of the *electromagnetic spectrum. The array will consist of 50 movable dishes, each of twelve metres diameter, situated at an altitude of 5000 metres in Zona de Chajnantor, in Chile's Atacama Desert (Fig. 1). This high-altitude site, which is also extremely dry, minimizes contamination from atmospheric water vapour. The receivers will be sensitive to frequencies between 30 and 950 GHz, corresponding to wavelengths between 1 cm and 0.3 mm. The array may be extended to cover an area of 14 km, allowing observations at a resolution of around 0.15 arcseconds even at the lowest frequencies. The array is expected to become operational in 2009.

The primary science goals of ALMA are to detect and study the earliest and most distant *galaxies, and to examine star and planet formation. Galaxies, stars, and planets are all formed within dust-obscured regions. The dust particles absorb *ultra-violet radiation from young objects and are consequently warmed to a temperature of around 3–100 degrees kelvin, whence they radiate energy in the sub-millimetre and far *infrared parts of the spectrum. By measuring the characteristic spectral lines of carbon monoxide (CO), ALMA will be able to determine galaxy *redshifts out to redshift $z = 10$.

http://www.eso.org/projects/alma/

alternative cosmologies
Most cosmologists today believe in the hot *big bang cosmology (HBBC; *see* 'Overview'), in which the *Universe began as an infinitely hot and dense initial *singularity which subsequently expanded and cooled and which is now governed by *Einstein's equation of gravitation. In HBBC, it is postulated that (1) the measured *redshifts of distant objects are due to expansion of the Universe, and (2) this expansion is governed by the general theory of *relativity. Alternative cosmologies disagree with one or both of these postulates, by trying to avoid the initial singularity and/or by modifying Einsteinian gravity.

Steady-state cosmology
See STEADY-STATE COSMOLOGY.

Spinning Universe
A spinning Universe was proposed in 1949 by Kurt Gödel, in which distant parts of the Universe rotate with respect to our local *inertial frame. Observations of distant

Fig. 1. Artist's impression of ALMA dishes in the Atacama Desert, Chile.

*galaxies and *quasars, however, show that distant sources are non-rotating to within 2.5×10^{-4} arcseconds per year. Observations of the *cosmic microwave background also put severe constraints on rotating Universes, and on anisotropic Universes in general.

Plasma cosmology
Plasma cosmology, developed in the 1960s by Hannes Alfvén, asserts that the *intergalactic medium is a plasma, an ionized gas in which electrons have become separated from the positively charged nuclei of their atoms. The large-scale structure of the Universe is then dominated by electromagnetic, rather than gravitational forces as in HBBC. Plasma cosmology also claims that the extended *galaxy rotation curves of spiral galaxies are due to electromagnetism, thus avoiding the need for *dark matter in galaxies. However, it is far from clear that plasma cosmology can explain away other evidence for dark matter.

While plasma cosmology can explain the temperature and near-isotropy of the cosmic microwave background (CMB), it cannot explain the distinctive features seen in the CMB power spectrum which were predicted by HBBC. Largely for this reason, few cosmologists today take plasma cosmology seriously.

Tired light
The tired light effect, suggested by Fritz *Zwicky in 1929, postulates that cosmological *redshift is due to photons 'tiring', or losing energy, as they travel through space. Since the energy E of a photon is given by the *Planck constant h times its frequency v, $E = hv$, light would get redder as it tires.

There are three main problems with this theory: (1) tired light does not obey the *time dilation observed in the light curves of distant *supernovae, (2) the cosmic microwave background radiation would not be expected to follow a *black-body spectrum, (3) distant images would tend to look more blurred than nearby images, contrary to observations by the *Hubble Space Telescope.

Intrinsic redshifts
The astronomer Halton Arp has claimed that the observed redshifts of galaxies and quasars may be intrinsic to the objects themselves, rather than due to expansion of the Universe. This claim is based largely on observations of pairs of objects which are close together

on the sky, but which have very different redshifts. Arp asserts that these pairs are physically associated, and thus at similar distances, meaning that redshift cannot be proportional to distance, and that quasars are relatively nearby and faint.

Arp's critics argue that the vast majority of quasars with measured redshifts are not correlated with nearby galaxies, and that a few *apparent* associations of distant quasars and nearby galaxies are likely to occur by chance. Strong evidence in support of the cosmological redshift of quasars comes from observations of the *Lyman alpha forest in the spectra of high-redshift quasars. This 'forest' of absorption lines in quasar spectra is explained in HBBC as being due to intervening clouds of neutral hydrogen. It has no natural explanation in Arp's theory.

modified Newtonian dynamics (MOND)
See MODIFIED NEWTONIAN DYNAMICS.

Varying gravitational constant
In 1937 the physicist Paul Dirac noted that the ratio of the size of the visible Universe (given by the speed of light c times the age of the Universe t) to the size of an *electron r is $ct/r \sim 10^{40}$. This is a dimensionless number since it is the ratio of two sizes, and is thus independent of whatever units one chooses to measure distance by. Dirac pointed out that this is remarkably similar to the ratio of the electrostatic to gravitational forces between an electron and a *proton $e^2/Gm_p m_e \sim 10^{40}$. Here e is the electron charge, G is Newton's gravitational constant and m_p and m_e are the masses of a proton and electron respectively. Both electrostatic and gravitational forces decline with the square of distance, and so this ratio is independent of the separation of the particles. Dirac thought it unlikely that two such large, dimensionless numbers should be approximately the same by chance, and he proposed a 'large numbers hypothesis' which said that Newton's gravitational 'constant' should actually vary with time in order to maintain the equality of these ratios. In other words, as the Universe grew and the first ratio increased, G in the denominator of the second ratio would simultaneously decrease.

Measurements from lunar laser ranging, experiments by the Viking Explorer satellite, and observations of pulsars (rapidly rotating neutron stars) have placed upper limits on the fractional rate of change of G of order one part in 10^{11}, ruling out Dirac's large numbers hypothesis. The coincidence of these large numbers may be explained by the *anthropic principle; if they were much different then stars would not be able to fuse hydrogen in their cores.

Narlikar, J.V. and Padmanabhan, T. *Annual Review of Astronomy and Astrophysics, 39, 211,* 2001.

angular-diameter distance

The angular-diameter distance is a way of expressing the distance to a cosmological object. It is *not* the true distance, but rather the distance that the object appears to have, based on its size, if one were ignorant of the cosmological effects of the expansion of the *Universe and of *spatial curvature.

The effect of the expansion is, counter-intuitively, to increase the apparent size of an object. This makes it appear closer and hence reduces the angular-diameter distance as compared to the true distance. This is because the angular-diameter distance refers to objects of a fixed *physical* size, and since at earlier times the Universe was smaller such an object is effectively larger as compared to the Universe at that time. Indeed, if you take an object of fixed physical size and move it so that it can be seen at further and further distances from you, at first it would appear to decrease in apparent size as expected, but eventually (typically beyond a *redshift of one or so) its apparent size would begin to increase again. However there are no objects in the Universe which have a fixed physical size and exist across a wide range of distances, so in practice this counter-intuitive behaviour is not observed.

Spatial curvature may either increase or decrease the angular-diameter distance, with a spherical geometry focusing light rays and making objects appear larger, and a hyperbolic geometry having the opposite effect.

Closely related to the angular-diameter distance is the *luminosity distance, which is the distance objects appear to have based on their observed brightness. The two relations are connected by the **reciprocity relation**, that their ratio is $D_L/D_A = (1 + z)^2$ where z is the redshift. This relation is independent of spatial curvature, and is a simple consequence of the conservation of *photon number.

The combination of luminosity and angular-diameter distance evolution leads to a strong reduction in the *surface brightness of distant objects; redshifting reduces the overall luminosity of an object, while the angular diameter effect spreads that luminosity over a larger area.

The most important current application of the angular-diameter distance is to *cosmic microwave background anisotropies, where it gives the characteristic scale of the peak structure seen in the *radiation angular power spectrum. This provides the best evidence that the Universe has a flat spatial geometry. In the future it is expected that the angular-diameter distance will be measured across a range of epochs using the phenomenon of *baryon oscillations in the galaxy distribution.

anisotropies

The word 'isotropy' means that things appear the same in every direction, and hence 'anisotropy' refers to a situation where they don't. The standard description of the global *Universe, the *Robertson–Walker metric, assumes both isotropy and *homogeneity, meaning that observable quantities should depend neither on our position in the Universe nor on the direction we look.

In practice isotropy does not hold precisely in our Universe (which would be a pretty boring place if it did), as we see stars and galaxies in particular positions on the sky. It holds only as an approximation valid on the largest scales. The distribution of material in the Universe can therefore be said to be anisotropic, and one of the main areas of cosmological investigation is in understanding the nature of the anisotropies and the physical processes behind their development.

By far the most widespread use of the word is *cosmic microwave background anisotropies, which play a key role in observational *cosmology. In fact, the cosmic microwave background radiation is very nearly isotropic (apart from a *cosmic dipole associated with the motion of the Earth relative to the microwave background), corresponding to radiation with an average temperature of 2.725 kelvin. However, superimposed on this uniform background are small deviations, of order tens of microkelvin in magnitude. These anisotropies are believed to indicate irregularities in the distribution of material in the Universe when the cosmic microwave background was formed, those irregularities later growing under the action of *gravitational instability to become *galaxies and other *cosmic structures.

anthropic principle

The anthropic principle comes in several forms, all centred around the idea that the *Universe has to have its observed properties (for instance the existence of stars and galaxies) because if it didn't then we wouldn't be here to observe it. It was first introduced by British physicist Brandon Carter in the early 1970s to explain certain coincidences in the Universe concerning large numbers. It has subsequently been used in several ways by astronomers, and is seen as of increasing importance in the field of *cosmology.

It comes in several forms, the most important of which are the weak anthropic principle and the strong anthropic principle. The weak anthropic principle is relatively uncontroversial, stating that whenever we consider the observed properties of the Universe, we should ask whether the outcome is conditioned by the requirement that it can host life. In their book *The Anthropic Cosmological Principle*, John Barrow and Frank Tipler define it as 'The observed values of all physical and cosmological quantities are not equally probable but they take on values restricted by the requirement that there exist sites where carbon-based life can evolve and by the requirement that the Universe be old enough for it to have already done so.' The stricture of 'carbon-based' is necessary because we want a Universe that supports our kind of life, not any old kind of life.

The strong anthropic principle makes the much bolder statement that the only types of Universe permitted to exist are those which do contain life during at least part of their existence. This idea is open to the criticism of being unscientific, as it is not falsifiable. Ordinarily the strong anthropic principle is avoided by cosmologists.

In fact, anthropic reasoning had already been used before it had a name, by Fred *Hoyle in the early 1950s in the study of creation of heavy elements in stars. Hoyle discovered that stars were simply unable to make heavy elements, and specifically the carbon atoms which he himself was made of, based on the understanding of atomic physics at the time, because the interactions were too inefficient. Since heavy elements clearly did exist (and were essential to life's existence),

Hoyle eventually decided that there must be an unknown reaction which was much more efficient in making heavy elements than those known. And indeed, atomic physicists discovered that such an interaction (the carbon resonance) did exist. Thus Hoyle was able to predict a property of physical interactions through his argument that heavy elements had to be created somehow.

The anthropic principle tends to create strong opinions; for a long time the astrophysics group at the Fermi National Accelerator Laboratory had a rule that if any visiting speaker mentioned 'the A word' the seminar would end immediately. There is no doubt that some of its stronger forms defy the normal rules of scientific process, but at least in its weak form it appears reasonable and amounts to little more than an essential *observational selection effect that needs to be taken into account in interpreting certain kinds of observations. Critics of the weak anthropic principle usually say that it is too obvious to merit such a grand name.

In cosmology, the anthropic principle has proved powerful in conjunction with theories of *cosmological inflation, particularly those of a self-reproducing type. In those inflation models, physical laws may vary within different regions of the very large scale Universe (scales much larger than our *observable Universe). For instance, in other distant regions parameters such as the *proton mass or the *cosmological constant may differ from those in our region of the Universe. Naturally, one is led to the idea that while all sorts of possible physical laws might happen around the Universe, the laws applying within our observable Universe are inevitably going to be of a type capable of supporting life.

The self-reproducing inflation scenario can be coupled to an important recent idea in *superstring theory, known as the *string landscape. This suggests that while string theory has a unique set of governing equations, the set of stable states of the system is extraordinarily large (estimates range from 10^{100} to 10^{1000}), with each possible state corresponding to different physical laws. This is sometimes drawn as a landscape with complex topography, with the deep valleys representing possible physical states. The self-reproducing inflationary cosmology is capable of moving regions of the Universe between these states, and so within the entire Universe, all possible states appear at different locations. Anthropic arguments can then potentially be used to select amongst them.

This leads to the key cosmological success of the anthropic principle, which is to explain the observed value of the cosmological constant, which is something that has proven baffling under other approaches. The argument was first made by Steven Weinberg (who previously won a *Nobel Prize for his part in establishing the *standard model of particle physics), and later refined by Alex Vilenkin, George Efstathiou and others. All of the different states in the string landscape correspond to different values of the cosmological constant, and there are so many of them that all possible values can be considered to exist somewhere in the Universe. But large values, much larger than that observed, cause the Universe to start accelerating before any galaxies have formed, causing *structure formation to cease before it has begun. And very low values, much smaller than that observed, are also disfavoured, since the observed value of the cosmological constant should be picked randomly amongst those values small enough to satisfy the structure formation constraint. The anthropic principle therefore can be used to justify both why the cosmological constant is non-zero, and why it is so small compared to natural expectations from particle physics. Nevertheless, this explanation remains controversial.

Barrow, J. D. and Tipler, F. *The Anthropic Cosmological Principle*, Oxford University Press, 1985.

http://arXiv.org/abs/astro-ph/0511774

anti-matter

According to fundamental particle theory, every particle has associated with it an *antiparticle, which can be thought of as its opposite. The particle properties are discussed under the anti-particle topic; this item concerns anti-matter—the overall distribution of anti-particles—in our *Universe.

Our Universe is, overwhelmingly, made from matter rather than anti-matter. We know that this is true because matter and anti-matter coming together annihilate, with all of the available energy converted into particles such as *photons of light which would readily be visible. Matter–anti-matter annihilations are extremely violent events. Even a nuclear explosion releases only around 0.1% to 1% of the available mass–energy of its

fuel, corresponding to energy released by a reconfiguration of the particles in the atomic nuclei (merging of hydrogen to helium in a fusion explosion, or splitting of uranium in a fission bomb). Matter–anti-matter annihilation has available the entire mass–energy of the colliding items; the accidental collision of a coffee cup with an anti-coffee cup could release comparable energy to a one tonne nuclear weapon. Thankfully there are no anti-coffee cups around to try this experiment with.

Another characteristic that ensures the ready detectability of such annihilations is that the decay products have an extremely well-defined characteristic energy. For instance, if an *electron and its anti-particle, a positron, were to annihilate, the standard route for this to occur is to produce two photons of light. However the simultaneous requirements of energy and momentum conservation dictate the final energies of each photon; for a non-relativistic collision (the electron and positron moving with relative speed much slower than the speed of light) each photon leaves with energy equal to the mass–energy of one of the original particles, equal to 511 kilo-electronvolts in *particle physics units. This corresponds to the gamma-ray part of the *electromagnetic spectrum, a very high energy indeed, and matter–anti-matter annihilations would give a characteristic signature of lots of photons with this particular energy. No such signal is seen in the largest mergers of objects taking place in our Universe, collisions of galaxy *clusters, indicating that they are made of the same type, presumably matter.

On even larger scales, there is no direct evidence that, for instance, some of the most distant galaxies we see are not made of anti-matter (what we see are the photons, but photons are an example of a particle which is its own anti-particle, so they cannot be used to decide). However galaxies collide and merge all the time, and there would have to be boundaries between regions of matter and anti-matter, where annihilations would take place, and these are not seen. Further, no one has been able to conceive of a mechanism which would allow a large-scale segregation of matter and anti-matter. The working hypothesis is therefore that our Universe is made from matter.

That is not to say that there is absolutely no anti-matter in the Universe. Particle accelerators such as CERN in Switzerland are able to manufacture anti-matter via high-energy particle collisions. However the amount they have been able to create so far is only a minuscule fraction of a gram. More interestingly, high-energy particles arrive at Earth from distant sources, known as *cosmic rays. These are particles which have been accelerated by the magnetic field of our Galaxy, and almost certainly are also arriving from distant *galaxies having been accelerated by their magnetic fields. For every ten thousand or so *protons that arrive at the Earth, there is around one anti-proton. However, these anti-protons are not taken as evidence that there is that much anti-matter in the Universe; rather, this modest abundance is compatible with the level of creation expected when normal cosmic rays impact on other particles within the Galaxy (called secondary production).

Presuming that the anti-matter density is negligible, that means that there is an imbalance between matter and anti-matter. Indeed, this has been measured quite accurately as we now have a good idea of how much normal matter there is, for instance from measurements of the *cosmic microwave background anisotropies. This is an observation that cosmologists would like to explain from first principles, and theories purporting to do this are called *baryogenesis. So far however they have been rather unsuccessful.

While the present Universe is largely bereft of anti-matter, that's not thought to have been true of the early Universe. In the hot young Universe particles such as electrons and protons are expected to have been in *thermal equilibrium. In this situation frequent particle collisions will create anti-particles, and a balance between creation and annihilation set up. Indeed, theories of the early Universe predict that there was almost, but not quite, a perfect balance between the amounts of matter and anti-matter when the Universe was young (much less than one second old). Electrons and positrons, for instance, would continually be annihilating into photons, and being recreated when those photons interact.

anti-particles

In the classification of fundamental particles embodied by the *standard model of particle physics, an important symmetry is that every particle has a corresponding anti-particle (except for a rare few particle types which are their own anti-particle). *Anti-matter means material made up from anti-particles.

The properties of anti-particles are essentially the same as the particles, in particular having the same mass, except that any charges they have are reversed. For instance, the anti-particle corresponding to the electron, known as the **positron**, has exactly the same mass as the electron but carries a positive electric charge rather than a negative one. *Quarks carry a rather more obscure charge known as their colour charge, and anti-quarks carry an anti-colour charge. Obviously, then, only uncharged particles could be their own anti-particle, and *photons of light are an example.

One of the most important properties of an anti-particle is that it can annihilate with its particle, destroying both. Because of energy conservation they cannot disappear completely; most commonly the annihilation creates high-energy photons of light which carry the energy away. For example, electron–positron annihilation creates two gamma-ray photons, whose energy matches the rest mass energy of the original particles.

Curiously, matter and anti-matter exhibit very similar properties but are *not* exactly equivalent. At a very subtle level, there are differences in the decay rates of some particles, as compared to their anti-particles, when mediated by the weak nuclear interaction (one of Nature's *fundamental forces). This is best established for a type of particle known as a K meson (or kaon), which is a light unstable particle formed from a quark and an anti-quark and easily produced in particle accelerators, whose different decay rates for particle and anti-particle was discovered in 1964. The technical name for this process is **charge-parity (CP) violation**, meaning that physical processes are not invariant under a simultaneous switch of the charges of particles and a reflection in space, a combination which swaps particles for anti-particles. Interestingly, adding an additional transformation to make CPT, where T stands for reversing the direction of time, is believed to be preserved in all fundamental interactions; the same processes that differentiate matter and anti-matter are therefore also sensitive to whether time is running backwards or forwards (i.e. processes which violate CP must also violate T).

That matter and anti-matter are different at a subtle level appears odd, but presumably is responsible for the Universe being overwhelmingly comprised only of matter. If there were no differences between the two, there would be no possible way to create this imbalance and it would just have to be said that the *Universe started out with the imbalance already in place. Theories which attempt to exploit CP violation to explain the observed matter–anti-matter asymmetry are known as *baryogenesis.

APM galaxy survey

The Automated Plate Measurement (APM) Galaxy Survey was one of two surveys (the other being the Edinburgh–Durham Southern Galaxy Catalogue) carried out in the late 1980s and early 1990s which digitized hundreds of sky survey plates in order to compile a catalogue of *galaxies over a very wide area of sky. Previous galaxy surveys had been based on either human inspection of survey plates (e.g. the *Zwicky Catalogue) or on machine scans of only a small number of plates.

The survey plates were taken with the United Kingdom Schmidt Telescope (UKST) in Australia. Each plate covers an area of six by six degrees on the sky and the plates are spaced on a uniform grid at five degrees apart. The one degree of overlap between neighbouring plates proved vital for calibrating the individual plates onto a uniform *magnitude system.

The plates were individually scanned using the Automated Plate Measuring (APM) facility, a high-speed laser microdensitometer, in Cambridge, England. Essentially, the machine scans a laser beam over the plate and measures the amount of light transmitted as a function of position. Astronomical sources are detected in real time as the plate is scanned and the position, integrated density, size and shape of each source is recorded. Galaxies and stars are distinguished by their shapes: galaxies are fuzzy, extended sources whereas stars are much more concentrated.

Altogether, 269 blue-sensitive (b_J) photographic plates were scanned in this way and the overlap between neighbouring plates used to put all plates on a common photometric system. The completed survey contains roughly three million galaxies with magnitude in the range $17 < b_J < 20.5$, covering more than 6 000 square degrees of sky. When the survey was completed in the mid 1990s, this was by far the largest volume of the *Universe that had been surveyed.

Fig. 1 shows the APM galaxy distribution as a density map in equal area projection on the sky, centred on the South Galactic pole. Each pixel covers a small patch of sky 0.1 degrees on a side, and is shaded according to the number of galaxies within the area: where there are more galaxies, the pixels are brighter. Galaxy clusters, containing hundreds of galaxies closely packed together, are seen as small bright patches. The larger

Fig. 1. Distribution of galaxies in the APM Galaxy Survey.

The APM Galaxy Survey
Maddox et al

elongated bright areas are *superclusters and filaments. These surround darker *voids where there are fewer galaxies. The small empty patches in the map are regions that have been excluded around very bright stars, satellite trails, and plate defects.

A colour version of this map, available from the website below, is colour-coded according to the apparent magnitude of the galaxies in each pixel: fainter galaxies are shown as red, intermediate are shown as green, and bright are shown as blue. The more distant galaxies tend to be fainter, and also show less clustering, and so the map has a generally uniform reddish background. The more nearby galaxies tend to be bright, and are more clustered, so the more prominent clusters of galaxies in the map tend to show up as blue.

The large area surveyed enabled the first reliable measurement of the clustering of galaxies on angular scales of two degrees and larger. It was found that the clustering strength on these large scales was significantly higher than predicted by the then-popular *critical density *cold dark matter model, in which the mass of the Universe is dominated by slow-moving, non-*baryonic dark matter. This was one of the first results to rule out this model and to hint that the matter *density parameter of the Universe is less than unity.

The survey also enabled measurements of the number–magnitude counts of galaxies over a wide range of magnitudes, which showed that galaxies have evolved significantly in luminosity since redshift $z \approx 0.2$.

The APM Galaxy Survey also formed the source catalogue for two *redshift surveys: the 1 in 20 sparse-sampled Stromlo-APM Redshift Survey and the *two degree field galaxy redshift survey.

http://www.nottingham.ac.uk/~ppzsjm/apm/apm.html

arXiv (pron. archive, despite the spelling)
The arXiv is a massive repository of research papers, mainly on physics topics including astrophysics and *cosmology, accesible freely over the World Wide Web (WWW). It was established as an automated system in 1991 by Paul Ginsparg, originally focused on a small community of researchers working in *superstring theory who had been interchanging research results via an e-mail distribution list set up by Joanne Cohn. It rapidly expanded into new subject

areas, with the astrophysics archive *astro-ph being established in early 1992. It currently contains eleven separate physics categories, some including many subcategories, plus a few other categories in mathematical sciences.

The main site http://arxiv.org is presently hosted by Cornell University, and has at least 18 mirror sites around the world. It operates largely automatically but with some elements of human intervention. Papers are submitted by approved users via a WWW interface, and each weekday the new batch of papers becomes visible as abstracts on the WWW. The full papers can then be downloaded in various formats including postscript and PDF. The arXiv is searchable in a wide variety of ways, including author and keyword searches and cited reference searches.

The arXiv revolutionized the way research scientists operate, by vastly increasing the speed of dissemination of results. Almost all working cosmologists routinely submit their papers to the arXiv as well as to refereed journals, in many cases doing so in advance of formal publication. Papers can be updated by their authors, for instance to match any changes suggested by peer review journals. Original fears that the arXiv would undermine peer review and lead to an uncontrolled free-for-all with consequent loss of confidence in paper quality have proven ungrounded, with essentially all arXiv papers continuing to be separately published via peer-reviewed journals.

ArXiv main site: http://arxiv.org

astroparticle physics
See PARTICLE ASTROPHYSICS.

astro-ph (pron. astro-pee-aitch)
Identification by an astro-ph number indicates that a paper is available within a massive free archive of physical science papers, known as the *arXiv. Astro-ph is one of many subject classifications within this archive. Presently around 700 papers are added to the astro-ph archive each month by scientists worldwide, usually in advance of them appearing in refereed journals.

Papers submitted in years up to 2006 were identified by a subject classification followed by a seven-digit number, such as 'astro-ph/0302207'; the first two digits indicate the year, the second set of two the month, and the final three a sequential number in that month. This particular astro-ph paper was the first announcement of results by the *WMAP satellite, subsequently published in the journal *Astrophysical Journal Supplements*. From April 2007 onwards, due to the volume of submissions threatening the 1000 papers per month limitation of that scheme, the scheme was altered to read arXiv:0704.0001, being year, month, and a four-digit identifier. The new scheme does not explicitly identify the subject class, which is held as separate data, with papers in all classifications lying within a single sequential numbering each month. It is anticipated based on current growth that a fifth digit may become necessary around 2020.

Some *cosmology articles, at the authors' choice, appear under a separate subject category gr-qc, standing for General Relativity and Quantum Cosmology, or under high-energy physics headings hep-ph and hep-th (phenomenology and theory).

Papers can be updated by the authors at any time, indicated by a version number 'v1', 'v2' etc. appended to the identifier. All versions remain archived and accessible, in order that updates cannot be used to wrongly establish priority.

Atacama Large Millimetre Array
See ALMA.

axions
Axions are hypothetical fundamental particles which are a candidate to be the *cold dark matter in the *Universe. Their original motivation came from particle physics, where they were introduced by Robert Peccei and Helen Quinn in 1977 to explain why particles and their corresponding *anti-particles have such similar properties (technically known as the strong CP problem). It was later realized that if their mass lay within a particular range, then they would be predicted to have a significant cosmological *density, perhaps greater than that of all the *baryons, making them a *dark matter candidate.

Particle physics theory gives little guidance as to the expected axion mass, if they do exist, but astrophysics and *cosmology strongly constrains it. To be a viable dark matter candidate, the mass of an individual axion must be very light, around one billionth of the mass of an *electron. Despite this light mass, axions are produced with negligible kinetic energy and constitute cold dark matter, rather than hot dark matter as would usually be expected of such a light particle.

Methods have been proposed to detect axions in the laboratory using magnetized microwave cavities, but are excruciatingly difficult to carry out and have yet to show any evidence for the existence of axions. On purely theoretical grounds, it is thought that it would be a considerable coincidence for axions to have the right mass to be dark matter, and so typically they are less favoured as a dark matter candidate than more massive particles known as *WIMPs.

baryogenesis

Our *Universe possesses a *matter–anti-matter asymmetry, containing essentially no *anti-matter. Attempts to explain this are collectively described as theories of baryogenesis, literally the creation of *baryons in the Universe (the main baryons being *protons and *neutrons). Baryogenesis is therefore the creation of the basic building blocks from which we are made. At present there are no compelling theories of baryogenesis, and the ambitions of theorists are restricted to attempting to estimate the order of magnitude of the baryon density in the Universe.

The Sakharov conditions

While it is perfectly possible that the Universe was born with the matter–anti-matter asymmetry in place, cosmologists prefer to believe that it arose as a consequence of physical processes taking place in the Universe, which they can then attempt to understand. In 1967, Soviet physicist Andrei Sakharov formulated the three conditions necessary to permit baryogenesis, now known as the Sakharov conditions. They are

- Violation of baryon number.
- C and CP (charge and charge–parity) violation.
- Departure from *thermal equilibrium.

The first condition is obviously necessary; if baryon number is conserved by all possible interactions then it cannot change during the lifetime of the Universe. The *standard model of particle physics does conserve baryon number (with one subtle but important exception discussed below), so that we can be confident that baryon number does not change except during the Universe's earliest stages. But various extensions, including *grand unification scenarios, do not conserve baryon number. Experimentally, however, baryon number violation has never yet been observed.

The second condition is expressed in a rather technical way, but in essence means that the physical laws applying to matter and anti-matter cannot be precisely the same. If they were, then any change in baryon number due to interactions of matter would be cancelled by an equal and opposite effect for anti-matter interactions. Even the standard model of particle physics violates C and CP, but only in a very subtle and small way.

Thermal equilibrium is characterized by a physical system in balance, where any interactions between particles (e.g. a particle and its anti-particle annihilating to form two photons) are matched by interactions going in the opposite direction (in this case two photons colliding and forming the particle–anti-particle pair). Systems typically evolve towards thermal equilibrium, and the Universe is believed to have been in thermal equilibrium for most of its early evolution. Baryon number cannot be generated in thermal equilibrium, because even if there are interactions generating baryon number, they will be cancelled out by the equal number of interactions proceeding in the reverse direction. The main way in which the Universe departs from thermal equilibrium is during *phase transitions, and it is usually assumed that the baryon number must be generated at one of these in the *early Universe.

Grand-unified baryogenesis

The simplest models of baryogenesis assume that the baryon asymmetry was created extremely early in the Universe's history, at the grand unification phase transition. This is a popular choice because grand unified theories automatically include baryon number violation, can readily satisfy Sakharov's second condition, and the phase transition ending the unification phase naturally leads to some departure from thermal equilibrium.

At such an early epoch, baryons would be in thermal equilibrium with photons, and both particles and anti-particles would be present with roughly the same numbers as photons. Only later in the Universe's evolution will it cool sufficiently that those particles and anti-particles will annihilate. Presently the *baryon-to-photon ratio is roughly one baryon for every billion photons. The aim of baryogenesis is to set in place a small imbalance during the early Universe, so that for every billion photons, there are a billion anti-baryons but a billion-and-one baryons. Later, the billion anti-baryons annihilate with a billion baryons, and leave over the residual one baryon per billion photons that we want.

Many theories of this type have been devised, but none are yet seen as compelling and there is certainly no understanding of why the outcome is one-in-a-billion, rather than say one-in-a-million or even one-in-a-hundred.

Electro-weak baryogenesis

An alternative approach uses a discovery that the standard model of particle physics does have one way of violating baryon number, via a type of interaction known as a **sphaleron** interaction. The physics is too complicated to explain here, but the upshot is that baryon number violation is, in principle, possible at the *electro-weak phase transition because of these interactions. They are however so weak that the observed asymmetry can only be created if the electro-weak phase transition leads to violent departures from thermal equilibrium, and it is not believed that it does.

Kolb, E. W. and Turner, M. S. *The Early Universe*, Addison–Wesley, 1990 [Technical].

baryon oscillations

'Baryon oscillations' refers to the behaviour of *baryons during the early stages of gravitational collapse. 'Baryons' is shorthand for *protons and *neutrons, the particles we are predominantly made of. Baryons are an important component of the *Universe, presently comprising roughly 4% of the total *density.

Before the epoch of *decoupling, baryons interact strongly with the *photons of radiation in the Universe, this interaction generating a *pressure force. In regions of high *dark matter density, gravitational attraction tries to draw the baryons in, but the pressure offers a restoring force and the baryons oscillate, alternately compressing and rarefacting in the gravitational potential well formed by the dark matter. These oscillations are directly analogous to sound waves, and so are often called acoustic oscillations.

While the radiation dominates, the sound speed in this medium is very high, being $c/\sqrt{3}$ where c is the speed of light. *Density perturbations on larger scales lead to oscillations on a longer timescale, the oscillation period being proportional to the length scale. Perturbations on the scale of the *Hubble length will have period equal to the age of the Universe at that time, and so will just be beginning to oscillate.

The oscillations continue until decoupling takes place, after which the photons no longer interact with the baryons. This removes the pressure support, and gravitational collapse of the baryons can begin in earnest, ultimately ending with the formation of *galaxies, galaxy *clusters and so on.

The oscillation phase has important observational consequences. They are most prominent in the *cosmic microwave background (CMB) anisotropies; decoupling in effect takes a photograph of the pattern of oscillations at that instant, which we can view today by receiving them. The *power spectrum of the CMB shows strong oscillatory features whose origin is the baryon oscillations, known as *acoustic peaks.

A smaller effect can also be seen in the galaxy power spectrum; regions where the oscillations were at maximum compression at the time the pressure support was removed enhance structure formation, and those at a rarefaction suppress it. The effect in the galaxy power spectrum is also sometimes referred to as **baryon wiggles**. The signal is very small, but was detected for the first time in 2005 by the *Sloan Digital Sky Survey and the *two degree field galaxy redshift survey. An important challenge for future surveys will be to measure the characteristic scale of the oscillations in the distant Universe as well as the nearby Universe, as comparing the two measures the expansion rate of the Universe and hence potentially constrains the properties of *dark energy.

baryon-to-photon ratio

Two of the constituents of the present *Universe are *baryons and *photons of light. The number of baryons has been unchanged at least since the Universe was one second old, and perhaps from even earlier, the last plausible epoch at which it might have changed being the *electro-weak phase transition. The number of photons is dominated by those forming the *cosmic microwave background (CMB), and has not changed significantly since its creation at *decoupling. The ratio of the number of baryons to the number of photons is therefore a universal constant, provided we average over a large enough volume to fairly probe the large-scale structure of the Universe.

The value of this constant is measured quite accurately in the *standard cosmological model, observations having strongly constrained both the baryon and photon densities. The ratio is that there is only one baryon for approximately every two billion photons. This is independently confirmed, with somewhat greater uncertainty, by cosmic *nucleosynthesis.

Explaining the tiny value of this ratio, which quantifies the *matter–anti-matter asymmetry in the Universe, has not been possible so far. Theories which attempt to do so are known as *baryogenesis scenarios.

Despite the numerical domination of the photons, in terms of energy *density it is the baryons which presently dominate, as the mass–energy of each individual *proton or *neutron is so much higher than the energy of each CMB photon.

baryons

Baryon is a collective noun referring to any particle made up of three *quarks. Quarks are one of the fundamental building blocks of Nature, and the physics of fundamental interactions favours groupings of three quarks; for instance, both a *proton and a *neutron are, at a more fundamental level, made up of three quarks. Baryons are therefore of considerable interest in astrophysics and *cosmology, as they are the main ingredients from which stars, planets, and indeed human beings are made. While other baryons exist, and can be created in particle accelerators, only the proton and neutron can be stable and so they are the only baryons which arise naturally in the *Universe. Baryons have corresponding *anti-particles sometimes known as anti-baryons, of which the anti-proton and anti-neutron are examples. However there is very little *anti-matter in the Universe.

The *density of baryons in the Universe can be measured in two main ways. One is from the abundance of the different light elements formed during *nucleosynthesis when the Universe was about one second old, and the second is from the pattern of *cosmic microwave background (CMB) anisotropies. Both indicate that baryonic matter makes up only about 4% of the total matter in the Universe, being subdominant to both *dark matter and *dark energy.

Within the *standard model of particle physics, believed to be valid in the present Universe, there is a quantum number called **baryon number B** which is a conserved quantity, and equals the number of baryons minus the number of anti-baryons. This implies that the lightest baryon, the proton, must be absolutely stable, as there is no lighter particle it could decay to while preserving baryon number. It is thought that baryon number is not preserved by interactions beyond the standard model, such as those in *grand unification theories. These permit protons to decay, albeit with an extremely long lifetime.

Cosmologists would like to have a theory which predicts why the baryon density in the Universe is 4% of the total, and name such theories *baryogenesis. Sadly, such theories are highly speculative and uncertain, and do not amount to even an order-of-magnitude understanding of the observed density.

Cosmologists sometimes abuse the word 'baryon' by also including *electrons within it, to the dismay of particle physicists. This is convenient because the Universe is charge neutral, so that there is one electron for every proton. With electrons having a mass of only around a two-thousandth that of the proton, if the total mass is all one is interested in then the electrons are usually not worth bothering about separately.

Bayesian inference

Bayesian inference is a system of logical deduction, enabling conclusions to be drawn about rival theories from observational data. Its basic precept is the assignment of probabilities to all quantities of interest, which are then updated each time new observational data become available. The probabilities are manipulated by various tools, the most important being a theorem known as Bayes' theorem. It was originally derived by the Reverend Thomas Bayes in 1763, after whom the entire inference system is now named. However the system of Bayesian inference was established mainly by work of Sir Harold Jeffreys in the middle of the 20th century, culminating in his classic monograph *Theory of Probability*.

The main rival viewpoint to Bayesian inference is sampling theory (also known as frequentist statistics), which relies on the concept of an infinitely repeatable experiment. There being only one *Universe, this seems particularly inappropriate in a cosmological context, and the majority of cosmological analyses are carried out from a Bayesian viewpoint. The main branches of Bayesian analysis used in *cosmology are *parameter estimation, in which probabilities are assigned to particular values of parameters, and *model selection, where probabilities are assigned to choices of the set of parameters to be varied in a fit to data.

While elegant in its mathematical formulation, Bayesian inference is often complicated in its actual implementation, requiring large-scale computations to evaluate the probabilities generated by what may be quite complex models and datasets. It is only relatively recently that computing facilities have become powerful enough to make general calculations possible, usually using what are called *Monte Carlo methods whereby the probability distributions are sampled in a random manner. Typical cosmological calculations are supercomputer-class problems.

Jeffreys, H. *Theory of Probability*, Oxford University Press, 1961 (3rd edition) [Technical].

Gregory, P. *Bayesian Logical Data Analysis for the Physical Sciences*, Cambridge University Press, 2005 [Technical].

Bianchi classification

The usual versions of the hot big bang cosmology assume that the *Universe is on average both homogeneous (i.e. all points are equivalent) and isotropic (i.e. all directions are equivalent). This assumption is compatible with all current observations on large scales. Nevertheless, it is interesting to consider more general models. For models which are homogeneous but anisotropic, a complete classification of possible relativistic models exists and is known as the Bianchi classification, after Italian mathematician Luigi Bianchi who devised it around the late 1890s.

*Isotropy can be violated while retaining homogeneity. Most simply, the Universe might be expanding at different rates in different directions, so that there is a separate *scale factor associated with each of the three principal coordinate directions. Another possibility is that there might be a global *rotation of the universe. In all, the

Bianchi classification consists of nine different types, usually indicated by Roman numerals Bianchi I, Bianchi II, ... Bianchi IX. However some of these classes have sub-classes, and some of the classes are contained as special cases of the others. Note that Bianchi's classification predated Einstein's discovery of *general relativity in 1915, and came from a branch of mathematics known as Lie algebras. The modern version of the Bianchi classification as applied to cosmology was first considered by Abraham Taub and later fully formalized by Engelbert Schücking and Christoph Behr, in the middle of last century.

A particular application of the Bianchi models is to study the general approach to a homogeneous cosmological singularity, without making the assumption of isotropy. The Bianchi IX case has seen particularly extensive study, and is sometimes known as the **Mixmaster Universe**. In an expanding Universe it always has two dimensions expanding and one contracting, but from time to time the dimensions switch roles so that each takes a turn as contracting against the general expansion. It has been shown that this model behaves chaotically in the vicinity of a past or future *singularity.

bias parameter

The bias parameter is a way of quantifying how strongly a particular class of object is *clustered relative to mass *density perturbations. In most simple models of *biased galaxy formation, it is assumed that the bias is both linear and scale-independent, in other words that the amplitude of the galaxy *correlation function $\xi_g(r)$ is a simple multiple of the mass density correlation function, $\xi_m(r)$, and that this multiple is independent of scale r.

Formally, the bias parameter b relates the fluctuations in galaxy density δ_g to fluctuations in mass density δ_m, where $\delta_g = b\,\delta_m$. Here the mass density fluctuation δ_m is the relative deviation of the local density ρ at a particular location from the mean density $\bar{\rho}$, $\delta_m = (\rho - \bar{\rho})/\bar{\rho}$. The galaxy density fluctuation δ_g is defined similarly as the relative deviation from mean galaxy density. Since correlation functions depend on *pairs* of quantities, for example the excess number of galaxy pairs at separation r, the amplitude of the correlation function scales as the *square* of the bias parameter. We thus have $\xi_g(r) = b^2\xi_m(r)$ for a class of galaxy with bias parameter b. For most galaxies, bias is thought to be larger than one on small scales (so that galaxies are 'positively biased') and to tend towards a constant value close to one on large scales.

The bias parameter b may be estimated by measuring the amplitude of redshift-space distortions due to galaxy *peculiar velocities. One can show that on large scales, where *cosmological perturbation theory is valid, the amplitude of the distortion is proportional to $\Omega_{\mathrm{m}}^{0.6}/b$, where Ω_{m} is the matter *density parameter. If the value of Ω_{m} is determined from independent observations, such as *cosmic microwave background anisotropies, then b can be calculated. The difficulty with this procedure is that non-linear effects dominate on small scales, thus requiring one to estimate the galaxy correlation on very large scales ($r \gg 10h^{-1}$Mpc), where the signal-to-noise ratio is small.

The *relative* bias between different types of objects is easy to determine by comparing their correlation functions (see Fig. 1). Elliptical, red, and luminous galaxies are more strongly clustered, and so have a larger relative bias parameter, than spiral, blue, and faint galaxies.

It is now becoming possible to estimate the bias parameter more directly, by comparing the galaxy correlation function $\xi_{\mathrm{g}}(r)$ directly with the mass correlation function $\xi_{\mathrm{m}}(r)$ determined from *gravitational lensing.

biased galaxy formation

Astronomers commonly use *galaxies to map out the large-scale structure of the *Universe, much as one might map out the human population from aerial photographs of the Earth's surface taken at night. However, there is no guarantee that galaxy light is directly correlated with mass, just as a large concentration of streetlights does not necessarily follow a high population density.

The possibility that galaxies do not faithfully trace the distribution of mass in the Universe is referred to by astronomers as a *biased* galaxy distribution. If one assumes that bias is both linear and scale-independent, it may be represented by a single number b known as the *bias parameter. In this case, the fluctuations in galaxy density δ_{g}, and fluctuations in mass density δ_{m}, are related by $\delta_{\mathrm{g}} = b\,\delta_{\mathrm{m}}$. Here the mass density fluctuation δ_{m} is the relative deviation of the local density ρ at a particular location from the mean density $\bar{\rho}$, $\delta_{\mathrm{m}} = (\rho - \bar{\rho})/\bar{\rho}$. The galaxy density fluctuation δ_{g} is defined similarly as the relative

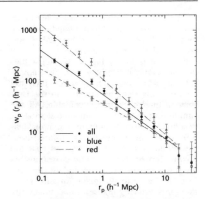

Fig. 1. The galaxy correlation function measured from the Sloan Digital Sky Survey. Note that the correlation function for red (typically elliptical) galaxies is larger on small scales but declines more quickly than that for blue (typically spiral) galaxies. Clearly, blue and red galaxies are biased in different ways, and the bias depends on scale.

deviation from mean galaxy density. Note that a local density is determined by averaging over a chosen volume: generally the smaller the volume (or scale) one chooses, the larger the amplitude of fluctuations from mean density. The bias parameter b is thus in general scale-dependent, although the scale-independent assumption appears to hold well on large scales, $r \gg 10h^{-1}$Mpc.

It is known that elliptical galaxies tend to reside in *clusters of galaxies, whereas spiral galaxies tend to be more isolated. Elliptical galaxies are thus more strongly biased than spirals. The relative clustering of different types of galaxies may be inferred by comparing their *correlation functions (see Fig. 1).

There is an ongoing debate as to whether elliptical galaxies preferentially form in high-density regions, or whether evolutionary processes such as galaxy mergers and tidal interactions cause galaxies in these high-density environments to appear as ellipticals. Is it 'nature or nurture' that leads to the high relative bias of ellipticals? The answer, while still unknown, is likely to incorporate both intrinsic galaxy formation processes ('nature') as well as evolution ('nurture').

big bang

The big bang, also known as the **initial singularity**, refers to the instant of formation of the *Universe. There is not actually a whole lot to say about it, as no one has the slightest idea what might have happened at the big bang, nor whether one actually happened at all. The term 'big bang' was coined by Fred *Hoyle and intended to be disparaging.

If, and it is a very big 'if', Einstein's theory of *general relativity were to remain valid to arbitrarily high densities, then the idea of a big bang as the start to the Universe is pretty much unavoidable. All the simple homogeneous cosmological models, described by the *Friedmann equation, feature an instant of creation at infinite density. While it might be suspected that the assumption of *homogeneity is responsible for this, Stephen Hawking and Roger Penrose were able to prove several **singularity theorems** showing that an initial singularity was inevitable under a much wider range of conditions (though the conditions allowing *cosmological inflation violate the assumptions of the theorems, rendering them invalid, an extension of them shows that inflation cannot have continued indefinitely into the past). Their theorems also demonstrated the existence of a singularity at the centre of a *black hole.

A common confusion is to ask 'what happened before the big bang?'. This question actually makes no logical sense, because both space *and* time were created at the big bang. In fact the question is essentially equivalent to the question 'what is north of the North Pole?', which everyone agrees is a question that doesn't actually have an answer.

Nevertheless, modern physics holds that it is very unlikely that general relativity is valid to infinite density; at high densities it should instead be replaced by a *quantum gravity theory such as *superstring theory. Whether or not these theories might still predict a big bang is unknown, though one of their motivations is the hope that they will avoid the singularities that general relativity suffers from. Various studies of *non-singular cosmologies have been made.

The possibility that our Universe might have *extra dimensions beyond the familiar three spatial ones further clouds the waters as concerns whether there was or wasn't a big bang. For example, the *ekpyrotic Universe theory suggests that what we see as a big bang is actually a collision of higher-dimensional objects known as *branes, and doesn't correspond to a singularity at all.

The phrase 'big bang cosmology', or 'hot big bang cosmology', is typically used to indicate the aftermath of the putative big bang, not the

bang itself. There is no doubt that the present Universe does indeed look as if it underwent such an explosion in the past, even if the existence of an actual bang is in doubt. *See also* 'Overview: the hot big bang cosmology'.

big crunch

The big crunch is one possible endpoint of the Universe's evolution. If the *density of material in the *Universe is sufficiently high, its gravitational attraction is enough to overcome the expansion of the Universe, and a recollapse phase is then inevitable. In the simplest models (those without a *cosmological constant), the collapse is simply the expansion in reverse. According to the equations of *general relativity, after a finite time the Universe recollapses completely, giving a *singularity of the same form as the original *big bang. Such a singularity corresponds to the end of *space-time itself, and everything in the Universe is unavoidably forced together (a science fiction fallacy would be to stay in a spaceship watching the final big crunch take place, but that would only be possible if one could pilot the spaceship out of space-time itself).

It is not presently known whether or not our Universe will end in a big crunch; the possible fates of the Universe are discussed in *future of the Universe. If it does, it is unclear whether it would actually reach a singularity, because the well-established laws of physics are believed to become inapplicable before the singularity is reached.

One possibility that has emerged recently from ideas in fundamental physics is that the big crunch may not end the Universe after all. Theories such as *superstrings predict that space-time may have more spatial dimensions than the three we are familiar with. Ordinarily the extra dimensions are very small compared to our three and so far have proven unobservable, but they may become important as the Universe approaches the singularity. Indeed, what we perceive to be a singularity may be an illusion of us not considering all the space-time dimensions, in which case the Universe may be able to continue evolving beyond the apparent singularity. One implementation of this is the *cyclic Universes model, in which the singularity corresponds to the collision of two *branes, one of which corresponds to our Universe. Once the two branes have passed through each other, the collapse is converted to expansion with the collision corresponding to a new big bang.

In this scenario there might be an infinite sequence of expansion and recollapse, each punctuated by a big crunch/big bang.

big rip

The big rip is one possible ultimate evolution of the *Universe. Rather more dramatic even than the *big crunch, it occurs if the Universe undergoes an infinite amount of expansion in a finite time. That sounds unlikely, but would be a consequence of the Universe being dominated by a particular type of *dark energy known as *phantom dark energy, which has the property that its *density increases as the Universe expands. As the density grows, it causes an increase in the expansion rate of the Universe, which in turn forces a faster increase in the dark energy density. This feedback is so strong that the expansion rate and dark energy density both become infinite after a finite time.

At present it is not known whether phantom dark energy is a possibility consistent with all physical laws, but we can say that it is permitted by current observations. The simplest model of dark energy is Einstein's *cosmological constant, whose density remains constant as the Universe expands. This gives predictions consistent with observations, and since observations can only ever have a finite accuracy that implies that models where the dark energy density is either increasing or decreasing can also be compatible with observations, provided the variation is sufficiently slow.

The current dark energy density is low enough that its impact can be felt only on the largest astronomical scales. However if phantom dark energy exists then, as its density grows, the associated repulsive gravity effect will make itself felt on shorter and shorter length scales. Eventually, it will overcome the attractive gravitational force that holds the stars within galaxies together, sending them flying apart. Our own *Milky Way Galaxy will be torn apart about 60 million years before the Universe's eventual demise. Within a few months of the end, planets will be stripped away from stars destroying the Solar System, and in the final 10^{-19} of a second even atoms will be torn apart.

The big rip is quite an intimidating prospect, especially as a suitably advanced civilization would be able to recognize well in advance that it was underway, but would be powerless to do anything to stop it. However, even if the physical laws in our Universe are indeed those which lead to a big rip, we can relax in the knowledge that it won't happen for at least twenty billion years!

black-body spectrum

A black-body is a perfect absorber and emitter of radiation. When placed in a bath of radiation, a black-body alters its temperature until an equilibrium is reached between the radiation absorbed and emitted. The radiation emitted (or absorbed) by a black-body is known as black-body radiation, and its properties are characterized by a single parameter, the temperature of the black-body. It is therefore an idealized concept, but nevertheless a useful one because many objects do act more or less as perfect black-bodies. In particular, the *cosmic microwave background radiation pervading the *Universe has a spectrum very close to a black-body with a temperature of 2.725 degrees above absolute zero.

The term black-body spectrum refers to how the energy of the black-body radiation is distributed across the *electromagnetic spectrum. This has a universal form shown in Fig. 1. The axes are somewhat complicated as the basic quantities being plotted, frequency f along the x-axis and energy *density ϵ up the y-axis, have been multiplied by factors of the temperature T and by various physical constants to make them dimensionless. [For completeness, c is the speed of light, h is Planck's constant, and k_B is Boltzmann's constant. The temperature T must be measured in kelvin.] The important points to note are firstly that there is a well-defined peak, meaning that most of the energy is emitted

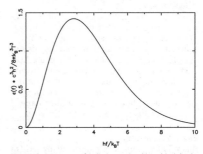

Fig. 1. The spectrum of radiation emitted by a black-body has a universal form known as the Planck spectrum. The x-axis shows the frequency of radiation f, and the y-axis the corresponding energy per unit volume ϵ. In each case the axes are normalized according to various fundamental constants of nature and to the temperature.

around a particular characteristic frequency $f \sim k_B T/h$, and secondly that the peak frequency is proportional to the temperature, so that hotter black-bodies emit radiation at higher frequencies.

An example of an emitter close to a black-body is the Sun. Its surface temperature is around 5 800 kelvin, and it emits its light mainly in the visible part of the electromagnetic spectrum. Colder objects, such as a central-heating radiator at perhaps 60 degrees Celsius, emit in the *infrared and we perceive the radiation as heat. The Universe itself, much colder still, has radiation in the *microwave part of the spectrum.

As well as the distribution of energy at different frequencies, the total energy of the black-body radiation is of interest. Adding up (integrating) the energy at each frequency, one finds that the energy density of the radiation is proportional to the temperature to the fourth power; double the temperature and the energy density will increase by a factor $2^4 = 16$.

As the Universe expands, the black-body form of the spectrum is preserved but the temperature decreases in inverse proportion to the size of the Universe, given by the *scale factor. At a microscopic level, this can be viewed as the *redshifting of individual *photons. The energy density of radiation therefore reduces as the fourth power of the scale factor. Three of these powers can be identified as the reduction in density due to the volume becoming larger, and the final power due to the reduction in energy of the photons due to redshifting.

black holes

Black holes are amongst the most enigmatic objects predicted by modern physics. They are regions of *space-time where the *curvature is so strong that even light is not able to escape. The edge of the black hole is known as the *event horizon, and anyone foolish enough to cross it can never return. Black holes are implicated in a wide range of astrophysical phenomena, including many of the most energetic events in the *Universe.

Black hole history

Black holes were first conceived by British mathematician John Michell in 1783. Using Newton's theory of *gravity, he reasoned that a sufficiently dense object would have an escape velocity greater than the *speed of light, which was by then known to be finite.

Such objects were also studied by French polymath Pierre-Simon Laplace soon afterwards.

However proper understanding had to wait for Albert *Einstein's greatest discovery—that gravity could be described as the curvature of space-time induced by the mass of objects. His theory of *general relativity, introduced in 1915, enabled the first proper descriptions of black holes to be made. Remarkably, it was only one year later that Karl Schwarzschild (who was shortly to die in the trenches during the First World War) produced the first mathematical description, which we now call the Schwarzschild black hole. It describes a black hole which is perfectly spherical and non-rotating (technically known as a 'static' black hole).

However it was to be many decades before the meaning of Schwarzschild's calculation would be properly understood. Indeed, the term 'black hole' was not introduced until 1969, by American physicist John Wheeler. (The contemporary Soviet astrophysics school instead used the terminology 'frozen star'.) The known black hole types were extended by the discovery of solutions to *Einstein's equation describing charged black holes—known as Reissner–Nordstrom black holes—and rotating black holes which are known as Kerr black holes after Roy Kerr who derived their mathematical properties in 1963.

From the late 1960s onwards, it became apparent that black holes, far from being a purely mathematical device, might actually have a role to play in a range of astrophysical phenomena. While by definition the black hole itself cannot be seen, the strong gravitational field in the vicinity of the black hole can lead to a range of energetic phenomena.

Black hole physics

According to the equations derived by Schwarzschild, the event horizon of a black hole has a radius r given by

$$r = \frac{2GM}{c^2} \,.$$

Here G is Newton's gravitational constant, c is the speed of light, and M is the mass of the black hole. This radius is known as the Schwarzschild radius. To get a feel for what this means, we can try out some sample values.

Mass	Characteristic object of this mass	Radius
6×10^{24} kg	Earth	1 cm
2×10^{30} kg	Sun	3 km
2×10^{36} kg	A million Suns	3 million km

We can see immediately why black holes are rather removed from everyday experience. In order to have a black hole the mass of the Earth, we would have to compress it to be only two centimetres across, a fearsome density far beyond our ability to create. To do the same with the Sun, it has to collapse to become just a few kilometres across.

However, the Schwarzschild formula shows that the radius of a black hole is directly proportional to the mass. This is at odds with what one normally expects—if you have a material of constant density then to double its radius you would need to have eight (i.e. 2^3) times as much mass. Because, instead, we need only twice as much mass to double the radius, this means that the larger a black hole is, the smaller its density.

In the vicinity of the event horizon, objects experience a powerful type of gravitational force known as a **tidal force**. Say you are orbiting just outside a black hole with your feet pointing towards it. Because gravity obeys an inverse square law, the strength of gravity felt by your feet is actually larger than that of your head, as your feet are closer to the hole. On the Earth such a difference is totally negligible, but near a black hole this difference in force can be so strong as to pull materials apart.

The tidal forces are one way in which you might work out that you are near a black hole. The event horizon is quite a sinister location, because there is nothing special about the space-time at that particular place and you could quite easily blunder across it. Strong tidal forces might alert you to the existence of a nearby black hole, though the larger the black hole the smaller the tidal effect (because of the lower density) you would need to look out for.

Within the black hole, at its centre, lurks a *singularity. This is a place where the curvature of space-time becomes infinite and the laws of physics, at least as given by general relativity, break down. Once you cross the event horizon, a collision with the singularity becomes inevitable and will happen after a finite amount of time. (If it's any consolation,

you will be shredded by tidal forces before being crushed to infinite density at the centre.)

The breakdown of physical laws at the singularity has worried many physicists, and led to Roger Penrose proposing in 1969 the **cosmic censorship hypothesis** which states that all singularities must be hidden behind event horizons, so that their effects cannot escape to influence the entire Universe. This hypothesis remains unproven after more than 30 years. A rival strand of thought is that modern fundamental physics theories such as *superstring theory may modify the predictions of general relativity in the high-density regime, so as to avoid the formation of the singularity.

If you are far enough from a black hole, its gravitational field is no different from that of any other body. For instance, if the Sun was replaced tomorrow by a black hole of the same mass, the Earth would quite happily continue around its orbit as if nothing had happened (it would soon get pretty cold though). The idea that black holes 'suck things in' is a common misconception; in fact black holes are so small that it would be quite a challenge to manage to fall into one.

However, the strong gravitational field very near a black hole does lead to dramatic effects; for instance light passing very close to the event horizon follows a strongly curved trajectory, an effect known as *gravitational lensing. This is a version of the light bending that Arthur *Eddington used to verify general relativity, but while at the surface of the Sun the bending is so small as to be only just measurable, near an event horizon light can be sharply bent and indeed can even spin around the black hole a few times before emerging in a completely unexpected direction. Spotting these distortions to the directions of light rays coming from distant stars would be another way of diagnosing the existence of a nearby black hole.

Quantum black holes

According to general relativity, nothing can ever escape a black hole. Remarkably, this conclusion is overturned when the effects of *quantum mechanics are included. In 1975, British physicist Stephen Hawking showed that, according to quantum theory, black holes will radiate energy and evaporate. This phenomenon is known as **Hawking radiation** or **Hawking evaporation**.

A simple picture of why evaporation can take place comes from the quantum vacuum. Whereas in non-quantum theories a vacuum is simply nothing, according to quantum theory the vacuum is a seething mass of particles coming into and out of existence, violating energy conservation for the brief periods allowed by Werner Heisenberg's uncertainty principle. Suppose a particle–*anti-particle pair comes into existence. In the vicinity of a black hole, one may cross inside the event horizon, and hence be unable to find its partner to annihilate. Instead, the other particle can escape to infinity as radiation. Energy conservation can be satisfied by extracting gravitational potential energy to create the rest mass–energy of the escaping particle.

Hawking showed that the emitted radiation matched that of a perfect black-body radiator, and that the temperature is related to the black hole mass by

$$T = \frac{\hbar c^3}{8\pi G k_B} \frac{1}{M}.$$

This utterly remarkable formula contains all four of the most fundamental constants of Nature—the speed of light c, the gravitational constant G, Planck's constant $\hbar$ measuring quantum effects, and Boltzmann's constant k_B relating temperature to energy.

This formula shows that the temperature is higher the *lighter* the black hole is. Indeed, for any large black hole mass, e.g. the mass of the Sun, the temperature is so low that the amount of radiated energy is minuscule (in fact, less than the black hole will absorb from the *cosmic microwave background). Only for miniature black holes, known as *primordial black holes, might the Hawking evaporation process operate at a significant level.

The endpoint of black hole evaporation remains controversial, with dispute as to whether the eventual evaporation will reveal the central singularity to the outside world, or whether the singularity will also disappear in the final burst of evaporation. A further possibility is that the evaporation process might leave behind a stable relic particle.

Black hole astrophysics
There are two main types of black holes whose actual existence in the *Universe is more or less certain.

The first is black holes formed at the end of the lifecycle of very massive stars, perhaps upwards of ten solar masses. These end

their lives in dramatic *supernovae, where the core collapses under gravitational attraction and the released gravitational potential energy violently expels the outer layers of the star. Stars of mass greater than about two solar masses undergo such core collapse, the lighter ones leaving behind a *neutron star. However there is a maximum permitted mass for a neutron star of around three to four solar masses, and those stars whose collapsing core exceeds this have no alternative other than complete collapse to a black hole. The existence of such black holes can be inferred from cases where the black hole is in a binary star system with a conventional star, from which it draws material. Cygnus X-1, the first such black hole to be discovered, is the most famous example, emitting strongly in the *X-ray part of the *electromagnetic spectrum.

The second class is *supermassive black holes, explored in detail under that heading. For example, our own *Milky Way Galaxy has a massive black hole at its centre (known as Sagittarius A*) with a mass of several million solar masses; modern *infrared observations can directly observe stars orbiting around the black hole. Supermassive black holes are also believed to be the central energy source for *active galactic nuclei and *quasars.

Black hole cosmology
It is not known whether black holes play a role in cosmology. In principle black holes are a very good *cold dark matter candidate, as they fulfil all the requirements and would remove the need to speculate about new types of particles such as *WIMPs. However no convincing explanation has ever been offered as to how those black holes would be formed.

Black holes which form during the early stages of the Universe, and which might have a much lower mass than the astrophysical black holes discussed above, are known as primordial black holes. There is presently no evidence firmly indicating that primordial black holes formed.

Stannard, R. *Black Holes and Uncle Albert*, Faber and Faber, 2005 [Popular].

Shapiro, S. L. and Teukolsky, S. A. *Black Holes, White Dwarfs and Neutron Stars*, John Wiley and Sons, 1983 [Technical].

Chandrasekhar, S. *The Mathematical Theory of Black Holes*, Oxford University Press, 1998 (first published 1983) [Very technical].

Boomerang experiment

The Boomerang experiment, a rather contrived acronym standing for Balloon Observations Of Millimetric Extragalactic Radiation And Geophysics, was the most important of the second-generation experiments to study *cosmic microwave background (CMB) anisotropies in the era in between their discovery by the COBE satellite and the precision measures by the *WMAP satellite. It was the first experiment, in 1998, to map out the structure of the anisotropies on angular scales from several degrees down to 10 arcminutes, and in doing so provided the first convincing single-experiment evidence that the *geometry of the Universe was close to spatial flatness. A subsequent flight of the experiment also measured the *polarization of the CMB.

Boomerang was a predominantly Italian–North American collaboration, which flew sensitive CMB detectors on long-duration balloon flights at Antarctica. Microwave observations are difficult at ground level due to microwave emission within the atmosphere (known as water vapour lines), and accurate experiments need to circumvent this. Satellite experiments such as COBE and WMAP completely solve this problem by leaving the Earth's atmosphere entirely, but at great expense. The alternatives are to seek high-altitude sites or to use balloons or rockets to get above most of the atmosphere. Antarctica is a particularly attractive site as it has high-altitude plains and the cold conditions minimize atmospheric water vapour.

Boomerang successfully exploited weather patterns at Antarctica, where natural wind currents can carry a balloon on a wide 'orbit' around the pole taking roughly two weeks to return to the starting point. This long-duration ballooning strategy enabled far more data to be taken than any other contemporary experiment. After an initial test flight in North America, two full flights were undertaken at Antarctica, one in 1998 targeting CMB anisotropies, and one in 2003 also measuring CMB polarization.

Fig. 1 shows the experiment shortly before the 1998 launch. The mission performed

Fig. 1. The Boomerang experiment shortly prior to launch at Antarctica in 1998. The experiment itself is on the right-hand side of the image, and the balloon on the left is already partially inflated. A dormant volcano, Mt Erebus, is seen in the background.

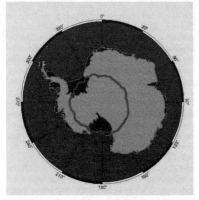

Fig. 2. The path followed by the balloon in the 1998 flight. The complete circuit took about 10½ days.

optimally, maintaining a height of about 35 kilometres (above all but 1% of the atmosphere) and circling Antarctica over a period of 10½ days. The path followed is shown in Fig. 2.

The Boomerang map of the microwave sky from the 1998 data is shown in Fig. 3. It covers a few percent of the sky, at an angular resolution of around 10 arcminutes. To uncover the full power of such a map a statistical analysis of the irregularities is needed, via the *radiation angular power spectrum as described in the CMB anisotropies topic. However one can perhaps already see by eye

that there is a characteristic size of the small spots, whose width is significantly larger than the instrument resolution and hence a true property of the microwave sky. This size corresponds to an angular scale where the anisotropies are particularly strong (in power spectrum terms, this corresponds to a peak, known as the first *acoustic peak), and is measured to be about one degree. This measurement directly probes the spatial geometry of the *Universe, which determines the angular size of distant objects through the *angular-diameter distance. The Boomerang team were able to use the data from the 1998 flight to show, in results published in 2000, that the Universe was close to spatial flatness, with the total density within about 10% of the *critical density.

A second Antarctic Boomerang flight took place in 2003, with results released in 2005. As well as improving the measurements of the microwave temperature anisotropies, the instrument had been upgraded to measure polarization of the microwave background, and their results were the most precise available up to that point.

boson stars

Boson stars are a hypothetical type of compact stellar object. Fundamental particles are commonly broken up into two classes, *bosons whose spin (a fundamental property possessed by all particles) is an integer, and *fermions whose spin is a half-integer (i.e. one-half, three-halves, etc.). The spin determines the statistical properties of the

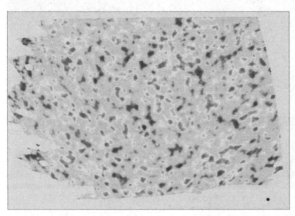

Fig. 3. The cosmic microwave background as imaged by Boomerang, covering a sky area of approximately 100 degrees across and 30 degrees high. The black dot on the lower right indicates the resolution of the map. The most prominent small features have size approximately one degree.

particles; for instance only fermions obey the Pauli exclusion principle, which forbids fermions from occupying the same quantum state. *Electrons, *protons, and *neutrons, which are the primary constituents of normal stars, are all fermions. *Photons of light are the simplest example of a boson, others being the famous Higgs particle of the *standard model of particle physics, and the particle thought to have driven a period of *cosmological inflation.

While all known types of star, including white dwarfs and neutron stars, are made of fermions, there has been speculation about whether it is possible to form stellar-like objects from bosonic particles. By far the most interesting possibility would be to construct stars from the *dark matter; the favoured assumption is that the dark matter is made of fundamental particles, and given that it seems as likely as not that those particles might be bosons. Whether they can in practice form collapsed objects is thought unlikely but not completely excluded. Direct detection of dark matter boson stars would be quite difficult, though if they lie in the right mass range they might be observable via *gravitational microlensing.

The properties of boson stars can be studied using a similar set of stellar structure equations to those used for more conventional stars; in particular the equations are quite similar to those describing neutron stars, and result in similar properties. Boson stars would be highly dense, and like neutron stars possess a maximum mass beyond which they collapse into *black holes. Their characteristic mass depends on the mass and interactions of the bosonic particles, and could be vastly different from the typical masses of conventional stars.

A related, and highly speculative, idea is that the dark matter in our Galaxy could be in the form of a single giant boson star. If the bosons are extremely light, then the 'star' could have a huge radius, comparable to the size of a *galaxy rather than a star, while being extremely diffuse. In this model, the visible galactic disk would be embedded within the giant boson star, but the structure of the dark matter halo would be determined by the stellar structure equations rather than by infall and *virialization as normally envisaged. It is however very unclear how such bosonic halos might form and especially how they would merge in a *hierarchical structure formation scenario.

bosons

Fundamental particles can be divided into two classes, known as bosons and *fermions, according to their spins. All particles possess a spin, which in the case of a boson is given by a positive integer, and for a fermion is given by a half-integer. If the spin is zero the particles are known as scalars, if it is 1 they are vector bosons, and if higher they are tensor bosons. Bosons are named after Indian physicist Satyendra Nath Bose.

Whether particles are bosons or fermions determines how they behave *quantum mechanically. Bosons behave as indistinguishable particles, meaning that if the locations, velocities etc. of two particles are swapped, the quantum state is unchanged—you can't tell which particle is which. This is known as Bose–Einstein statistics, from which the term bosons derives. Fermions, by contrast, are distinguishable particles.

The classification of known particles made by the *standard model of particle physics separates them into these two classes, and features many particles of spin-1, which are responsible for mediating the *fundamental forces between particles. These so-called gauge bosons are the *photons which cause electromagnetism, the W and Z bosons that give the weak nuclear force, and the *gluons which give the strong nuclear force. Additionally, the standard model has a spin-0 boson known as the Higgs particle, responsible for giving mass to other particles. *Gravity is not included in the standard model, but can be considered to be mediated by bosonic particles called *gravitons, which have spin-2.

Once physics beyond the standard model is considered, there are many more proposed bosonic particles, and attempts to exploit these for *cosmology are commonplace. In particular, spin-zero bosonic particles, also known as *scalar fields, possess very interesting properties discussed under that heading. Amongst other things, they are a favoured candidate both to explain *cosmological inflation in the *early Universe and the nature of *dark energy in the present *Universe. Bosons are also a candidate to be the *dark matter in the Universe, and have even been postulated to condense into stellar-like objects known as *boson stars.

bottom-up universe
See HIERARCHICAL STRUCTURE FORMATION.

branes
Branes, short for 'membranes', are the fundamental objects of a scenario for high-energy physics known as the *braneworld.

braneworld
The braneworld is a contemporary picture of our *Universe, which speculates that our visible Universe may be confined to a three-dimensional volume which resides in a higher-dimensional space. This picture is motivated by *superstring theory and *M-theory, and if proven right constitutes a major revision of our understanding of the Universe. There is as yet no observational support for the scenario.

The brane and the bulk
The word 'brane' is short for membrane, and first appeared in the late 1980s amongst researchers working at the theoretical end of string theory. It had become apparent that as well as one-dimensional strings, it was possible to consider fundamental objects with higher dimensionality. A two-dimensional surface was known as a membrane, but since string theory postulates that there are nine spatial dimensions there is no reason to stop at two dimensions; one can have three, four, five and more dimensional objects residing in this space. Probably initiated as a rather weak pun, the jargon for such objects became p-branes, where p indicates the dimension of the object—superstrings would be 1-branes. Nowadays they are commonly just called branes. In the braneworld scenario, we are particularly interested in 3-branes, being objects with three spatial dimensions just as our Universe appears to have.

In the simplest superstring scenarios, the strings are not allowed to end and so must form loops. However this changes in the presence of a brane; the strings are still not permitted to end in empty space, but they may end on the brane, where their endpoints represent fundamental particles. Because the string ends are not permitted to detach from the brane, the particles are confined to it.

From there it is a short step to the braneworld concept. The particles represented by the string ends are not some mysterious hypothetical particles—they are the particles in the Universe that we are made of. Because they are confined to the brane, our Universe appears to us to have three space dimensions, whereas in reality there are more. The full space, in which the brane is embedded, is known as the *bulk*. This is schematically illustrated in Fig. 1, for the simplified situation of a two-dimensional Universe embedded in a three-dimensional bulk.

Not all forces are tied to the brane, however. In superstring theories of the kind where strings can end on branes, there are also loops of string which, having no ends, need not stay with the brane but rather can travel freely through the bulk. These string loops are responsible for the gravitational force. Gravity, then, is able to penetrate the bulk, and only under restricted circumstances will behave as if it too were confined to the brane. This offers the best prospect for experimental verification of the braneworld idea.

The Randall–Sundrum braneworld
The full M-theory motivated braneworld picture, sometimes called Hořava–Witten theory after its originators Petr Hořava and Edward Witten, is an extremely complicated scenario. Within this theory, based on the full eleven-dimensional M-theory, six of the space dimensions are *compactified to very small sizes in the same manner as in superstring theory, but the final extra dimension has a different status and may be larger (this is the dimension used to relate eleven-dimensional M-theory to ten-dimensional superstring theory). In order to understand its implications, a lot of work has been done in the context of a much simplified version (what physicists often call a toy model) invented by Lisa Randall and Raman Sundrum in 1999, and called the Randall–Sundrum braneworld.

In the Randall–Sundrum scenario, it is assumed that there is only a single extra dimension, giving a bulk with five *spacetime dimensions and a brane with three space and one time dimension embedded within it (just as in Fig. 1, with one extra dimension added to each). Then one can consider the distance from the brane into the extra dimension. Randall and Sundrum discovered a remarkable feature. If the brane contains mass (very reasonable since it is full of stuff), then this mass can cause a curvature in the extra dimension; this is just the basic result of Einstein's *general relativity theory that the

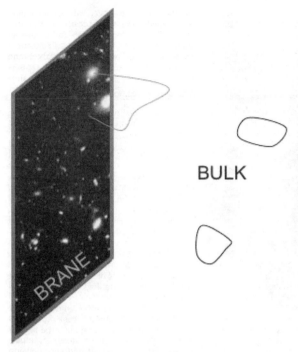

Fig. 1. A schematic of the braneworld. Our Universe (here displayed as two- rather than three-dimensional) is confined to a surface known as the 'brane'. Only the gravitational force sees the full higher-dimensional space known as the 'bulk'. The gravitational force is mediated by loops of string which travel freely in the bulk. Segments of string have their ends fixed to the brane, the endpoints representing the fundamental particles we see. The brane is infinite in extent along its two dimensions, rather than bounded as it appears in the figure.

presence of matter curves space giving grav- ity. In this case, the extra dimension devel- ops a strong curvature, so strong in fact that there is an *event horizon which means that even gravitational forces can only act a finite distance away from the brane. This naturally means that even the gravitational force is effectively confined to the brane, as required if the model is to have a chance of matching observations.

Multiple branes

The Randall–Sundrum scenario is elegantly simple, but more complex scenarios exist which contain multiple branes. A common set-up is to have two parallel branes with the bulk wedged between them like a sandwich (see Fig. 2); the full Hořava–Witten theory is of this form. In this type of scenario there is no meaning to the region outside the two branes; the entire space-time is the central bulk bounded by a brane at each edge. We would live on one of these boundary branes.

This scenario raises the possibility that interesting things, including other civiliza- tions, might reside on the second brane. If so, they could in principle be extremely near, even just tiny fractions of a metre, but as the separation is in the untraversable extra dimension it is a distance we cannot cross. It would be extremely difficult to detect their presence as only gravitational signals can be sent across the bulk. Either generating or detecting such gravitational waves is well beyond our current technical capability.

An interesting scenario is that the *dark matter in the Universe, whose presence has only been detected through its gravitational influence, might reside on a different brane. The formation of *galaxies in our Universe may then be due to the clumping of material in an entirely separate Universe. If so, then by definition the gravitational effect of dark matter is all we will ever feel, and direct dark matter detection experiments are doomed to failure.

BULK

Fig. 2. A Universe with two branes. The bulk is the slab between the two branes. There is no meaning to the region outside the branes. The branes extend infinitely along their two dimensions.

Large extra dimensions

If extra dimensions can exist, how large can they be? The usual braneworld scenarios presume that they are extremely small indeed, given for instance by the characteristic length scale of fundamental physics, the *Planck length, or at least the scale characteristic of *grand unification. But there is no reason of principle why they should be so small, and indeed it was proposed in 1998 by Nima Arkani-Hamed, Savas Dimopoulos, and Gia Dvali that their size could be as large as a millimetre. Their reasoning was that the only signature of large extra dimensions is a modification to the gravitational force, and because gravity is so weak no-one knew whether or not Newton's law of gravity was still valid on the millimetre scale. Subsequently delicate experiments have confirmed Newton's theory down to length scales rather smaller than that, but it still remains possible that the extra dimensions are significantly larger than microscopic scales.

Braneworld cosmology

Braneworld cosmology is a subject in its infancy, but one which proposes a substantial rewrite of our view of cosmology, particularly in the early Universe. Amongst the new ingredients the braneworld brings to cosmology are:

- Gravitational forces can penetrate the bulk. At low energies the structure of the theory must prevent this to avoid violation of our everyday experience of gravity, but at very high energies there can be an effect. Indeed, the Randall–Sundrum scenario predicts a high-energy modification to the *Friedmann equation describing the expansion of the Universe, which may for instance alter how *cosmological inflation takes place.
- The influence of other branes may be important. If a second brane moves relative to ours, that can modify the physics on our own brane. Indeed, Dvali and Henry Tye proposed in 1998 that such inter-brane dynamics might be responsible for inflation.
- Branes might collide, as exploited in the *ekpyrotic Universe and *cyclic Universe models. Such collisions may correspond to the big bang as experienced by observers on the brane.

All of these topics, and more, are under intense investigation at present.

A final curiosity worth mentioning is that in one viewpoint the expansion of the Universe can be interpreted as due to the motion of our brane through the bulk. One of the first lessons people are taught in cosmology is that the question 'What is the Universe expanding into?' has no answer; it is space itself which is expanding. According to the braneworld the question might not have been so stupid after all, and the answer is that the Universe is expanding into an extra dimension.

Randall, L. *Warped Passages: Unravelling the Universe's Hidden Dimensions*, Allen Lane, 2005.

bulk motions

As well as moving apart from each other as the *Universe expands, *galaxies have *peculiar velocities relative to this *Hubble expansion. If there is a large overdensity in a particular region of the Universe, due for instance to a large *cluster or *supercluster of galaxies, then nearby galaxies will tend to have a net average peculiar velocity toward the overdense region. These galaxies are then said to exhibit a bulk motion, also known as a bulk flow. Since peculiar velocities and bulk motions are caused by gravity, they can be used to trace out the distribution of *all* mass in the Universe, including *dark matter. This makes the study of bulk motions useful as a cosmological tool.

Overdense regions are known as 'attractors' and pull in nearby galaxies. Underdense regions are known as *voids and tend to repel nearby galaxies.

The earliest studied bulk motion was the 'Virgo infall', the peculiar motion of local galaxies, including the *Milky Way, toward the *Virgo cluster. It is important to emphasize that this 'infall' is only relative to the Hubble expansion, and in fact we are moving away from the Virgo cluster, only more slowly than from other points in space at a similar distance away from us. In fact, the local bulk motion is now recognized to be more complex than simple Virgo infall, and is due to a number of nearby attractors and voids, not just the Virgo cluster.

One can constrain a combination $\Omega_m^{0.6}/b$ of the matter *density parameter, Ω_m, and the factor by which galaxies are biased tracers of mass, b, by comparing observed bulk motions with those predicted from the observed galaxy density field. Observations in the early 1990s suggested a density parameter $\Omega_m \approx 1$ (in agreement with theoretical prejudice at the time), but more recent observations imply $\Omega_m \approx 0.3$, consistent with other estimates from, for example, *cosmic microwave background anisotropies.

On larger scales, it was thought until recently that galaxies within *redshifts as large as 15 000 km/s exhibited bulk motions of order 600 km/s. If true, this would have had serious implications for cosmology in a supposedly *homogeneous Universe. However, more recent measurements have shown that the net bulk motion within 6 000 km/s is consistent with zero. The cause of the discrepancy between the earlier and more recent measurements is not entirely clear, but is doubtless related to the difficulty of measuring accurate redshift-independent *distances to galaxies.

http://arXiv.org/abs/astro-ph/0003232

causality

Causality refers to the process of cause and effect in the *Universe. Because the *speed of light, acting as a universal speed limit, is finite, regions of the Universe which are well separated are unable to communicate with each other. They are said to be **causally disconnected**. During its early stages, our *observable Universe would have been made up of a large number of separate causally disconnected regions.

That the Universe is well-ordered on very large scales, for instance being very close to *homogeneity on average, is rather mysterious given that causality indicates that separate regions are unaware of each other's existence. This is known as the *horizon problem. *Cosmological inflation provides an explanation for this otherwise baffling situation.

An important application of causality is the evolution of *density perturbations during early stages of the Universe's evolution, when their scale may be much larger than the *Hubble length which acts as the limit of causal interactions. The different parts of such a perturbation are then unaware of each other's existence, and must evolve separately according to physical laws applied to the local environment. This can be used to show, for instance, that *adiabatic perturbations do not evolve while their scale is larger than the Hubble length. More generally, the evolution of large-scale perturbations can be followed using the idea that separate parts of the perturbation evolve independently, an approach known as the **separate Universes** picture since the different parts of the perturbation are each treated as if they were an isolated homogeneous Universe.

CDM

See COLD DARK MATTER.

CDM model

See COLD DARK MATTER MODEL.

Cepheid variables

Cepheid variables are a class of pulsating variable star, whose light output varies periodically in a characteristic way: a rapid brightening phase is followed by a more gradual fading: see Fig. 1 (over page). They are named after the star δ Cephei, discovered in 1784 by the 19-year-old English amateur astronomer, John Goodricke.

Cepheids are amongst the best understood variable stars. They have used up their hydrogen supply and have started fusing helium in their cores into heavier elements. The pressure of *gravity ionizes the helium by removing one of its *electrons. When ionized, helium is opaque to light, trapping the heat generated from fusion, causing the star to expand. This expansion cools the outer layers of the star (as air escaping rapidly from a bicycle tyre is cooled), and the helium ions recombine with electrons and become transparent to light, allowing the trapped energy to escape. The outer layers of the star then fall inwards under gravity, recompressing and reionizing the helium, and so the cycle repeats. The stellar surface is hotter and thus more luminous as it expands, cooler and fainter as it recontracts.

Cepheids are important in *cosmology as they possess two properties that allow us to determine distances to other *galaxies. First, they are intrinsically very luminous, ranging from a few hundred to tens of thousands times more luminous than the Sun, so that they can be seen to large distances. Second, the period of variability correlates tightly with mean luminosity, the period–luminosity relation, discovered in 1912 by Henrietta Leavitt. The least luminous Cepheids have periods of 1–2 days, while the most luminous have periods of order 100 days. The intrinsic luminosity of a Cepheid may thus be derived by measuring its period of variability. By comparing this with the apparent brightness, the distance to the Cepheid may be determined using the inverse square law.

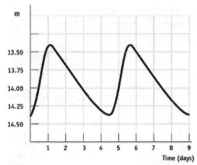

Fig. 1. The light curve for a typical Cepheid variable star, showing the star's brightness in magnitudes against time in days.

Edwin *Hubble used observations of Cepheid variables in the *nebulae M31 and M33 in the 1920s to show that they were at much larger distances than stars in our own Galaxy, and thus were 'Island Universes' in their own right. Cepheid distances were also important in deriving the *Hubble law, which states that, on average, the recession velocity of a galaxy is linearly proportional to its distance.

It is now known that there are two types of Cepheid with distinct period–luminosity relations. Type I Cepheids have a high proportion of heavy elements ('metals') in their outer layers, and are about four times more luminous for a given period than the metal-poor Type II Cepheids. The type of a Cepheid star may be determined from its spectrum; Type I Cepheids show a large number of absorption lines compared with Type II.

Although they are very luminous, and can be detected to large distances, Cepheid variables need to be resolved from other stars in order to measure their periodicity. Blurring due to the Earth's atmosphere had limited the use of Cepheids as distance indicators to relatively nearby galaxies, such as M31, which is at a distance of roughly 750 kpc (2.5 million light years). With the launch of the *Hubble Space Telescope, astronomers have been able to observe from above the atmosphere, and the Hubble Space Telescope Key Project to measure the Hubble constant has measured Cepheids in galaxies as far away as 21 Mpc (70 million light years).

See also DISTANCE LADDER.

CfA Redshift Survey

The Center for Astrophysics (CfA) Redshift Survey was started in 1977 and the first part was completed in 1982. It contained *redshifts for 2 401 *galaxies brighter than *magnitude 14.5 at high Galactic latitude in the merged catalogues of *Zwicky and Nilson, and produced the first large area and moderately deep three-dimensional maps of *cosmic structure in the nearby Universe.

The second CfA survey (CfA2), carried out between 1985 and 1995, extended the original CfA survey one magnitude fainter and contains redshifts for about 18 000 bright galaxies in the northern sky. Although it was already known from statistical analyses that galaxies are clustered, a plot of redshift against right ascension for the first strip of galaxies in the CfA2 survey (Fig. 1) gave the striking impression that galaxies appear to be distributed

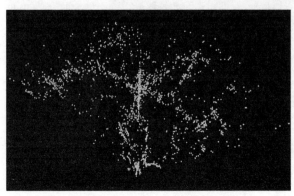

Fig. 1. A plot of right ascension against recession velocity (proportional to distance) for about 1 100 galaxies in the first strip of the CfA2 Redshift Survey. The Coma cluster forms the torso of the 'stick man' near the centre of this plot.

on surfaces, almost bubble like, surrounding large empty regions, or *voids.

These surveys allowed the first large-scale, quantitative analyses of the clustering of galaxies in three dimensions.

http://cfa-www.harvard.edu/~huchra/zcat/

Chandra satellite

The Chandra satellite (shown in Fig. 1) is a NASA satellite designed to measure *X-ray emission from astrophysical objects. It was known as AXAF (Advanced X-ray Astronomy Facility) before launch and subsequently renamed after Indian-born astronomer Subrahmanyan Chandrasekhar, famous for wide-ranging contributions to astrophysics including the discovery of the Chandrasekhar limit for gravitational collapse of white dwarf stars. As X-rays do not penetrate the Earth's atmosphere, satellite observations are the only way of exploring this part of the *electromagnetic spectrum. X-rays are towards the high-energy end of the spectrum, and primarily probe very energetic physical environments, including stellar and galactic *black holes, *supernovae, and the hot gas within galaxy *clusters.

Fig. 1. An artist's impression of the Chandra satellite.

Chandra is one of NASA's four 'great observatories' missions, along with the *Hubble Space Telescope, the *Spitzer Space Telescope operating in the *infrared, and the now-defunct Compton gamma-ray observatory. It

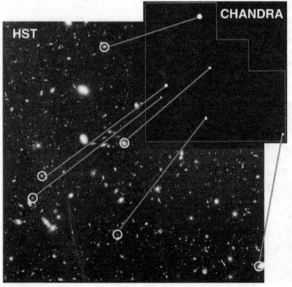

Fig. 2. The main image is the Hubble Deep Field, showing distant galaxies imaged in visible light. The upper right insert shows Chandra observations of the same field. Only a small number of the galaxies seen by Hubble emit in the X-ray part of the spectrum.

was launched into Earth orbit by a space shuttle in 1999. Chandra was designed principally to achieve high imaging resolution, complementing the European Space Agency's *XMM-Newton satellite whose main aims were to achieve high sensitivity and spectroscopic resolution.

Chandra's main science goals are based around non-cosmological topics, particularly study of supernova remnants and *active galactic nuclei, but it has also made significant contributions to cosmology. One example is Chandra observations of the *Hubble Deep Field, Fig. 2, which show that very few of the galaxies seen in visible light observations also emit strongly in the X-ray, and further that the brightness of galaxies seen at visible wavelengths is not a useful indicator of how strong their X-ray emission might be.

The most powerful images Chandra has taken so far comprise the two Chandra Deep Fields, one in the northern sky and one in the southern. For instance, the northern field has at the time of writing accumulated around two million seconds (about 25 days) of combined observations in one region of the sky in order to pick up distant low-luminosity sources. The faintest objects found in this survey emit so little light that there are roughly four days in between the arrival of successive individual *photons of X-ray light.

Chandra Home Page: http://chandra.nasa.gov

closed Universe

A closed *Universe is one which possesses spherical *geometry, meaning that the density of material within it is greater than the *critical density. Such a Universe has a finite volume. In the simplest cosmological models introduced by *Friedmann, which did not possess a *cosmological constant, the expansion of a closed Universe would eventually halt, to be followed by recollapse into a *big crunch. During the complete lifecycle of such a Universe, from *big bang to big crunch, there would just be time for something moving at the speed of light to make a complete circuit of the Universe and return to its starting point.

Modern cosmological models do include a cosmological constant, or more generally *dark energy, which widens the possibilities. If the cosmological constant is sufficiently positive, a closed Universe can expand forever. In that case, it is possible, given enough time, to completely traverse the Universe and

return to the starting point, just as a trip around the equator achieves on Earth. Current observations indicate that the Universe either has precisely the flat geometry or is rather close to it, and if indeed it is marginally closed, it will be of the type of closed Universe which expands forever. One should however beware of making predictions of what might happen in the distant future (*see* FUTURE OF THE UNIVERSE).

cluster baryon fraction

The cluster baryon fraction, as the name suggests, is the fraction of the mass of a *cluster of galaxies which is comprised of *baryons. Baryonic matter exists in two main forms within a cluster. One is as stars within the constituent *galaxies, and the other is in the form of gas which has fallen into the cluster and been heated in the process. The latter is heated so drastically (typically to millions or tens of millions of degrees) as it falls in that it is unable to condense to form stars, instead remaining as diffuse hot gas held in place by the balance of its own *pressure against the cluster's gravitational attraction.

In fact, the gas is the dominant type of baryonic matter in clusters; in a typical large cluster the gas mass is about a factor of ten larger than that in galaxies. That is nicely in agreement with observations showing that stars contribute only about 0.5% of the *critical density of the *Universe, while baryons are believed to comprise around 4%. That baryonic gas is ten times as prevalent as stars is presumably true throughout the Universe, but only in galaxy clusters does the gas become hot enough to be readily observed via its *X-ray emission.

Observations of the gas alone are sufficient to allow the baryon fraction to be estimated, provided it is assumed that the gas is in hydrostatic equilibrium, whereby the gas pressure forces balance the gravitational attraction which is due to the total mass of the clusters. Early treatments would additionally model the cluster as possessing a single gas temperature, but more recent observations have shown that this is not always a good approximation and that better results can be obtained by measuring and allowing for temperature variations across clusters.

The basic premise in using the cluster baryon fraction is that the composition of a galaxy cluster matches that of the Universe as a whole. As galaxy clusters are the largest gravitationally collapsed objects in the Universe and retain essentially all the material that falls into them, they should indeed be good samples. Detailed numerical simulations suggest that perhaps 10% of the baryons are lost in the collapse process, an effect that observations are usually corrected for.

The cluster baryon fraction technique reached maturity in 1993 with a paper by Simon White and collaborators. They derived a value of around 1/6, which was inconsistent with the then-popular critical-density Universe (requiring 96% of the critical density as dark matter) and/or the 4% baryon density derived from *nucleosynthesis. They named this the 'baryon catastrophe', but this emotive term has fallen from favour as the *standard cosmological model has evolved to its present form. Their result has in fact stood up very well, and the current values of the parameters are in good agreement with their number.

Allied to observations of the *cosmic microwave background (CMB) indicating that the Universe is exactly or nearly spatially flat, plus the nucleosynthesis measurement of a 4% baryon density which is now reinforced by CMB measurements, the cluster baryon fraction can be seen as an important independent indicator of the need for *dark energy in the Universe, as it shows that dark matter alone cannot provide enough density to give a spatially flat Universe.

clustering

One of the most useful properties of *galaxies and *quasars from the cosmologist's point of view is that they cluster together, rather than being randomly distributed throughout space. The large-scale clustering pattern of galaxies and quasars provides information about the distribution of *density perturbations in the Universe, and the smaller-scale clustering pattern yields important clues as to the amount and type of *dark matter in the Universe. The recent detection of *baryon oscillations in the galaxy distribution on scales of 140 Mpc confirms our picture of *structure formation by gravitational collapse of small density perturbations.

Clustering may be statistically quantified by a hierarchy of *correlation functions. The simplest of these, the two-point correlation function, gives the excess probability, above random, of finding two objects at a given separation from each other. If the distribution obeys *Gaussian statistics, then it is fully characterized by the two-point function. In general, however, the distribution may be further characterized by the three-point function, the probability of finding three objects in a specified triangular configuration, the four-point function, and so on.

It is currently believed that the initial density perturbations (as seen in *cosmic microwave background anisotropies) do obey Gaussian statistics. However, amplification of the initially small perturbations via *gravitational instability will produce non-Gaussian statistics, particularly in the small-scale clustering of galaxies and quasars. The reason for this is that the distribution of large density perturbations (with a standard deviation comparable to or larger than the mean) will be skewed towards higher values since negative densities cannot be realized, and so cannot be Gaussian, which is necessarily symmetric.

The three-point and higher order correlation functions are time-consuming to calculate and non-trivial to visualize, and so an alternative statistic is to calculate the moments of *counts in cells. Here one places cells of specified size and shape (normally cubic or spherical) either at random or on a regular grid, and counts the number of objects that fall in each cell. The second moment, or variance, of the cell counts then yields a quantity equivalent to the integral of the two-point function over the volume of one cell. The third moment yields the integral of the three-point function and so on.

Complementary measures of clustering include the *power spectrum, which is the *Fourier transform (FT) of the two-point correlation function, the bi-spectrum (the FT of the three-point function) and the tri-spectrum (the FT of the four-point function). These statistics are particularly easy to apply to whole-sky samples, and so have proved popular with CMB analyses, but are less straightforward to apply to real galaxy and *quasar surveys which suffer from incomplete sky coverage and radial variations of the mean density.

clusters of galaxies

A cluster of *galaxies is a gravitationally bound collection of at least several hundred, and possibly many thousands of, galaxies within a region a few megaparsecs across. This extremely high concentration of matter forms a deep gravitational 'potential well', which traps and heats a large amount of gas, causing most galaxy clusters to emit strongly in the *X-ray part of the *electromagnetic spectrum.

History

The first cluster was unwittingly discovered by Charles Messier, whose famous *Messier catalogue recorded 11 nebulae within the *Virgo cluster. This concentration was remarked upon by William Herschel in 1811 the *Coma and Perseus clusters were studied in the early years of the 20th century by Max Wolf in Heidelberg. Once the extra-Galactic nature of nebulae was confirmed, many further clusters were discovered, notably by Harlow Shapley and Edwin *Hubble.

The first comprehensive cluster catalogues were compiled in the 1950s by Fritz *Zwicky and by George Abell, whose *Abell cluster catalogue is still widely used today. The compilation of cluster surveys from optical observations of their constituent galaxies is plagued by the problem of **projection effects**. This effect comes about due to chance projections of galaxies along the line-of-sight to the cluster: the observer may mistake these galaxies, at very different distances, for members of a single physical cluster. This problem is particularly acute when searching for distant clusters, when there will be more unassociated intervening galaxies along the line-of-sight. Use of colour, as well as association on the sky, can help in cluster-finding, since clusters are dominated by *elliptical galaxies of a similar reddish colour. Only by obtaining *redshifts of the galaxies can one be certain that they belong to the same physical system. Alternative ways of surveying clusters, unaffected by projection effects, are discussed below.

One may estimate the masses of galaxy clusters by assuming that they are *virialized systems, meaning that the motions of galaxies within them have been randomized over time by gravitational interactions with other cluster members. By measuring the *peculiar velocities of galaxies within clusters, one can show that the typical time taken for a galaxy to cross from one side of a cluster to the other (the 'crossing time') is of order 1 Gyr. Since this is much shorter than the *Hubble time, it seems that the virial approximation is a rea-

sonable one. Under this assumption, one may show that the mass of a galaxy cluster is given approximately by

$$ M \approx \frac{3R_e \sigma^2}{G}, $$

where R_e is the effective radius of the cluster (i.e. the radius that contains half of the total galaxies), σ^2 is the line-of-sight *velocity dispersion of galaxies within the cluster, and G is Newton's gravitational constant. One finds masses $\sim 10^{14}$ solar masses for a relatively poor cluster such as Virgo and $\sim 10^{15}$ solar masses for the richer Coma cluster, see Fig. 1. Both of these masses are significantly larger than is obtained by summing the masses of the constituent galaxies ($\approx 1.3 \times 10^{12}$ and $\approx 6 \times 10^{12}$ solar masses respectively). This would imply a mass-to-light (M/L) ratio of around 300, far larger than for individual galaxies, even allowing for their high M/L *galactic halos. This problem was noticed by Zwicky in the 1930s, and was the first evidence for 'missing' or *dark matter.

Usually clusters feature one very large central galaxy of elliptical type, often called the **brightest cluster galaxy** or BCG. This central galaxy is made from mergers of smaller galaxies and also accretion of cooling gas flowing towards the cluster centre. In some cases, where clusters have recently undergone major merger events, there may be two or more such galaxies which have not yet had time to merge into a new larger central galaxy, as for example in the case of the Coma cluster.

Cluster gas

At least part of the missing mass is due to intergalactic gas, which is heated in cluster environments to temperatures of 10^7–10^8 K and which thus emits in the X-ray part of the spectrum. This gas is then far too hot to undergo collapse into galaxies and stars. Its X-ray emission provides an alternative method for compiling surveys of galaxy clusters, which is potentially less biased than using optical surveys. The first X-ray cluster sample came from the Einstein satellite and was known as the Einstein Medium Sensitivity Survey (EMSS) catalogue. Subsequent surveys were made by the *ROSAT satellite, and new surveys are ongoing with the *XMM-Newton and *Chandra satellites.

The hot gas in clusters may also be detected indirectly by the effect it has on photons

Fig. 1. Two images of the nearby Coma cluster, on the same angular scale. The left image shows optical light, where the main visible features are the constituent galaxies. The right image shows X-rays, emitted from the hot intra-cluster gas; only the brightest galaxies can be seen in this image. A much more detailed X-ray image of Coma can be seen at the entry for XMM-Newton satellite, and an alternative optical image in the Coma cluster topic.

from the *cosmic microwave background (*see* SUNYAEV-ZEL'DOVICH EFFECT).

Cluster dark matter

Even more important than the hot cluster gas is dark matter, whose *gravity is responsible for holding the cluster together. Without it, the galaxies, moving at their high observed velocities, would simply fly apart. Further, the hot cluster gas needs this extra gravity if it is to be held in hydrostatic equilibrium. Considering the amount of gravity needed to confine the hot gas provides a simple way to estimate the mass of the cluster, and in particular to compare the amount of *baryonic material (the galaxies and the gas) with the dark matter. This ratio, known as the *cluster baryon fraction, is an important probe of the material composition of the *Universe.

Cluster formation and evolution

Being the largest objects to have undergone gravitational collapse in our Universe to date, clusters are the most recent objects to form as part of the *hierarchical structure formation process that has taken place in our Universe. Indeed, one should say that the epoch of cluster formation is still underway and will continue for some time; only a few percent of galaxies are presently in galaxy clusters, but more are being accumulated into these structures as time goes by.

It is difficult therefore to say precisely when a cluster forms. A grouping of galaxies continues to evolve in two ways, one by continual accretion of gas and smaller galaxies from the surrounding areas (thought to take place predominantly down the filaments at whose vertices clusters tend to form, as part of the *cosmic web shown in Fig. 2), and the other by major mergers where two similarly-sized groups or clusters merge into a single system. While the former process leads to a fairly steady evolution of the cluster properties, such as a slowly increasing X-ray temperature as the gravitational potential well deepens, the latter can lead to drastic changes, including intense brightening, which may last a billion years or so before the cluster settles down.

Equally, the dividing line between galaxy clusters and *galaxy groups is rather arbitrary, and may not be consistently made between different detection methods. For those who detect clusters via their X-ray emission, an arbitrary dividing line is usually set at a temperature of about 25 million kelvin, corresponding to an equivalent particle energy of $k_B T = 2$ keV, where k_B is the Boltzmann constant which relates temperatures to energies.

The general environment of the cluster, particularly its gas, is known as the *intracluster medium (ICM). There is considerable interest in understanding the physical

Fig. 2. A galaxy cluster identified in a large-scale *N*-body simulation of cosmic structure formation. The galaxy cluster lies at the centre, and is located at a convergence of long filaments of galaxies. Material rains down these filaments onto the cluster. This image shows a slice through the Millennium Simulation.

processes that affect the ICM, and in how the ICM environment affects the cluster galaxies, which are discussed in the ICM entry.

An observed evolutionary effect in the galaxy population is the Butcher–Oemler effect, first described in 1978 by American astronomers Harvey Butcher and Augustus Oemler. They observed that the galaxies within clusters at intermediate redshift (around $z = 0.4$) tended to have more blue galaxies (mainly spirals) than those in the present Universe.

Clusters as cosmological probes
Galaxy clusters are an important probe of the *cosmological model. As well as the cluster baryon fraction mentioned above, which probes the mixture of material in the *Universe, the observed number density of clusters depends on several *cosmological parameters. The number of galaxy clusters in the local Universe primarily measures the size of *density perturbations on a scale corresponding to the cluster mass. Having used such measurements to fix the size of density perturbations, the way in which the number of clusters changes with redshift (i.e. looking at distant clusters observed at a time when the Universe was younger) probes the rate of change of those density perturbations. This growth rate depends primarily on the total density of matter in the Universe, but also to some extent on the existence and properties of *dark energy.

Present cluster observations measure the size of density perturbations quite accurately, and give good agreement with alternative measures such as *CMB anisotropies. They favour a matter density well below the *critical density, in agreement with other measurements but with less accuracy. As yet, it has not been possible to examine dark energy properties using galaxy clusters.

CMB
See COSMIC MICROWAVE BACKGROUND; COSMIC MICROWAVE BACKGROUND ANISOTROPIES.

cmbfast
Cmbfast is a computer program which calculates the *cosmic microwave background (CMB) anisotropies and the galaxy *power spectrum for *cosmological models, to enable them to be compared with observations. The program was created in 1996 by Uroš Seljak and Matias Zaldarriaga and made freely available to the research community (see the WWW address below).

The CMB anisotropies were discovered by the *COBE satellite in the early 1990s, and soon afterwards anisotropies were detected by other experiments too. Progress in understanding the observations was very slow however, due to the difficulty of predicting what was expected in different cosmological models, and in particular what happened if the assumed *cosmological parameters were varied. Once those assumptions were made, the observable quantities could only be predicted by solving a complex set of equations, which required simulation on a computer. Various researchers wrote such computer programs, known as Boltzmann codes, but they typically took many hours to compute the predictions for a single set of cosmological parameters. Further, only researchers with access to the closely guarded Boltzmann codes could exploit them.

Seljak and Zaldarriaga overcame these problems at a stroke. Firstly, they devised a new way of solving the equations governing the evolution of *perturbations in the Universe, which reduced the computer time required for a single choice of parameters from hours to minutes (subsequent increases in computer power have reduced this to typically tens of seconds). Secondly, they made their program freely available to the research community, and it swiftly became the most widely used program in cosmology, a large fraction of working cosmologists having used it at some time or other.

Since the original version was released, various upgrades have taken place, for instance including *polarization of the CMB, *gravitational lensing, and *dark energy models. Another computer code, called CAMB, was devised at Cambridge to independently verify cmbfast's predictions, relying heavily on heritage from cmbfast.

Despite the massive speed-up achieved by cmbfast, and advances in computing power since it was first written, the speed with which model predictions can be made remains the limiting factor in trying to constrain cosmological models using observational data. Carrying out such comparisons at research level remains a supercomputer class problem, but without cmbfast and its successors, it wouldn't be feasible at all.

cmbfast home page: http://www.cmbfast.org
CAMB home page: http://camb.info

COBE satellite

See COSMIC BACKGROUND EXPLORER (COBE) SATELLITE.

coincidence problem

This refers to the puzzle that the densities of *dark matter and *dark energy are very similar at the epoch when we happen to be measuring them, whereas at all other epochs one or other is totally dominant. See *cosmological constant and *quintessence for more details.

cold dark matter (CDM)

Cold dark matter is a form of *dark matter, distinguished from other forms of dark matter by its constituent particles having a non-relativistic velocity throughout the *Universe's history. Cold dark matter is the form favoured observationally, and is a key component of the *standard cosmological model, which is also sometimes known as the *cold dark matter model. Although its *density is quite well determined by astrophysical measurements, it is not known what form the dark matter takes.

Cold dark matter properties

Like any dark matter candidate, the role of cold dark matter is to provide the gravitational attraction needed to explain a range of astrophysical and cosmological phenomena. In order to have so far escaped direct (i.e. non-gravitational) detection, its interactions with normal matter must be either extremely feeble or completely non-existent. This excludes the possibility of interactions via the electromagnetic or strong nuclear *fundamental forces, while leaving open the possibility of weak nuclear interactions.

Cold dark matter is normally assumed to be in the form of some new type of fundamental particles, perhaps motivated by *supersymmetry. Such particles would have first come into existence in the *thermal equilibrium state created after *cosmological inflation, which marks the beginning of the hot phase of the Universe's evolution. At this time they may interact frequently with other particles, but at some point their interactions would become weak enough for them to decouple from the thermal bath (the simplest way that this could happen would be in direct analogy to *neutrino decoupling, discussed under that topic). In order to be cold dark matter, their motion should be non-relativistic at this time, meaning that their mass–energy is already much greater than the ambient thermal energy.

The distribution of cold dark matter will feature *inhomogeneities, perhaps inherited from the inflationary period, whose *gravitational instability leads to *structure formation. Because of their negligible initial velocities, the cold dark matter particles move entirely due to this gravitational force. Such gravitational motions are independent of the particle masses (the equivalence principle embodied in *general relativity). Accordingly, the only free parameter of a cold dark matter cosmology is the total density of dark matter, making it the simplest possible hypothesis. For hot dark matter, by contrast, the initial distribution of particle velocities would also have to be specified.

Most experiments seeking to detect dark matter directly, as described in the *dark matter topic, are seeking cold dark matter, usually under the assumption that the cold dark matter is comprised of *WIMPs. For such experiments, the actual mass of the particles and their interaction rates are also important as well as the density.

Cold dark matter candidates
There are many candidates to be the cold dark matter, including the following.

WIMPs—WIMPs stands for Weakly-Interacting Massive Particles, where the 'weak' refers to the weak nuclear force. They are motivated by supersymmetry, which although not yet confirmed is regarded as a fairly inevitable property of fundamental particle theories.

Axions—Axions are particles proposed to explain why particles and *anti-particles have such similar properties. Although extremely light, their formation mechanism ensures they behave as cold dark matter. Dedicated experiments have been designed to test for the existence of axions but so far have not seen anything.

WIMPzillas—These particles had the misfortune to be proposed in the same year as the not-at-all-good Hollywood remake *Godzilla* was released in the cinemas, landing them with a name which has made them difficult to take seriously ever since. WIMPzillas are extremely massive particles, perhaps with masses characteristic of *grand unification theories (individual particles having masses as big as a nanogram). Being so heavy, there are far fewer of them required to supply the observed dark matter density, making them extraordinarily hard to detect directly.

Primordial black holes—*Primordial black holes, being low-mass black holes formed within the first second of the Universe's existence, can act as cold dark matter. They are the main candidate which is not a type of fundamental particle; however it is hard to see how such black holes could form with a suitable abundance.

cold dark matter model (CDM model)
A cold dark matter model, typically abbreviated as a CDM model, is one in which *cold dark matter plays the dominant role in *structure formation, providing the gravitational attraction that causes *galaxies and other structures to form. The *standard cosmological model is a cold dark matter model in this sense; even though *dark energy is the most abundant material in that model, it does not participate in the structure formation process. This model is often described as a ΛCDM model (pronounced *lam-da see dee em*), the Greek capital lambda being the standard symbol for the *cosmological constant which is the simplest form of dark energy.

Historically, cold dark matter models stood against other choices—including hot *dark matter models (HDM), warm dark matter models (WDM), and cold plus hot dark matter models (CHDM)—as a possible theory of structure formation, but by the late 1990s it had become by far the dominant option. This is now true to such an extent that it is usually taken as read that cosmological models will be based around the assumption of cold dark matter as a major constituent.

Coma cluster
The Coma cluster (Abell 1656) is the second-closest *cluster of galaxies to us after the *Virgo cluster. It is situated in the constellation of Coma Berenices at a distance of roughly 100 Mpc (330 million light years) and probably contains more than 10 000 *galaxies, a large number of which are elliptical or lenticular in morphology (*see* GALAXY CLASSIFICATION). It covers a region of sky more than 4 degrees in diameter, with a physical size of at least 20 million light years. While the optical image of Coma, Fig. 1, shows two cores, centred on the giant elliptical galaxies NGC 4874 and NGC 4889, the *X-ray image of Coma (*see* XMM-NEWTON SATELLITE, Fig. 3) shows that the hot X-ray emitting gas is distributed in an approximately spherically-symmetric way, indicating that the cluster is relatively well *virialized.

The Coma cluster has proved very useful in studies to estimate the value of the *Hubble constant, since it provides a large number of luminous galaxies at a distance close enough that several different *distance estimates may be made, but far enough away that *peculiar velocities relative to the *Hubble expansion are expected to be small. A distance estimate

Fig. 1. The Coma cluster of galaxies. This optical image is dominated by the two giant elliptical galaxies NGC 4874 and NGC 4889, with hundreds of fainter galaxies also visible.

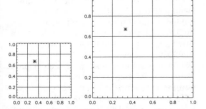

Fig. 1. Two-dimensional illustration of comoving coordinates. The asterisk maintains the same comoving coordinates (0.33, 0.67) even though its physical distance from the origin increases as the coordinate system expands from left to right.

to Coma of 100 ± 10 Mpc from surface brightness fluctuations implies a value for the Hubble constant of 71 ± 8 km/s/Mpc, consistent with other recent measurements.

William Herschel described the Coma cluster in 1785 as 'the nebulous stratum of Coma Berenices, almost everywhere equally rich in fine nebulae.' However, it was not realized at the time that these *nebulae were other galaxies, millions of light years away from our own.

comoving coordinates

A comoving coordinate system is a reference frame that expands in tandem with the expansion of the *Universe, thus factoring out the effect of the *Hubble expansion. Any object with a zero *peculiar velocity will then have fixed comoving coordinates, and any pair of objects each with zero peculiar velocity will have a fixed separation in comoving coordinates.

A two-dimensional example of a comoving coordinate system is a grid of lines drawn on the surface of a balloon. As the balloon is inflated, the grid lines themselves move further apart, and so the coordinates of a stationary ant would be fixed, see Fig. 1.

Comoving coordinates are particularly useful when investigating the evolution of cosmic structures, since the mean density of matter in the Universe is fixed when measured using comoving coordinates, in contrast to the case when *proper coordinates are used, in which

case density scales as the inverse cube of the *scale factor of the Universe.

compactification

Some modern theories of fundamental physics, such as *superstring theory and *M-theory, postulate that the Universe has more than three spatial dimensions, without which the theories are not consistent. This is apparently in gross contradiction to the actual *Universe, with the three dimensions up–down, back–front and left–right. Compactification is the process by which the extra dimensions are hidden from our view in order that the theories remain viable.

The origin of the word comes from 'compact', which, as well as conveying a sense of smallness in everyday usage, has the technical meaning of possessing a finite length/volume. A circle is an example of a compact one-dimensional line. The idea, then, is that the extra dimensions are very small, not just in everyday terms but small even compared to microscopic particles. If so, then we simply have so far been unable to probe the structure of space with enough resolution to see the effects of the extra dimensions.

Superstring theorists devote a lot of time to understanding compactification, because the properties of the extra-dimensional space determine the physical laws at the low energy scales applicable to our present Universe. The goal is to find a compactification mechanism which gives the low-energy physics of the *standard model of particle physics. Unfortunately there are a very large number of possible compactifications, and as yet no ideas as to how Nature might have selected amongst them.

The term 'compactification' carries implications of a dynamical process, that those dimensions were initially not compact and then some process made them so. That implication is probably not correct; if the dimensions are compact now they almost certainly always were. If anything, the picture would be that all the dimensions were initially small and, for some reason or other, only our familiar three dimensions became large. Incidentally, it may well be that the three large dimensions we experience are compact too (*see* TOPOLOGY OF THE UNIVERSE). The important issue is really just whether the dimensions are large or small, not whether they are finite or infinite.

Although the extra dimensions are unlikely to have switched between compactness and non-compactness, their size may well vary with time; indeed given that the three large dimensions are most definitely expanding, it would be strange if the small ones were not able to evolve. However, such evolution has the interesting consequence that it would lead to a variation in the values of what are known as the fundamental constants of Nature, such as the strength of gravity G. Observational searches for such *varying fundamental constants are one possible probe of the compactified dimensions.

concordance cosmology
See STANDARD COSMOLOGICAL MODEL.

cone plot
Also known as a wedge plot, this is a way of illustrating the distribution of sources in a *redshift survey. Using polar coordinates (r, θ), with r being distance from the origin and θ (Greek theta) azimuthal angle, r generally represents *redshift and θ represents one of the angular coordinates on the sky, normally right ascension. (*See* Fig. 1 of CfA REDSHIFT SURVEY for one of the first, and probably most famous, cone plots to be made.)

confidence limits
Observational measurements are never exact, instead always carrying some level of uncertainty due to the combined effects of experimental uncertainties, randomness, and approximation in theoretical modelling. In order to assess the statistical *significance of a result, measurements are usually assigned confidence levels, specifying a range within which the measured quantities are believed to lie with a certain probability.

In many situations, the experimental errors follow a *Gaussian distribution, the width of which can be measured by the standard deviation which is almost invariably indicated by σ (Greek sigma). For a Gaussian distribution, 68% of the probability lies within a distance of σ from the mean and 95% lies within a distance of 2σ. These ranges are referred to as the one-sigma and two-sigma confidence ranges respectively.

For the specific case of *cosmological parameters, it is quite common for the probability distributions of parameters to deviate significantly from the Gaussian approximation. In such cases, it remains conventional to quote confidence ranges containing 68% or 95% of the possible values, and even to still refer to them as one-sigma and two-sigma though they are no longer directly identifiable with the standard deviation. In the case of non-Gaussian uncertainties, one may see the positive and negative uncertainties quoted with different values, reflecting the asymmetry of the distribution.

The standard convention is that when a number is quoted with an uncertainty, for instance $p = 3.4 \pm 0.3$ or $q = 2.7^{+0.2}_{-0.3}$, the quoted range is the 68% confidence range, meaning that there is a 68% chance of the true value lying within that range. Clearly this leaves a significant probability that the true value is outside the stated range, so if one wants to be confident that the true value is enclosed one should consider at least the two-sigma and perhaps even the three-sigma limits (the latter corresponding to 99.7% for Gaussian distributions, though for non-Gaussian ones the corresponding percentage is likely to be less).

The convention in the case of limits, such as $r < 0.52$, is that the limit is the 95% confidence one, i.e. there is a 95% probability that the value is indeed less than the one stated.

Copernican principle
See COSMOLOGICAL PRINCIPLE.

correlation function
*Galaxies, and many other astronomical sources, are not randomly distributed in space, but cluster together in high-density regions. This clustering may be quantified using the correlation function. There are in fact a whole hierarchy of correlation functions: two-point, three-point, four-point and so on. The two-point correlation function

is by far the most widely used, and the term 'correlation function' used on its own may be assumed to refer to the two-point function.

The galaxy two-point correlation function gives the excess probability above random of finding two galaxies with a given separation. It is traditionally written as the Greek letter ξ (xi) and may be defined as

$$P = [1 + \xi(r)]\,\bar{\rho}dV_1dV_2.$$

Here, P is the probability of there being a galaxy in each of two infinitesimal volumes dV_1 and dV_2 separated by a distance r, where the mean galaxy density is $\bar{\rho}$. Note that if galaxies are unclustered (randomly distributed in space), then $\xi(r) = 0$ on all scales r (dV_1 and dV_2 are assumed small enough that there is a negligible probability of either containing more than one galaxy). If galaxies are clustered, then ξ is positive. If galaxies were to avoid each other at a certain scale (i.e. anticlustered), then ξ would be negative.

In principle, the correlation function is simply determined by counting the numbers of pairs of galaxies in a sample as a function of separation, and by comparing these counts with what would be expected if galaxies were randomly distributed. In practice, this is complicated by having a survey *selection function that declines with distance and a survey of incomplete sky coverage, so that edge effects have to be taken into account, i.e. the fraction of a spherical shell around each galaxy that is included within the survey volume. This is most easily accomplished by distributing a large number of points at random within the survey volume and with the same selection function as the survey. One then counts the number of galaxy-galaxy $GG(r)$ and galaxy-random $GR(r)$ pairs as a function of separation r. The estimated correlation function is then given by

$$\xi(r) = \frac{N_r}{N_g}\frac{GG(r)}{GR(r)},$$

where N_g and N_r are the total number of galaxies and random points respectively. In practice, more complicated estimators are used which provide a better correction for survey edge effects.

The spatial correlation function ξ is always measured from *redshift surveys, so that the distance of each galaxy is derived from its *redshift. Due to *peculiar velocities, these inferred distances are not exactly equal to the

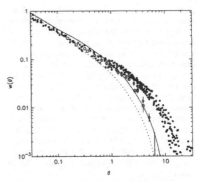

Fig. 1. The galaxy angular correlation function, $w(\theta)$, measured from the APM galaxy survey (solid symbols) and from the earlier Lick galaxy catalogue (open symbols). The solid and dotted lines show $w(\theta)$ predicted by critical density cold dark matter models assuming a Hubble parameter of 40 and 50 km/s/Mpc respectively. It is clear that neither model is compatible with the APM data points.

true distances, and so the correlation function suffers from *redshift space distortions. These distortions may be isolated by measuring the correlation function $\xi(\sigma, \pi)$ as a function of two components of separation parallel and perpendicular to the line of sight, denoted π and σ respectively. When this is done, only the line-of-sight separation π is affected by peculiar velocities, and so $\xi(\sigma, \pi)$ is asymmetric (see PECULIAR VELOCITIES, Fig. 1). One may integrate the two-dimensional correlation function $\xi(\sigma, \pi)$ over the line-of-sight direction to obtain an estimate of the projected correlation function $w_p(\sigma)$, also commonly written as $w_p(r_p)$, where $r_p \equiv \sigma$ is the projected separation. This projected correlation function is unaffected by peculiar velocities, but still incorporates distance information, unlike the angular correlation function $w(\theta)$.

Before the era of large redshift surveys, galaxy clustering was quantified by measuring the angular correlation function, $w(\theta)$, which describes the apparent clustering of galaxies as seen projected on the sky. Here, physical separation r is replaced by angular separation θ. The angular correlation function measured from the *APM galaxy survey is shown in Fig. 1. This measurement showed that the amplitude of $w(\theta)$ was much higher on angular scales of 2–20 degrees than was predicted by the then-popular *critical density *cold dark matter model, in which the mass of the Universe is dominated by slow-moving, non-*baryonic dark matter. This was one of the first results to rule out this model and to hint that the *density parameter of the Universe is less than unity.

The three-point correlation function gives the excess probability of finding one object in each of three volumes of specified separation and geometry. The three-point function is a function of three separations (or two separations and an angle) and so is significantly more time-consuming to estimate, and also much harder to visualize than the two-point function. For this reason, it has not been widely used.

Cosmic Background Explorer satellite (COBE) (pron. *koh-bee*)

The Cosmic Background Explorer satellite, known as COBE, was one of the landmark cosmological projects operated by NASA, and widely regarded as the most successful cosmological satellite mission ever. Launched in 1989, it discovered the *cosmic microwave background anisotropies that had been predicted almost 25 years earlier, and which have subsequently proved the key to developing the *standard cosmological model. It also made a measurement of the thermal spectrum of the *cosmic microwave background (CMB) which has yet to be bettered, and made the first precision all-sky maps of the diffuse

*cosmic infrared background. Scientists John Mather and George Smoot were awarded the 2006 *Nobel Prize in Physics for their achievements with the COBE satellite.

The satellite

The COBE satellite was conceived in the 1970s as part of the thesis project of Mather, who would later become the overall project leader of the mission and Principal Investigator of one of its instruments. Studies of the cosmic microwave background were being hampered by strong emission of microwaves from within the Earth's atmosphere, leading to ambitious rocket-based experiments which sought to make measurements from high in the Earth's atmosphere. It was realized that the only effective way to circumvent this problem would be to observe from above the Earth's atmosphere, which required a satellite observatory. The mission was designed to carry three separate instruments with different science goals, described below.

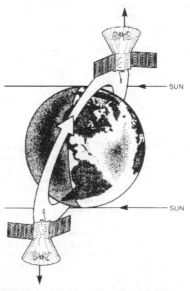

Fig. 2. A schematic of the COBE satellite's orbit. As the Earth went round the Sun, the orbit precessed so as always to be perpendicular to the Earth–Sun direction, limiting the number of solar microwaves entering the instruments.

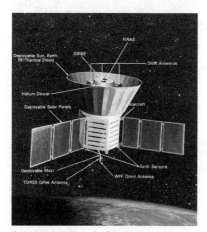

Fig. 1. An artist's schematic of the COBE satellite, launched in 1989.

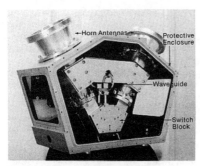

Fig. 4. A DIRBE image of the bulge region of the Milky Way.

Fig. 3. A pre-launch photo of one of the DMR detectors. It compared the intensity of radiation coming from two regions of the sky 60 degrees apart. Its angular resolution was approximately 7 degrees. The location of the DMR instruments can be seen in Fig. 1.

After a series of delays, one associated with the Challenger space shuttle disaster in 1986, the satellite was finally launched on a Delta rocket into a sun-synchronous low-Earth polar orbit, shown in Fig. 2, in November 1989. This orbit was chosen to minimize the noise into the detectors from the Sun and Earth, being the two brightest microwave sources in the sky, and to offer a stable thermal environment where the satellite would not be passing between Sun and shade. It eventually was able to take four years of science data.

Many of the COBE team later went on to work on the *Wilkinson Microwave Anisotropy Probe (WMAP) satellite.

COBE:DMR

The Differential Microwave Radiometer (DMR) experiment, led by Smoot, sought to measure cosmic microwave anisotropies, a task which had eluded ground-based experiments for over two decades. Its strategy was not to measure the temperature in a given direction exactly, but rather to compare the temperatures in two directions 60 degrees apart. One of the detectors can be seen in Fig. 3. This strategy avoided the need for a calibration source on board. During the course of the mission, each region of the sky was compared in temperature with many different points 60 degrees away. Eventually, all this data could be used to construct a map of the sky. The maps were made at an angular resolution of 2.8 degrees, with the formal

resolution of the instrument being seven degrees. By modern standards this is a very poor resolution (an effective area 200 times that of the full moon), but perfectly adequate to measure variations on scales larger than that.

There were three separate DMR instruments operating at slightly different microwave frequencies. Because the CMB has a *black-body spectrum, the known frequency dependence of the signal can be used to minimize the effects of contamination such as galactic microwave emission. Maps were made at each frequency, and then could be decomposed into a CMB map and maps of other sources.

(For the results of the COBE DMR experiment, *see* COSMIC MICROWAVE BACKGROUND ANISOTROPIES). Subsequent to COBE, the study of CMB anisotropies has developed into one of the most important branches of observational cosmology.

COBE:FIRAS

The Far InfraRed Absolute Spectrometer (FIRAS) was led by Mather, and aimed to measure the frequency spectrum of the CMB (i.e. the variation of intensity with frequency) for a wide range of frequencies on both sides of the maximum intensity, and to compare with a reference black-body spectrum source carried on board the satellite. It indeed confirmed that the spectrum was of black-body form with no measurable deviations, and determined the temperature to be 2.725 ± 0.001 kelvin. Its results are described in more detail under the item on the cosmic microwave background, and have yet to be bettered.

COBE:DIRBE

The third experiment on COBE was the Diffuse InfraRed Background Experiment (DIRBE), led by Mike Hauser. It made the first accurate all-sky maps of the cosmic infrared background, highlighting in particular *infrared radiation from dust in the plane of the *Milky Way (see Fig. 4) and from other galaxies. (For its main cosmological results, *see* COSMIC INFRARED BACKGROUND).

http://lambda.gsfc.nasa.gov/product/cobe/

cosmic dipole

The *cosmic microwave background (CMB) radiation, the remnant of the *big bang fireball, has a very nearly uniform temperature of 2.725 degrees above absolute zero, also known as degrees kelvin, or simply kelvin, and written 2.725 K. The largest departure from isotropy in the CMB temperature has a dipole pattern (that is a pair of 'hot' and 'cold' spots in opposite directions on the sky, with a smooth variation of temperature between them, see Fig. 1). This dipole pattern is naturally explained as being due to our motion with respect to the restframe of the CMB radiation. The relative temperature change $\Delta T/T$ is simply given by the *Doppler shift, $\Delta T/T = v/c$, where v is our velocity relative to the CMB frame and c is the *speed of light. A maximum temperature fluctuation of $\Delta T = 3.36$ mK (thousandths of a kelvin)

implies a velocity of 370 km/s in the direction of $(l, b) = (264°, +48°)$ in Galactic coordinates. After correcting for the motion of the Earth around the Sun, this motion has two major components:

- The motion of the Sun around the centre of the Galaxy at ≈ 220 km/s
- The motion of our Galaxy relative to the CMB frame.

This latter *bulk motion is due to the gravitational pull of several nearby *superclusters of galaxies.

cosmic infrared background (CIRB)

The cosmic infrared background refers to all diffuse *infrared radiation external to the *Milky Way. It arises from the cumulative radiation from stars and *galaxies dating back to the epoch when these objects first began to form. Measurements of the CIRB constrain models of the *star-formation history of the *Universe and the build-up over time of *dust and elements heavier than hydrogen, including those of which living organisms are composed.

The Diffuse InfraRed Background Experiment (DIRBE) on the *Cosmic Background Explorer (COBE) satellite made whole-sky maps of the infrared sky in the late 1990s. These maps are dominated by emission from the Milky Way galaxy and the Solar System,

Fig. 1. An all-sky plot, in *Mollweide projection, of the cosmic microwave background temperature observed by the *Wilkinson Microwave Anisotropy Probe (WMAP). The dark shading in the lower-left quadrant of the image corresponds to a temperature about three thousandths (0.003) of a degree hotter than average, and the light shading in the upper-right quadrant to a temperature 0.003 degrees cooler than average. This is the cosmic dipole, and is due to our motion relative to the microwave background. Also visible in this image is microwave emission from our own Milky Way Galaxy, whose plane lies horizontally along the centre of the plot.

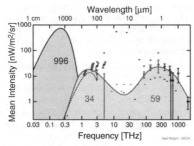

Fig. 1. A plot of the mean intensity of three components of background radiation. On the left, with a total intensity of 996 nW/m²/sr, is the cosmic microwave background (CMB). In the centre, at 34 nW/m²/sr, is the far-IR cosmic background and on the right at 59 nW/m²/sr is the near-IR cosmic background. Whereas the CMB originates from redshifted photons from the hot plasma at recombination, both infrared backgrounds come from unresolved, individual stars and galaxies.

since these sources are significantly closer to us than extra-Galactic sources. However, given DIRBE's observations of the infrared sky at a number of different wavelengths between 1.25 and 240 microns, one may model the Galactic and Solar System components and subtract them from the maps, leaving the extra-Galactic component.

The CIRB as observed by DIRBE may be broken down into a near-infrared component (around 1–10 microns in wavelength) and a far-infrared component (around 10–1000 microns), see Fig. 1. The near-IR component consists mostly of diffuse starlight *redshifted into the infrared. The far-IR component is mostly due to starlight that has been absorbed by dust and then re-radiated at longer wavelengths. (Stellar radiation heats up the dust particles, which then radiate with a *black-body spectrum.) The total CIRB has an intensity of around one ten-millionth of a watt per square metre per steradian.

Since the far-IR CIRB was detected by COBE, astronomers have searched for sources that might contribute to the infrared background. One of the most successful searches has been that done using the *SCUBA instrument at the James Clerk Maxwell Telescope in Hawaii. This instrument, operating at sub-millimetre wavelengths, has identified a population of high-redshift, luminous, dusty objects at 850 microns wavelength. The inte-

grated light from these objects is comparable to the CIRB intensity at 850 microns, suggesting that most of the 'background' light at these wavelengths has actually been resolved into individual objects.

As stars form and fuse hydrogen into helium and heavier elements, they emit radiation which will contribute to the CIRB either directly, or indirectly via dust heating. Observations of the CIRB can thus constrain the star-formation history of the Universe. However, since one observes the integrated emission over cosmic history, one cannot deduce a *unique* star-formation history.

Hauser, M. G. and Dwek, E. *Annual Review of Astronomy and Astrophysics, Vol. 39, pp. 249–307* (2001).

http://www.astro.ucla.edu/~wright/CIBR/

cosmic microwave background (CMB)

The cosmic microwave background is relic radiation left over from earlier stages of the hot *big bang (*see* 'Overview'), and is currently the most important way of understanding the properties of the young *Universe. It is comprised of standard radiation, but from a part of the *electromagnetic spectrum with a much longer wavelength than visible light, placing it in the *microwave part of the spectrum. The Universe is filled with this radiation, which bathes the Earth from all directions. It is, in fact, the dominant form of radiation in the present Universe. Its long wavelength however makes it quite difficult to measure accurately.

Properties of the CMB

The cosmic microwave background is extremely close to *isotropy, meaning that the radiation received from each direction is more or less identical. Indeed, for almost 30 years after its discovery, observations were not able to find any evidence of *anisotropy. Nowadays the departures from perfect isotropy are the most important property of the CMB; those departures will be discussed in the next topic *cosmic microwave background anisotropies. The remainder of this topic will take the CMB to be isotropic.

The radiation received corresponds to perfect emission of radiation from an object at temperature 2.725 kelvin. Perfect emitters radiate what is called a *black-body spectrum: they don't emit radiation just at a single frequency, but there is a well-defined peak, whose frequency is given by Wien's Law as

$$f_{\text{peak}} \approx 10^{11} \times T \text{ Hz/K}$$

where T is the temperature measured in kelvin. The higher the temperature, the higher the frequency of radiation emitted. A temperature of a few thousand degrees, such as the surface of the Sun, emits light in the visible part of the electromagnetic spectrum, while objects at room temperature emit in the *infrared. Radiation in the near infrared can be felt as heat. The peak intensity of the cosmic background falls at around 3×10^{11} Hz which is in the microwave part of the spectrum, hence the term 'cosmic microwave background'. We will see later that the CMB is very accurately of black-body form; indeed so far no deviations have been detected though they are predicted to exist at some level.

Although the CMB is very low intensity radiation, the Universe is filled with it, and if the total energy is added up a surprisingly large result is obtained. In fact, the CMB is by far the dominant form of radiation in the Universe, with its total energy easily exceeding that of all the light which has ever been emitted by stars (the stars are easier to detect because the radiation comes from a compact source and is at a higher frequency). In total it contains around one ten-thousandth of the present total energy of the Universe (this comparison assumes that any mass in the Universe is converted into an equivalent energy using Albert *Einstein's famous $E = mc^2$ relation).

Discovery of the CMB

The discovery of the CMB is one of the more entertaining, and often-told, hit-and-miss stories of cosmological progress. With the benefit of generous hindsight, its discovery can be traced all the way back to a 1941 paper by Andrew McKellar. McKellar observed transitions between rotational energy levels of interstellar molecules, specifically cyanogen (in the form CN), and found that the observed absorption was consistent with the gas being bathed in radiation at a temperature around 2.3 kelvin (with a range encompassing the value now known to be correct). The relation to *cosmology was not made, understandably as the hot big bang model was yet to be discovered. [Incidentally, this technique can now be used to

show that the background radiation was hotter in the past, by studying distant gas clouds seen when the Universe was significantly smaller.]

The first theoretical prediction of the CMB was made in 1948 by Ralph Alpher and Robert Herman. Following on from earlier work in collaboration with George *Gamow on cosmic *nucleosynthesis, they realized that there would be leftover radiation from the big bang, and predicted a temperature around 5 kelvin. They did not realise at the time however that the radiation might be detectable (and indeed had been detected by McKellar). Over the following decade, several other observers using radio telescopes noted the presence of a thermal background around 3 kelvin, without making a connection to cosmology. Those include Tigran Shmaonov in Russia, the French astronomer Emile Le Roux, and Ed Ohm working with the Bell Telephone Laboratories telescope in Holmdel, New Jersey. During this time, the pioneering work of Alpher and Herman received some publicity, particularly from Gamow, but was largely forgotten.

In the mid 1960s theorists once more came to realize that a cosmic microwave background was expected, and would be key evidence in support of the hot big bang. The realization came independently from a powerful Soviet research group led by Yakov *Zel'dovich, and from a Princeton University group led by Robert Dicke. Bizarrely, members of the Soviet group were aware of Ohm's work on the Holmdel telescope, while the Princeton group were not despite themselves being located in New Jersey. Unfortunately though the Soviets misinterpreted Ohm's results as implying no cosmic background. Dicke, in the meantime, proposed that the Princeton group build a radiometer capable of detecting the predicted background, while forgetting that he himself had actually designed and used a radiometer to place an upper limit (of around 20 kelvin) on an isotropic cosmic background way back in 1946. Indeed, the detector on the Holmdel telescope was exactly of the design invented by Dicke.

While all this was going on, Arno Penzias and Robert Wilson were working on the Holmdel telescope (see Fig. 1) and trying to get to the bottom of the mysterious excess signal. Increasing levels of desperation, including the removal of large quantities of pigeon

Fig. 1. Penzias (right) and Wilson with the Holmdel telescope in the background.

droppings, did nothing to eliminate the signal. Eventually, they learned of the Princeton group's plans to detect a cosmic microwave background, and when the teams got together they rapidly concluded that the Holmdel telescope was most likely detecting the cosmic microwave background, at a temperature of approximately 3.5 kelvin. Penzias and Wilson's paper appeared in the *Astrophysical Journal* in 1965, adjacent to a paper from the Princeton group interpreting the result as cosmological. Penzias and Wilson went on to receive the *Nobel Prize in Physics in 1978 for their discovery.

Origin of the CMB

The origin of the cosmic microwave background lies in the hot dense early stages of the Universe's evolution. Let us wind back the evolution of the Universe to a time when it was say one-millionth of its present size. At that time, the characteristic temperature of the Universe would have been about three million degrees, creating a harsh environment within which atoms could not survive. Instead, any time that an electron attempted to bind with a nucleus to form an atom, it would immediately be blasted away again by the high-energy radiation surrounding it. The Universe would therefore be in a plasma state, and any radiation present would strongly interact with the free electrons and bounce around the Universe in a random walk. In short, the Universe would be an opaque fog.

The Universe would continue to expand and cool, with the radiation losing energy through the *redshifting effect. The cooler the

radiation becomes, the less able it is to disrupt atoms, and eventually by a temperature of around 3 000 kelvin, atoms were first able to form in the Universe, the electrons becoming bound to the nuclei to form mostly hydrogen and helium atoms. At this epoch, the Universe made a rapid transition from opaque to transparent, since at this stage the individual *photons are by definition too feeble to interact with the atoms. Thus freed, the radiation propagates uninterrupted through the Universe.

Strictly speaking, two separate physical processes are taking place around the time the microwave background forms. *Recombination refers to the process whereby electrons combine with nuclei to form atoms, and the time of recombination is typically defined to be when this process is 90% complete. *Decoupling refers to the time after which photons can propagate freely, and happens shortly after recombination. More details on those can be found under their separate headings.

The formation process described above explains why the CMB has a thermal (blackbody) spectrum. Such spectra are characteristic of situations where there are frequent interactions in order to establish a thermal equilibrium, which appears puzzling for the CMB since its photons are not able to interact significantly. But in the hot big bang model the photons *were* in a strongly interacting equilibrium state, in the early Universe when they were much hotter. This neat explanation of the thermal property of the CMB was the clinching evidence for the hot big bang and against the *steady-state cosmology; in the latter case no convincing mechanism for thermalization has ever been established.

After the cosmic microwave background forms, the Universe continues to expand and so the photons of light continue to redshift. At the time of decoupling their characteristic temperature was around 3 000 kelvin, which is about half the temperature of the surface of the Sun and would mean that most photons had near infrared wavelengths. The subsequent expansion of the Universe increases their wavelength about one thousand times, shifting them into the microwave part of the electromagnetic spectrum where they are presently detected. The redshifting of light preserves the thermal spectrum, while reducing the corresponding temperature. In fact, the temperature of the Universe is

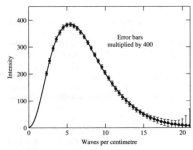

Fig. 2. The frequency spectrum of the CMB as measured by the FIRAS instrument on board the COBE satellite. The x-axis shows the frequency of the radiation and the y-axis the intensity of light at that frequency. The error bars are so small that they have been multiplied by 400 in order to make them visible on this plot. The spectrum corresponds to a perfect black-body spectrum at a temperature of 2.725 kelvin, and the peak at five waves per centimetre corresponds to the microwave part of the electromagnetic spectrum.

just inversely proportional to its size, which is one of the most important relations in cosmology.

The microwave photons have been travelling uninterrupted since the epoch of decoupling, and hence have been travelling for almost the entire age of the Universe. Those we detect at the Earth have therefore come from the far reaches of the observable Universe. The point of origin of all the photons we detect is known as the *last-scattering surface.

The thermal spectrum

The best measurement to date of the thermal spectrum of the microwave background is that by the Far InfraRed Absolute Spectrometer (FIRAS) experiment on board the *COBE satellite. This experiment compared the emission from the Universe against that of a reference black-body, in order to map out the intensity at different wavelengths. The result is shown in Fig. 2, a remarkable figure because the errors in the measurement are so small that they have had to be multiplied by 400 to make them visible in the plot. The measurements are perfectly consistent with a black-body spectrum, showing no evidence of any deviations and limiting any such deviations to be very small. The temperature is measured to be 2.725 kelvin with an uncertainty of 0.001 kelvin, making it the most accurately determined cosmological parameter by far.

Nevertheless, deviations are predicted at some level, though it will be challenging to detect them. There are several types predicted.

Compton μ distortion—To exhibit a perfect thermal spectrum the photons actually need two properties. The photons have to have the right distribution in energy (thought to occur before decoupling through frequent scattering from free electrons), known as **kinetic equilibrium**, but there also have to be the right number of photons with regard to their total energy, known as **chemical equilibrium**. The latter is harder to arrange, as it requires interactions which change the actual number of photons, which simple scattering from electrons does not do.

In the high energies of the very early Universe, interactions changing photon number are commonplace and full kinetic and chemical equilibrium is reached. However Rashid Sunyaev and Zel'dovich showed that once the Universe is more than one millionth of its present size, these number-changing processes become inefficient. If energy is injected into the photon distribution after this time but before decoupling, chemical equilibrium cannot be obtained, only kinetic equilibrium. This gives a characteristic distortion to the spectrum known as a μ-distortion (because μ is the symbol used to indicate a so-called chemical potential, measuring the difference between the actual and equilibrium numbers). The COBE-FIRAS observations place a tight upper limit on the μ-distortion, but do not detect it. Most theoretical models do predict that it is negligible, though some more exotic possibilities such as evaporating *primordial black holes are constrained by it.

Compton y-distortion—A more interesting type of distortion is known as the *y-distortion, again first discussed by Sunyaev and Zel'dovich. This arises after decoupling. If the material in the Universe is heated sufficiently (for example by energy from young stars, supernovae, or accreting black holes) it can become ionized, with the free electrons once more able to scatter the microwave photons. As the ionized gas will be hotter than the photons (unlike the situation before decoupling when electrons and photons are in thermal equilibrium), this scattering on average heats the photons, again giving a spectral distortion of a characteristic form.

This distortion can be seen in the direction of galaxy *clusters, where it is known as the *Sunyaev–Zel'dovich effect and will be discussed separately under that heading. It is also of interest to try to measure the distortion averaged over all directions, which indicates the total amount of energy generated in the Universe by the sources above. Current experiments have not been able to detect that signal, with COBE-FIRAS setting a tight upper limit $y < 1.5 \times 10^{-5}$. In this case, theoretical models do predict that there will be a detectable signal, at a level perhaps three to ten times less than the COBE limit. At present however no experiment has been conceived capable of providing that extra sensitivity.

Recombination lines—The vast majority of the photons are part of the thermal spectrum. However, right at the end of recombination a final set of photons are created when the last electrons drop into their lowest energy levels, and since there are no free electrons left these photons cannot scatter to attain a thermal spectrum. Such photons form what are called recombination lines, because they have very well defined energies corresponding to the energy level differences of the electrons in the atoms. However the feature is spread out in wavelength because the photons are emitted over a range of epochs, actually leading to quite broad spectral features which are located on the right-hand side of the thermal maximum seen in Fig. 2. Detecting them would be an interesting probe of the process of recombination, but at present doing so seems extremely difficult, as they correspond to such a tiny fraction of the photon population. Since the ratio of atoms to photons is roughly one in a billion (10^9), there is only about one recombination photon per billion thermal ones. Moreover, they live in a part of the spectrum where the intensity is dominated by the *cosmic infrared background radiation associated with emission from dusty galaxies.

Radiation in the young Universe

As we saw earlier, in the present Universe the CMB contributes only a small fraction of the total energy density of the Universe, around one ten-thousandth. However, while the Universe has been expanding, the density of photons has been reducing not just because the volume has been increasing, but also because the wavelength of the individual photons has been stretched. In combination, this means that the radiation energy density has been falling as the fourth power of the scale factor, whereas non-relativistic matter has been losing its energy density just according to the volume expansion. Accordingly, if we track back in time, the radiation will have contributed a larger fraction of the total. Indeed, when the Universe was around one ten-thousandth of its present size, the radiation and matter densities would have been equal, known as *matter–radiation equality.

Prior to that, when the Universe was younger than about fifty thousand years, radiation was the dominant form of material in the Universe and it is said to have been *radiation dominated.

Kragh, H. *Cosmology and Controversy*, Princeton University Press 1996.

Alpher, R. A. and Herman, R. *Genesis of the Big Bang*, Oxford University Press, 2001.

cosmic microwave background anisotropies (CMB anisotropies)

The most important feature of the *cosmic microwave background (CMB) is that the temperature of radiation coming from different directions is not perfectly uniform, even if one ignores the *cosmic dipole caused by the Earth's motion relative to the microwave background. The cosmic microwave background is *anisotropic. While part of this anisotropy can be attributed to astrophysical objects, particularly *microwave emission from our galaxy and others nearby and microwave scattering by hot gas in galaxy *clusters, a substantial part of these irregularities is believed to be intrinsic to the radiation and already present at the time the microwave background formed.

Comparison of the observed CMB anisotropies with theoretical predictions gives the best evidence in favour of the *standard cosmological model.

Discovery

The discovery of the cosmic microwave background in 1965 initiated a long search for the anisotropies, which were predicted to exist in the late 1960s, for instance via the *Sachs–Wolfe effect. Progressively more sophisticated

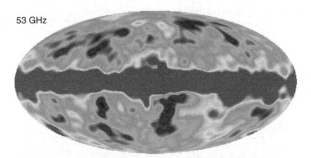

Fig. 1. The COBE satellite was the first to discover cosmic microwave anisotropies, in 1992. This image, shown in Mollweide Projection, shows the final data from four years of observation, released in 1996. The central horizontal band is emission from our own galaxy, but the other features are believed to be primordial. The minimum angular resolution is approximately seven degrees. The main features seen in this map have been confirmed by subsequent experiments, particularly the WMAP satellite—*see* Fig. 5.

experiments were able to impose more and more stringent upper limits on the level of the anisotropies. Such searches received a blow in the 1980s with the widespread acceptance of *dark matter as a key component of *cosmological models: dark matter cosmologies predicted a much lower level of anisotropy than models where *baryons were the dominant gravitating material. By the end of the 1980s, strong upper limits had been set by experiments probing a wide range of different angular scales, with no indication that the CMB was anything other than uniform.

The anisotropies were finally detected in 1992 by the DMR instrument on the *Cosmic Background Explorer (COBE) satellite, the instrument having principal investigator George Smoot. DMR had a very poor angular resolution of around seven degrees, but was able to make all-sky maps enabling a statistical detection of the anisotropies, which has been confirmed by subsequent experiments. ['Statistical detection' means that the signal was not strong enough, as compared to instrument noise, that any particular features could confidently be said to be real. However they were able to say that, on average, there were excess anisotropies that could not be attributed to the instrument but instead must be genuine features in the CMB.] In the final maps, released in 1996, the accumulated signal, as compared to instrument sensitivity, had become good enough to identify actual features in the CMB. The locations of those features have also subsequently been con-

firmed. One of their final maps is shown in Fig. 1.

Foreground radiation

As well as the primordial anisotropies already present at last-scattering, the microwave radiation received by us contains contributions from various other sources, known collectively as **foregrounds** since they lie closer to us than the *last-scattering surface from which the CMB *photons emerge. The most prominent foreground is emission from material lying in the disk of our own Galaxy, which from our viewpoint wraps around a circle in the sky (the microwave version of the visible *Milky Way). This material emits radiation both through the synchrotron process of electrons spiralling in magnetic fields, important at low frequencies, and radiation from cool interstellar *dust which is important at higher frequencies. In addition to these, there is foreground emission from nearby *galaxies (referred to usually as point sources as they are typically of much smaller angular size than the beam of the instrument detecting them), and the *Sunyaev–Zel'dovich effect from galaxy clusters.

Removal of contamination by foregrounds is one of the major challenges of CMB studies. This decontamination is possible because the foregrounds have a different dependence on observing frequency than the CMB, which allows them to be modelled in the data if observations are made at different frequencies. For instance, the COBE satellite used three different frequency bands of detection, and the *Planck satellite will use nine. Template maps of galactic dust emission made at other frequencies can also be used to estimate the levels of foreground contamination.

Origin of the anisotropies

The physics underpinning the production of anisotropies is rather complex, and we can only give a flavour. A detailed calculation would exploit *cosmological perturbation theory, and include all interactions between the photons and the baryons. Such a calculation can only be done numerically; computer programs which do this are known as Boltzmann codes and the most widely used example is *cmbfast.

Amongst the effects contributing are

- Differences in *gravitational potential at last-scattering, known as the *Sachs–Wolfe effect.
- Variations in the *density, and hence temperature, from point-to-point on the last-scattering surface.
- Variations in the velocity of the fluid from point-to-point on the last-scattering surface, leading to a *Doppler effect.
- That the last-scattering surface has a finite thickness, leading to damping of perturbations on smaller scales.
- That a fraction of photons rescatter en route to us, measured by the *optical depth to rescattering following *reionization. Around 10% of photons are affected this way.
- Variations in the gravitational potential as the photons traverse the *Universe towards us (the integrated Sachs–Wolfe (ISW) effect).

The main quantity to be predicted is the *radiation angular power spectrum, known colloquially as the C_ℓ (pronounced *see ell* or *see sub ell*) after its standard notation. This quantifies the main statistical property of the anisotropies, namely how their typical size depends on angular scale. The angular scale of features is measured by the multipole number ℓ (whose technical origin is a spherical harmonic decomposition of the anisotropy pattern), with the rough correspondence of multipole value ℓ corresponding to angular size $\theta \approx 180^o/\ell$.

An example of theoretical prediction, that for the standard cosmological model, is shown in Fig. 2.

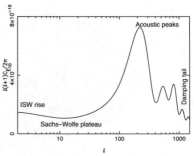

Fig. 2. The predicted power spectrum of temperature anisotropies in the standard cosmological model, with the four main features indicated. By convention, the *y*-axis is the combination $\ell(\ell + 1)C_\ell/2\pi$ rather than C_ℓ itself; this quantity relates most closely to the visual appearance of maps. The maximum anisotropies are at $\ell \sim 200$, corresponding to about one degree; this is the scale on which maps will show the strongest features.

There are four main features in the C_ℓ predicted by the usual cosmological models. They are

- The integrated Sachs–Wolfe (ISW) effect, operating only at very small values of ℓ, due to variations in the gravitational potential as the CMB propagates towards us.
- The Sachs–Wolfe plateau, a fairly flat region at smallish ℓ values due to variations in the gravitational potential on the last-scattering surface.
- The *acoustic peaks, a series of oscillations which are essentially a photograph of the complex fluid motions in the Universe as the CMB was created.
- The damping tail, where the anisotropies on small angular scales (large ℓ) are suppressed due to *Silk damping.

Modern CMB anisotropy experiments

Following the first detection by COBE, during the 1990s a large number of different CMB experiments came into operation, either ground-based at high-altitude sites or balloon-borne. A major contaminant of CMB studies within the atmosphere is microwave emission from water vapour, and as a strategy for reducing this the South Pole has been a favoured choice of location for experiments.

The next major breakthrough came in 1998 with the *Boomerang experiment, a long-duration balloon flight at the South Pole. This experiment made the first accurate mapping

Fig. 3. An all-sky map of the CMB anisotropies from three years of observations by the WMAP satellite, shown in Mollweide projection. Observations at several frequencies have been combined to subtract foregrounds as well as possible, to leave the primordial anisotropies; accordingly any features in the galactic plane are suppressed as compared to Fig. 1. The angular resolution is about fifteen arcminutes, around 30 times better than COBE, allowing small-scale features to be seen. The corresponding power spectrum is in excellent agreement with the standard cosmological model, and was the key piece of data establishing that model as the first precision cosmology.

of the first peak in the CMB power spectrum, and by locating it at approximately one degree it provided the first strong observational evidence that the *geometry of the Universe is spatially flat.

The state of the art in CMB observations are the maps made by the *Wilkinson Microwave Anisotropy Probe (WMAP) satellite, with data so far released in 2003 and 2006. WMAP

was the natural successor to COBE, with several science team members common to both projects, but offered much higher angular resolution and sensitivity. It provided exquisite maps of the microwave sky down to a resolution of about one quarter of a degree; see Figs. 3 and 4. WMAP confirmed the patterns first seen by COBE—see Fig. 5.

Fig. 4. An alternative rendering of the WMAP data, now shown on the surface of a sphere. We would be located at the centre of the sphere, whose radius is approximately the size of the observable Universe, about 14 000 megaparsecs. This is the largest scale map you will ever see! The disadvantage of this rendering, as opposed to the Mollweide one in Fig. 3, is that we can only see half the sky.

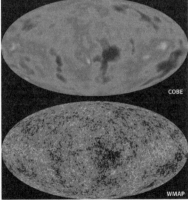

Fig. 5. A comparison of the maps from the COBE and WMAP satellites. While WMAP has considerably better angular resolution, the agreement on the largest-scale features is excellent. [Note that a minor cheat has been used in making the comparison: WMAP data on emission of microwaves from our galaxy has been used to subtract that contaminant from the COBE measurements.]

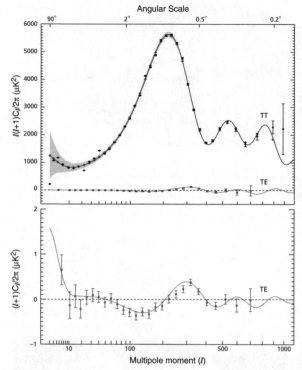

Fig. 6. The power spectra of CMB anisotropies obtained from the first three years of data from the WMAP satellite. The upper panel shows the temperature anisotropies (TT); the points have error bars, but except at large ℓ they are so small as to be almost invisible. The curve shows the best-fitting cosmological model. The shaded region indicates the uncertainty due to cosmic variance, which is large only at small ℓ. The lower panel shows the correlation between temperature and polarization anisotropies (TE, also shown at the bottom of the upper panel). The x-axis has been distorted to highlight the main features.

The power spectrum measured by WMAP is shown in Fig. 6, and shows stunning agreement with the theoretical predictions outlined above, which had been made decades previously. For the first time, a single experiment traced the microwave anisotropy structure all the way from the largest scales across several of the acoustic peaks, enabling cosmologists to constrain *cosmological parameters significantly. By the time the first three years of data were released in 2006, WMAP dominated the constraining power of cosmological data and was able to support the standard cosmological model on its own, though the most powerful constraints still come from combining WMAP with other datasets such as the galaxy *power spectrum.

Polarization

The current frontier of CMB anisotropy studies is *polarization. Theoretical models predict that the CMB should be polarized (due to anisotropies already existing at the time that photons last scatter), though at a level which is only a few per cent of the temperature anisotropies.

Polarization comes in two types, known as E-type and B-type. The E-type also exhibits a correlation to the temperature anisotropy, so that in total there are three extra quantities that can be measured. The B-type polarization is of special interest as it is only generated by primordial *gravitational waves, as for instance may have been generated by *cosmological inflation, but it is the weakest of the three signals. Theoretical predictions for each are shown in Fig. 7 (there being no B-type polarization predicted in the standard cosmological model).

CMB polarization was first convincingly detected by the *Degree Angular Scale Interferometer (DASI) experiment, led by John

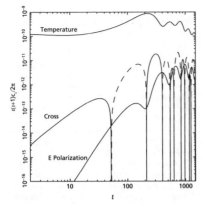

Fig. 7. Theoretical predictions for the temperature and polarization anisotropies in the standard cosmological model. These were obtained by running the cmbfast code. The temperature and *E*-polarization signals are always positive, while the correlation between them may be positive or negative, with the negative part shown as the dashed curve. We see that the temperature anisotropies are by far the dominant form.

Carlstrom, in 2002. It has subsequently been mapped in more detail by WMAP and by a later flight of Boomerang; an all-sky polarization map from WMAP is shown in Fig. 8. So far only the temperature–polarization cross-correlation and the E-type polarization have been measured. The correlation of the temperature with the E-type polarization as measured by WMAP is shown in Fig. 6.

Future polarization experiments are likely to be limited by the extent to which foregrounds are polarized, which is presently extremely uncertain. Upcoming attempts to improve knowledge of polarization come from the Planck satellite and from a variety of ground-based polarization experiments under construction as we write.

Extracting cosmology from the CMB

The CMB anisotropies are currently the single most powerful piece of observational support for the standard cosmological model. Their power arises from three properties:

- The anisotropies can be accurately measured.
- The anisotropies for a given cosmological model can be accurately predicted.
- Those predictions depend significantly on the cosmological parameters that we are seeking to measure.

The procedure for obtaining cosmological constraints is as follows. Observations are made, usually at several different microwave frequencies in order to enable foreground subtraction. The best possible map of the CMB is made, and then the C_ℓ estimated from that map. Finally, a *parameter estimation computation is carried out, where theoretical predictions are made with different values of the cosmological parameters, in order to assess which ones generate predictions for the power spectrum compatible with the observations. If different possible choices of parameters are also being considered, a *model selection analysis might also be made.

Current applications of this approach derive most of their constraining power from the WMAP satellite observations, and give

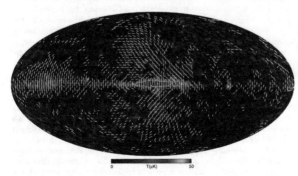

Fig. 8. An all-sky map of the large-scale polarization pattern from the WMAP satellite. Polarization has a direction, shown by the orientation of the bars, as well as a magnitude indicated by their length.

rise to the parameter values described in the standard cosmological model topic.

In addition to the power spectrum, attempts have been made to investigate whether there is any primordial *non-Gaussianity in the CMB anisotropies. Thus far none has been found.

cosmic rays

The term 'cosmic rays' does not appear to have a very precise meaning, but generally speaking it refers to any high-energy particles which are detected on Earth but whose origin is from outside the Earth. Their extra-planetary origin was first demonstrated by Victor Hess in 1911. Cosmic rays come in many varieties, and also span an extremely wide range of energies, with the most energetic greatly exceeding the energies which *particle accelerators are capable of.

The lowest-energy cosmic rays emanate from the Sun, and intermediate-energy ones are presumed to be created within our Galaxy, typically being connected with *supernovae. However the origin of the highest-energy cosmic rays is uncertain. Cosmic rays are interesting both as a way of learning about high-energy astrophysical environments, and of studying fundamental particles at extreme energies.

Types of cosmic ray

Cosmic rays are fundamental particles, but come in a wide variety of types. The majority are individual *protons, but almost the whole range of atomic nuclei are represented, with the *electrons stripped off. *Electrons also make up a component of the cosmic rays, and sometimes gamma rays, the highest-energy form of *photons, are included in the definition.

Cosmic ray detectors

Most types of cosmic ray are not detected directly at the surface of the Earth. On arriving at Earth, they collide with particles in the Earth's atmosphere, sparking a chain of reactions resulting in a large number of particles known as a cosmic ray air shower. By studying these showers, the properties and energy of the original cosmic ray can be reconstructed.

Modern detectors come in two main forms. One seeks to detect the air shower particles themselves at ground level, where they may be spread out over hundreds of metres. The air shower particles are highly *relativistic,

and while nothing can travel faster than the *speed of light in a vacuum, the particles may find themselves exceeding the speed of light within a particular medium, for example water. Such particles then emit a beam of radiation known as Cherenkov radiation, which is the optical equivalent of a sonic boom sent out when an aeroplane exceeds the sound speed in air.

An alternative method is to use the phenomenon of atmospheric fluorescence. A charged particle passing close to an atom in the atmosphere may temporarily excite one of its electrons to a higher energy level, and the photons created when the electrons return to the lowest energy state can be seen by sensitive detectors provided there is no stray background light. This technique enables the cosmic ray shower to be studied within the atmosphere.

The most impressive detector is the Pierre Auger Observatory, named after the French cosmic ray pioneer who first discovered air showers. Ultimately it will have components in both northern and southern hemispheres; the southern hemisphere site began operating in 2003 in Argentina and is the largest physics experiment in the world, with detectors spread over an area several thousand square kilometres. Two detector types are in operation; 1 600 large water tanks to directly detect air shower particles via the Cherenkov effect, and four detectors of fluorescence in the atmosphere. The geographical location of the array is shown in Fig. 1.

Cosmic ray energies

Cosmic rays span a very wide range of energies, as shown in Fig. 2 (which uses *particle physics units). The observed spectrum is for the most part well fitted by a power-law spectrum, though two features known as the 'knee' and 'ankle' of the spectrum can be seen relative to this. The cause of the knee, at around 10^{15} electron-volts, remains unclear but is thought to be created by the details of how particles are accelerated in an expanding shockwave around a supernova.

Ultra-high energy cosmic rays

A particular enigma in cosmic ray physics is the very highest-energy events. These are extraordinarily rare; the highest-energy event ever was seen by a detector in Utah known as the Fly's Eye detector in 1991, with a comparable event later seen at the

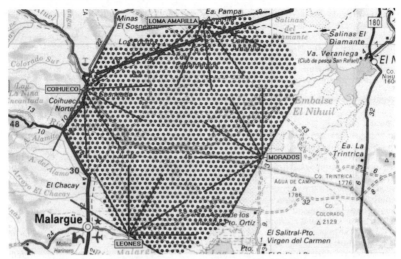

Fig. 1. A map of the location of the vast southern Pierre Auger Observatory. The dots indicate the location of water tank Cherenkov detectors, and the four groups of radial lines the position and viewing directions of the four atmospheric fluorescence detectors.

Japanese AGASA detector. Each shower is believed to have been induced by a cosmic ray with energy in excess of 10^{20} electron-volts. The trouble is that, rare

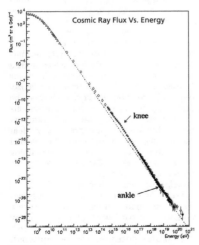

Fig. 2. The energy spectrum of cosmic rays.

though these events are, the predicted rate is precisely zero. This is because such energetic particles can interact directly with the *cosmic microwave background (CMB), and in 1966, very soon after the CMB's discovery, Kenneth Greisen, Georgi Zatsepin, and Vadem Kuzmin used this to show that no cosmic rays of energies above 5×10^{19} electron-volts should be capable of traversing intergalactic distances. This limit is known as the GZK cut-off, and at the time of writing tentative observational evidence is emerging that the cosmic ray spectrum does steepen sharply beyond this energy.

Solving the mystery of ultra-high energy cosmic rays is one of the key goals of the Pierre Auger Observatory.

Astrophysics from cosmic rays

The main astrophysics, as opposed to particle physics, to come from cosmic rays concerns supernovae, as they are believed to be the main origin of cosmic rays. The cosmic rays are not directly created in the explosions themselves, but are caught up in the tangle of magnetic fields created by the expanding shell of material around the explosion. As the particles move about within the complicated magnetic field, some gain energy and are eventually able to escape. After escaping their source, the paths of cosmic rays continue to be altered by the galactic magnetic field, which scrambles their routes so that they appear to come from all directions, rather than directly from their source.

Cosmic rays, despite their name, have not as yet been used to study *cosmology.

cosmic strings

Cosmic strings are a hypothetical type of object which may exist in our *Universe, having formed in a *phase transition. They are an example of a *topological defect; their formation is discussed under that heading. Cosmic strings would be extremely thin lines of high energy passing through our Universe. They are not to be confused with *superstrings, which are an unrelated concept.

Cosmic strings may have formed, for instance, during a *grand unification phase transition, or during the phase transition bringing a period of *cosmological inflation to an end. If a suitable transition takes place, their formation is guaranteed by a process known as the **Kibble mechanism** after physicist Tom Kibble. This states that regions separated by further than the distance that light could have travelled (known as causally separated regions) must experience the phase transition independently, and thus are unable to guarantee that there will not be defects between those regions. The result is a cosmic string network, such as that shown in Fig. 1.

A cosmic string network consists of very long (probably infinitely long) strings and also finite loops of string. A rule of topological defects such as cosmic strings is that they cannot come to an end, so they must either continue for ever or join their ends into a loop. There is some dispute between cosmologists as to the likely balance between infinite strings and loops.

Once formed, a cosmic string network does not remain static. The strings have considerable tension, which causes the strings to move at relativistic speeds. Loops undergo oscillations, and strings can collide and interact; for example if a piece of infinite string happens to intersect with itself it can chop off a small loop. Loops can also radiate energy. The net effect of these processes can be investigated using computer simulations, which indicate that the cosmic string network adopts a **scaling solution** which guarantees

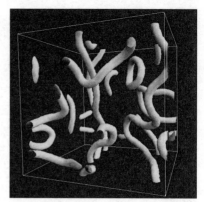

Fig. 1. A computer simulation of cosmic strings. The simulation box has been given periodic boundary conditions, so that a string exiting one side of the box is to be considered joined to the other side of the box.

that the total energy in the network remains a fixed fraction of the total energy of the Universe. Therefore, if a string network formed in the early Universe, there will still be strings in our present Universe. The prediction is that there will be perhaps one long string, but also many loops of string, within our *observable Universe.

No one has ever observed a cosmic string, but if they do exist they can have several important observational consequences. By their nature, they correspond to irregularities in the distribution of energy in the Universe, and those irregularities can be amplified by *gravitational instability to produce structures such as *galaxies. Indeed, during the 1980s cosmic strings were regarded as a strong candidate model for explaining the origin of structure, rivalling the inflationary cosmology. Eventually, however, the possibility that cosmic strings are solely responsible for the observed structures was excluded by precision observations of the *cosmic microwave background anisotropies, as cosmic string models predict quite different patterns of anisotropy than those observed. Nevertheless, it remains an interesting possibility that cosmic strings might contribute part of the perturbations leading to structure formation, which is particularly plausible if cosmic strings form in a phase transition which ends inflation. If so, increased observable precision may unveil their effects.

Cosmic strings also cause *gravitational lensing; light from a galaxy positioned behind a cosmic string could be bent round either side of the string to create a double image, and a cosmic string might create a series of such double images along its length. They can also gravitationally lens the cosmic microwave background radiation, leading to sharp discontinuities in the temperature on either side of a moving string. This is known as the Kaiser–Stebbins effect, after Nick Kaiser and Albert Stebbins who first predicted it. Observation of either of these lensing effects would be powerful evidence indicating that there are cosmic strings in our Universe.

cosmic structure

Cosmic structure, also known as the large-scale structure of the *Universe, describes the way that *galaxies and other tracers of mass in the Universe are distributed. Although on the largest scales we believe the Universe to be *homogeneous, the distribution of matter on smaller scales is far from random: galaxies tend to cluster together in groups or *clusters, and clusters themselves form *superclusters.

Observations of cosmic structure provide important constraints on *cosmological models, since this structure is believed to trace regions of above-average density in the *early Universe at the end of *cosmological inflation.

Origin of cosmic structure

It is now widely believed that cosmic structure has its origins in tiny *density perturbations created during cosmological inflation. One can show that the observed temperature variations in the *cosmic microwave background (CMB) radiation of $\Delta T/T \sim 10^{-5}$ correspond to density perturbations at the *last-scattering surface of $\Delta\rho/\rho \sim 10^{-4}$. This is because the CMB photons lose energy (are *redshifted) as they climb out of the *gravitational potential associated with overdense regions, the *Sachs-Wolfe effect. Conversely, a CMB photon emitted from an underdense region will gain in energy. Thus hotter (blueshifted) regions of the CMB correspond to underdensities, and cooler regions correspond to overdensities. Observations of the CMB thus provide an important verification of the inflationary origin of cosmic structure.

Growth of cosmic structure

Today, on the scale of galaxy clusters, around 8 Mpc, $\Delta\rho/\rho \sim 1$, thus there has been a factor $\sim 10^4$ growth in density fluctuations. These density fluctuations grow due to *gravitational instability: dense regions tend to collapse and become denser, underdense regions expand, relative to average density regions. Once the Universe enters the *matter-dominated era, density perturbations grow with the *scale factor of the Universe, i.e. $\Delta\rho/\rho \propto a(t) \propto 1/(1 + z)$.

For *baryonic matter (ordinary atoms), density perturbations can only start to grow when the Universe becomes transparent. Before this time, the *pressure from radiation resists the pull of gravity. Since last-scattering occurs at $z \sim 1\,000$, we would expect the initial density perturbations of $\Delta\rho/\rho \sim 10^{-4}$ to have only grown to ~ 0.1 by today, and so galaxies should never have formed! This problem is avoided by assuming that the matter content of the Universe is dominated by *cold dark matter. In this case, density perturbations start growing as soon as the density of dark matter exceeds that of radiation, at a redshift $z \approx 3\,000$ or so, and factors of $\sim 10^4$ growth in $\Delta\rho/\rho$ can be achieved. At $z = 3\,000$ the horizon scale is roughly 100 Mpc in *comoving coordinates (i.e. that is the size the horizon has expanded to today). This scale then sets the expected upper limit on the sizes of cosmic structures.

Observed structures

Starting on the smallest scales, stars are bound together by *gravity into galaxies. Galaxies themselves tend to be concentrated into groups or clusters of galaxies. Finally, galaxy clusters form loose associations known as superclusters. One thus observes a *hierarchy* of structures on different scales. As a general rule, the smaller structures show the largest density contrasts $\Delta\rho/\rho$, and also tend to be the most dynamically relaxed systems. This fits in with *hierarchical structure formation theories, in which small structures form first and then later coalesce to form larger and larger structures. Very few galaxy clusters appear to be dynamically relaxed, and it is not even clear whether all superclusters are gravitationally bound or not.

Measuring structure

There are several ways of quantifying cosmic structure. First one has to nominate a tracer of the structure. Tracers can be galaxies (selected in any of a number of ways, such as from optical, infrared or radio surveys), *quasars, or groups or clusters of galaxies, possibly selected from *X-ray surveys. One can even trace the distribution of mass, whether luminous or not, via *gravitational lensing surveys.

One then chooses a statistic to quantify the structure. Commonly used statistics include the *correlation function, the *power spectrum and the distribution of *counts in cells. Before one interprets the results of any statistical analysis, it is important to ascertain that the sample being analysed comprises a fair sample, that is that it forms a representative sample of the Universe. One can test for this by subdividing the sample into contiguous sub-volumes and analysing each sub-volume independently. If each sub-volume is a fair sample, then each should produce consistent results, in which case the full sample is certainly a fair sample. Note that whether a sample is fair or not depends on the scales at which one is quantifying structure (the largest scale must be much smaller than the sample size, see COSMIC VARIANCE), and on the statistic being applied. For example, a sample which is fair for measuring the two-point correlation function may not be fair when measuring higher-order statistics, such as the three-point correlation function, if it happens to contain a large supercluster.

cosmic variance

Cosmic variance arises from the fact that we can only make observations of the *observable Universe within our horizon, and so there is a fundamental limit on how accurately we can estimate *cosmological parameters. Cosmic variance becomes a serious problem as we attempt to make observations on scales approaching the horizon distance, since we can extract only a single sample of this size. As an example, the uncertainties in low-multipole estimates of *cosmic microwave background anisotropies are dominated by cosmic variance.

Cosmic variance may be estimated using *numerical simulations, which allow one to generate several realizations of the observed Universe, or by subdividing an observed sample into sub-samples, and estimating the required statistic for each sub-sample in turn.

cosmic web

The cosmic web is the name given to the filamentary, or web-like, distribution of matter in the Universe. It is apparent both in *redshift surveys of the galaxy distribution (see SLOAN DIGITAL SKY SURVEY, Fig. 1, and TWO DEGREE FIELD GALAXY REDSHIFT SURVEY, Fig. 1), and also in the results of *numerical simulations of structure formation (see N-BODY SIMULATIONS, Fig. 2). According to simulations, the cosmic web began to form from *dark matter, before ordinary *baryonic matter had a chance to collapse. After the epoch of *recombination, ordinary matter, primarily neutral hydrogen, was able to cluster along filamentary structures. *Galaxies and stars began to form from collapsing gas clouds in the highest-density regions, and today they occupy 'knots' in the cosmic web. The largest bound structures, *clusters of galaxies, formed at major junctions of the filaments. Today, material is pulled along these filaments into clusters by the force of *gravity.

cosmic X-ray background

Observations of the sky by the first *X-ray rockets and satellites in the 1960s and 1970s revealed the presence of a diffuse background. It was realized that the source of this high-energy radiation must come from outside the Solar System (or at least from beyond the Moon's orbit) since the X-ray background was occulted (blocked out) by the Moon.

The observed X-ray background may be divided into two components. The first, 'soft' component, corresponding to energies below 0.3 keV, arises mostly from sources within the *Milky Way Galaxy and is also known as the Galactic X-ray background. The observed Galactic X-ray background is produced largely by emission from hot gas within 100 parsecs of the Sun, and so is truly diffuse in nature.

The second, 'hard' component (above 0.3keV) arises mostly from unresolved sources outside of the Galaxy, and is known as the cosmic X-ray background. Recent X-ray telescopes, such as the Chandra X-ray observatory, have in fact resolved around 80% of the cosmic X-ray background to discrete extra-galactic X-ray sources, the bulk of which are *active galactic nuclei.

cosmogony

The study of the origin and creation of the *Universe is known as cosmogony. We can trace the current expansion of the Universe

back to a *big bang, a point in time when the
Universe had infinite density. At times earlier than 10^{-43} seconds, the *Planck time, the
currently understood laws of physics break
down, and so the cause of the big bang itself
remains a matter of conjecture. However, it is
thought that *quantum cosmology may one
day be able to explain the origin of the Universe as a tiny fluctuation on the quantum
scale.

Some cosmological models attempt to circumvent a creation event by postulating a
*cyclic Universe, which has existed indefi-
nitely and goes through a series of big bangs
and *big crunches, or by assuming that our
Universe is only one of an infinite number of
parallel universes comprising a *multiverse.

cosmography

Cosmography is the process of mapping the
*Universe. There are many ancient cosmogra-
phies, but, beyond what was observable to the
unaided human eye, these relied on conjecture and imagination, rather than on scientific observation.

*Galaxies provide the most obvious tracers
of the *cosmic structure, and so the modern
study of cosmography very much parallels the
compilation of *galaxy surveys. Before it was
realized that there exist galaxies beyond our
own *Milky Way Galaxy, the Universe and the
Galaxy were often regarded as synonymous.
Thomas Wright of Durham in 1750 correctly
interpreted the diffuse band of light from the
Milky Way as being due to a flattened distribution of stars (see Fig. 1).

Since then, there has been a long history of
constructing maps of the sky. Here we briefly
discuss attempts to survey significant areas of
sky (of order a quarter or more of the entire
sky).

The Carte du Ciel ('Map of the Sky') was an
international project to map the positions of
millions of stars brighter than 14th magnitude
using 22 000 photographic plates taken at 18
observatories around the world. The project
was started in 1887, but progressed much
more slowly than anticipated, and although a
partial catalogue was published in 1958, it was
never completed. By comparing the precise
locations of stars with more recent observations of the same areas of sky, the Carte du
Ciel photographic plates have proved useful
in measuring stellar motions.

The first complete survey of the Northern
Sky was the Palomar Observatory Sky Survey

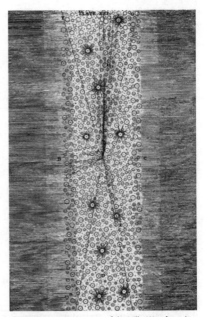

Fig. 1. Thomas Wright's map of the Milky Way, from *An
Original Theory or New Hypothesis of the Universe*,
published in 1750.

(POSS), carried out using the Oschin Schmidt
Telescope on Mount Palomar, California,
1950–57. This type of telescope, invented by
Bernhard Schmidt, is specially designed to
provide a very wide field of view, typically 6
by 6 degrees (compared with the 2 by 2 degree
Carte du Ciel plates), making it ideal for survey astronomy, where the aim is to cover
large areas of sky rather than study particular objects in detail. The POSS, which imaged
the sky above declination $\delta = -33$ degrees
using blue- and red-sensitive plates, has been
widely used to construct catalogues of galaxies (for example the Uppsala Galaxy Catalogue) and galaxy *clusters (notably the *Abell
cluster catalogue).

A second Palomar survey was carried
out in the 1980s and 1990s using more
sensitive photographic plates in blue, red
and near-infrared passbands and after
making improvements to the telescope
optics.

A Southern hemisphere equivalent of the
Palomar surveys was carried out with the

United Kingdom Schmidt Telescope (UKST) at Siding Spring, Australia, and the European Southern Observatory (ESO) Schmidt telescope on La Silla, Chile. Starting in 1973, the UKST and ESO Schmidt telescopes surveyed the sky south of declination $\delta = -17$ degrees using blue- (UKST) and red-sensitive (ESO) plates.

Between them, the POSS and UKST/ESO surveys cover the entire sky to a blue magnitude limit of $B_J \approx 22.5$. All of these survey plates have been digitized by the Space Telescope Science Institute and images of any part of the sky from these and other surveys may be accessed from the Digitized Sky Survey website below.

The *InfraRed Astronomical Satellite (IRAS) surveyed almost the entire sky in four mid to far infrared passbands in 1983, providing one of our first views behind the plane of the Milky Way Galaxy, which heavily absorbs optical light. A follow-up *redshift survey, the Point Source Catalogue Redshift (PSCz) Survey, measured redshifts for 15 000 galaxies over 83% of the sky.

The *Two Micron All Sky Survey (2MASS) provided the first complete sky coverage in the near-infrared part of the *electromagnetic spectrum, around 2 microns in wavelength, using two dedicated 1.3m telescopes, one in Arizona and one in Chile.

The largest sky survey to date, in terms of volume surveyed, is the *Sloan Digital Sky Survey. This is the first project to survey a substantial fraction of the sky in the optical using digital detectors rather than photographic plates.

In the X-ray part of the spectrum, the *ROSAT satellite has made all-sky maps in bands at 0.25, 0.75, and 1.5 keV (*see* PARTICLE PHYSICS UNITS). The X-ray emitters of greatest cosmological interest are *clusters of galaxies.

Maps of the sky have also been constructed at radio and *ultra-violet wavelengths.

All of the above surveys have used detected radiation to map the locations of astronomical sources. What we would really like to do is to map the distribution of all matter, including *dark matter. This is now becoming possible using the technique of *gravitational lensing. Here one infers the presence of possibly unseen mass by the gravitational lensing effect it has on light rays from sources behind the lensing mass.

http://archive.stsci.edu/cgi-bin/dss_form

cosmological constant

The cosmological constant is the simplest form of *dark energy, and according to the *standard cosmological model it is the dominant form of energy *density in the present *Universe. Understanding the origin of the dark energy, and in particular whether it is in the form of a cosmological constant, is one of the highest priorities in modern *cosmology.

History

The cosmological constant, always indicated by the Greek symbol Λ (capital lambda), was introduced by *Einstein into his theory of *general relativity. His motivation was to allow *cosmological models in which the Universe was static (i.e. not expanding), this being a decade or so before Edwin *Hubble's discovery of the expansion of the Universe. Once the expansion of the Universe became established, Einstein retracted the cosmological constant, famously denouncing it as '...the biggest blunder of my life'; he had missed a chance to predict the expansion of the Universe. [This phrase comes from an account of a conversation with Einstein by George *Gamow in his biography. Incidentally, Einstein is separately quoted as saying that 'the greatest mistake' of his life was a letter sent in his name (probably written by another with his blessing) to President Roosevelt in 1939 alerting him to the possibility of the development of nuclear weapons. This letter is seen by some as sparking the arms race, though in hindsight that race appears inevitable.]

Despite Einstein's retraction, the cosmological constant was intermittently invoked by cosmologists throughout the following decades, for diverse reasons such as explaining the *redshift distribution of *quasars (later found to be primarily due to evolution in the quasar population rather than a cosmological effect). It was never a popular choice however.

Its role in modern cosmology can be traced to a seminal paper by Princeton-based cosmologist James Peebles in 1984. The *inflationary cosmology had recently been introduced and predicted a flat Universe, while observations indicated that the density of matter, including *dark matter, fell well short of the *critical density needed to achieve this. The cosmological constant was able to plug that gap, and Peebles showed that it led to satisfactory models of *structure formation. In fact, his model was pretty much the *standard cosmological model in its modern form.

For more than a decade, however, acceptance of this model was not widespread, and it was seen as one of several competing models to explain the Universe. In 1998, however, observations of distant *supernovae effectively ruled out the competitors, and the existence of dark energy finally became accepted by the vast majority of cosmologists. Einstein's biggest blunder, in fact, was yet more prescience on his part.

Observational evidence
The cosmological constant is a specific kind of dark energy (*see* DARK ENERGY). Thus far, the cosmological constant has been fully successful in explaining observed data, faring at least as well as any other dark energy model which has been proposed.

The cosmological constant problem
From the viewpoint of general relativity theory, the cosmological constant is simply a parameter which is permitted by the structure of the theory, but which is otherwise undetermined. It measures the energy density of empty space, and hence is sometimes known as the **vacuum energy**.

From a particle physics perspective things are rather more interesting, as the vacuum energy is something which in principle can be predicted, being equivalent to the zero-point energy of all the particles. It is well known that, in *quantum mechanics, even at absolute zero some energy remains, and the same is true in particle physics theories. Unfortunately, the value predicted by these theories is too big to match observations. And not just 'too big' in the usual sense of ten or a hundred times too big; in particle theories with *supersymmetry the prediction is too big by a factor of about 10^{60}, and without supersymmetry this increases further to 10^{120}. This extraordinary discrepancy between theory and observation is known as the cosmological constant problem, thought by many to be the most vexing problem in all of fundamental physics.

Thus far, all attempts to understand the small value of the cosmological constant from physics alone, for instance attempting to cancel the quantum zero-point energy by a classical cosmological constant, have met with failure. Recently, however, a promising though controversial approach has been developed

using the *anthropic principle, which adds the extra ingredients that the value of the cosmological constant may vary on very large scales in the Universe, and that only regions where it is unnaturally small prove suitable for galaxy formation and hence ultimately life. In such a scenario, there may be no unique prediction of the cosmological constant from fundamental principles, but nevertheless all life in the Universe will arise in regions where it is suitably small and so that is what any observers see. This argument also motivates why its value should be non-zero, as all suitably small values are acceptable and arise in different locations in the Universe.

The coincidence problem
A second problem related to the cosmological constant is known as the coincidence problem, which asks why we should be living at an epoch when the dark energy and dark matter densities happen to be almost equal (to within a factor of three or so). At all significantly earlier epochs the dark energy density would be completely subdominant to matter and radiation, whereas only a short time (cosmologically speaking) into the future, the cosmological constant will become completely dominant, since its density is by definition constant, while that of matter reduces in inverse proportion to the volume. It seems that we have developed the ability to try to understand our Universe just as it passes through a rather special phase of its existence.

The coincidence problem is, arguably at least, also resolved by the anthropic principle argument that solves the cosmological constant problem. Unlike the cosmological constant problem, though, there are potential solutions based on physical arguments alone, particularly within *quintessence models.

Future probes
If the dark energy is indeed a cosmological constant, then its density is already quite well measured and all that future observations can hope to do is add further accuracy. Arguably this is unnecessary given the absence of any theory achieving better than an order-of-magnitude guess at its value. Future focus, therefore, is much more directed at seeking to demonstrate that the dark energy is *not* a cosmological constant, for instance by showing that its density varies with time. The ways in which this might be achieved are described at *dark energy.

cosmological inflation

Cosmological inflation, usually abbreviated to inflation, is the key idea of *early Universe cosmology. It postulates that the *Universe underwent a period of accelerated expansion during its earliest stages, during which its size increased by a huge factor. Such an expansion explains a series of otherwise puzzling features about the large-scale Universe, known as the *horizon problem, the *flatness problem, and *relic particle abundances. Inflation's most important role in modern cosmology is that it provides a quantitative theory for the origin of *density perturbations in the Universe, which ultimately induce *structure formation, with the predictions of the simplest inflationary models showing excellent agreement with observations.

Historical motivation

The birth of inflationary cosmology is usually recognized as being in a 1981 paper by American cosmologist Alan Guth, which is by now one of the most cited papers in all of physical science. In fact it was not the first paper to feature an accelerating expansion, now taken as the definition of inflation. Such models go all the way back to the *de Sitter space model of the Universe devised in 1917 by Willem de Sitter, corresponding to a Universe dominated by a *cosmological constant. The possibility that such evolution might be relevant to the very early Universe was first explored in detail in a cosmological model by Soviet cosmologist Alexei Starobinsky in 1980.

Guth's key contribution was to recognize that such an expansion could explain some otherwise mysterious features of the Universe, known as the horizon problem, the flatness problem, and the *monopole problem. The last of these was particularly influential in his thinking, and indeed according to his historical account in his book *The Inflationary Universe* it was only through publicizing his discovery as a way of solving the monopole problem that he learnt of the existence of the other problems.

Although Guth's paper also included a model of particle physics that might lead to inflation, Guth himself already realized that his model would not work as it was impossible to bring the inflationary era to a satisfactory end. This was resolved the following year in independent work by Andreas Albrecht and Paul Steinhardt and by Andrei Linde, with a model known as 'new inflation'. The follow-

ing year, Linde produced another simplification inventing a model known as 'chaotic inflation'. Since then, the creation of candidate inflation models has been a burgeoning industry, but even so a large fraction of inflation models are essentially specific realizations of either the new inflation or chaotic inflation scenarios.

Modelling the inflationary epoch

In order for the expansion of the Universe to accelerate, it must be dominated by a material which has negative *pressure. Such a material is a *scalar field which, in fundamental physics models, describes particles which have zero spin. Many such fields are predicted by particle physics models such as *supersymmetry.

The simplest inflation models, including the new inflation and chaotic inflation classes, are differentiated by choices for the *potential of the scalar field, which measures the energy *density of different configurations (describing for instance different numbers of particles and different interactions between them). Cosmologists have invented a large number of possible inflationary models by making different choices for the potential, sometimes motivated by considerations from particle theory and sometimes by simplicity. Ultimately, one hopes to use observations to distinguish between the different models.

It is not known what might have happened before inflation, but calculations usually assume that at some stage the inflationary scalar field, often called the *inflaton, comes to be the dominant form of matter in the Universe, and hence the driving term in the *Friedmann equation describing the expansion. Typically then the scalar field density evolves very slowly as the Universe expands, while all other materials are rapidly diluted to negligible quantities. The Universe has become completely scalar field dominated.

An example inflaton potential is shown in Fig. 1, with a possible value of the inflaton field during inflation shown schematically by the circle. The inflaton is normally indicated by the Greek letter ϕ, and its potential by the function $V(\phi)$.

During this phase the expansion of the Universe is approximately exponential in time, and in a short timespan it can expand by a huge factor, resolving the cosmological problems stated above in the process. As it does

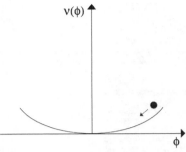

Fig. 1. A schematic drawing of the inflaton field ϕ sitting on its potential. As inflation proceeds, the inflaton will evolve towards the minimum, where inflation will end and reheating take place.

so, the inflaton slowly evolves towards its minimum energy state. Eventually this minimum is approached and inflation comes to an end.

Reheating after inflation

At the end of inflation, the inflaton particles remain the dominant constituent of the Universe. In order to regenerate the familiar particles of the hot *big bang model (*see* 'Overview')—including the *baryons we are made of, the *photons of radiation, and the *dark matter—the inflaton particles must decay. This process is known as *reheating, since in the process the cold Universe created by inflation is heated back to a high temperature through the creation of radiation. In some models, reheating is preceded by a particularly violent first phase of decay known as *preheating.

The details of reheating are the most poorly understood part of the inflationary cosmology, relying as they do on the way that a particle of uncertain identity, the inflaton, decays. However, provided the Universe fully thermalizes after reheating, the details of how it might have taken place are erased and have no particular observational consequences. It is important, however, that the reheating temperature is low enough that this thermalization process is unable to recreate the troublesome relic particles that inflation was designed to dispose of.

Perturbations from inflation

The most important consequence of inflation is that it leads to *density perturbations in the Universe. The mechanism is intrinsically quantum mechanical; according to Werner Heisenberg's famous uncertainty principle, quantum fluctuations are always present in any system. During a period of inflation, these fluctuations get caught up in the rapid expansion and are stretched to enormous length scales, comparable to or even much larger than our present *observable Universe. In the process, they are converted from quantum variations into real variations in the density of the Universe, referred to then as **classical** perturbations (in physics parlance, the word 'classical' means 'non-quantum'). Ultimately, these perturbations lead to the formation of *galaxies and other structures in the Universe.

In addition to the density perturbations, inflation predicts the existence of large-scale *gravitational waves, produced by quantum fluctuations in the *metric itself. No evidence for these has yet been found, but it is hoped that one day the gravitational waves will be discovered and act as a distinctive probe of how inflation took place.

See INFLATIONARY PERTURBATIONS.

Eternal inflation

An unusual feature of inflation is that it might be an eternal process, continuing in some regions of the Universe at the present and into the future. This cannot be true within our own observable Universe, but inflation predicts that the portion visible to us is only a tiny region of the full Universe.

The eternal inflation picture, also known as the self-reproducing inflationary Universe, was introduced by Alex Vilenkin in 1983 and further developed by Linde and collaborators. The idea is that during the energetic early stages of inflation, the quantum fluctuations in the motion of the inflaton—the same quantum fluctuations invoked to provide the density perturbations—might dominate over the classical rolling down the potential. In such a situation, the field might effectively roll up the potential rather than down. Moreover, in those regions where it rolled up, the expansion rate will then be faster as it is proportional to the total energy density. Thus more of the volume of the Universe ends up in regions where the field happened to fluctuate up than in the regions where it fluctuated down.

The typical history of a region of the Universe would be as follows. It would start at very high energy in the region where the

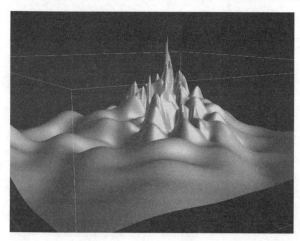

Fig. 2. A computer simulation of the eternal inflationary Universe. Our entire observable Universe would be a minuscule region in one of the foothills; at the spikes inflation is still continuing.

quantum fluctuations dominate the motion, with the Universe rapidly expanding. The energy would jiggle up and down in response to the quantum fluctuations, but eventually by chance it would hit a region of lower energy, where the quantum fluctuations are smaller. The classical rolling of the field down its potential would then start to dominate, pulling it away from the quantum regime. Much later, as the field heads towards the minimum of the potential, the density perturbations that we can observe would be generated, then eventually inflation in our region would come to an end.

Fig. 2 shows a computer simulation of the eternal inflationary Universe.

The eternal inflation scenario is often invoked in the context of the *anthropic principle, as it can lead to widely-separated regions of the Universe ending in different physical states.

Observational status

Inflation is presently very well supported by observations. Concerning the large-scale Universe, inflation predicts that it should be spatially flat and this has now been verified to approximately 1 percent accuracy. The Universe is also confirmed to be on average homogeneous and isotropic.

The most powerful tests come from study of the perturbations in the Universe, particularly *cosmic microwave background (CMB)

anisotropies and galaxy *clustering. These observations support the idea of initial irregularities which were *adiabatic perturbations, which is what the simplest inflation models predict, and furthermore that they are close to the *Harrison–Zel'dovich spectrum in their form, again as predicted by the simplest inflation models. There is also no evidence of primordial *non-Gaussianity of the perturbations, which is difficult to produce from inflation at a significant level.

The real 'smoking gun' for inflation would be detection of the large-scale gravitational waves that inflation produces. These would generate signatures in the CMB anisotropies, which have not yet been detected. Unfortunately inflation makes no clear prediction of the size of this signal; while some models lead to a significant effect, other models predict a level too small to ever be detected. There is therefore no guarantee that this test will ever be possible.

Inflation and particle physics

Inflation is very well motivated observationally, and so it would be nice to know whether it is actually a prediction of modern physics theories such as *superstring theory which purports to give a complete description of Nature. So far there is no definitive answer to this question; indeed it has proven hard so far to produce convincing scenarios. Whether this is a genuine difficulty, or is due to a lack of imagination so far on the part of theorists, remains to be seen. Once inflationary models become more constrained by observations it may become possible to make stronger progress.

Relation to dark energy
There are close parallels between inflation in the early Universe and studies of *dark energy in the present Universe. Indeed, if inflation is defined as an accelerating Universe, then dark energy can be said to be driving a period of inflation. Much dark energy work exploits the ideas developed to study early Universe inflation, particularly that the acceleration might be driven by a *scalar field, which in the dark energy context is known as *quintessence.

An even bolder suggestion is that early Universe inflation and quintessence are not just manifestations of the same physical mechanism, but might actually be caused by the same 'stuff'. Such theories were named 'quintessential inflation' by their inventors James Peebles and Vilenkin, but so far no compelling models have been devised.

Current inflationary research
Present research in inflationary cosmology can be divided into four main topics, though there are of course overlaps amongst them.

Particle physics model-building—The aim here is to make detailed models of inflation based on fundamental physics models, particularly *M-theory and the *braneworld. So far it has not proven possible to make truly compelling models which can be tested against observations.

Inflationary phenomenology—This area is motivated by fundamental physics, but the aim is usually to make simplified models from which observational predictions can readily be made, and which hopefully capture the main physics of the inflationary epoch.

Predicting inflationary perturbations—
Under this heading, researchers are developing techniques for making accurate predictions of the perturbations arising from inflation, so as to compare them with the precision cosmological data expected in the near future from projects such as the *Planck satellite. For the simplest inflation models this is essentially already achieved,

but not yet for more complex models such as braneworld inflation.

Comparison with observations—Taking detailed predictions from inflation models and confronting them with observational data will hopefully one day unveil inflationary dynamics. At the time of writing, observations have not reached the quality needed to seriously constrain such dynamics.

Guth, A. H. *The Inflationary Universe*, Jonathan Cape, 1997.

Liddle, A. R. and Lyth, D. H. *Cosmological Inflation and Large-scale Structure*, Cambridge University Press, 2000 [Very technical].

cosmological models
A cosmological model is a general term for any mathematical model seeking to capture some of the key observed features of the *Universe. A model is normally a collection of physical laws assumed to be true, plus a set of assumptions as to the environment in which those laws are to be applied. The laws then allow the physical situation to be evolved so as to produce predictions which can be set against observations to test the validity of the model assumptions being made.

An example of a simple cosmological model would be to assume the physical law is Einstein's *general relativity, and that the Universe contains a single material which satisfies the assumptions of *homogeneity and *isotropy. This leads to the *Friedmann equation governing the expansion of the Universe. Provided one sticks to global properties of the Universe, this can in fact give quite a good description.

As understanding of our Universe has developed, the range of physical laws considered has grown. A model of the homogeneous Universe would now include various separate types of material (typically *dark energy, *dark matter, *baryons, *photons, and *neutrinos), and would model the interactions between them to predict processes such as *nucleosynthesis, *recombination and *decoupling. This is something which has been done with considerable success.

Further complication then arises if one wishes to model the *structure formation process, which violates the homogeneity assumption. The initial *perturbations in the Universe must then be specified, and the physical equations enhanced to allow the inhomogeneities to be evolved according to *cosmological perturbation theory. At this level of complexity, powerful computers are essential to carry out *numerical simulations of how the Universe evolves, and highly detailed predictions can be made for observables such as the *cosmic microwave background anisotropies.

Ultimately, one thinks of a cosmological model as being defined by a set of *cosmological parameters indicating the different physical laws operating and the environmental variables describing the initial conditions. Typically these parameters are to be determined from observational data, a process known as *parameter estimation. For instance, one might aim to measure the relative amounts of different materials in the Universe or the present expansion rate, by testing for which values of those parameters a suitable match to observations can be achieved. More generally, one might be interested in testing different cosmological models against one another to see which fits the data best, which can for instance be achieved using the statistical technique of *model selection.

cosmological parameters

Cosmological parameters are a series of numbers which describe the detailed properties of our *Universe, indicating for instance the densities of the different constituents, global properties of the Universe such as its *geometry, and properties of the *density perturbations. One of the main goals of contemporary *cosmology is the accurate determination of these parameters, a process known as cosmological *parameter estimation.

Defining cosmological parameters

Defining cosmological parameters is the first step towards comparing a theoretical model with observational data. Faced with some data, we need to decide what physical processes are supposed to be responsible, which are assembled together to form a model of what goes on in the actual Universe. Typically a *cosmological model will be based on some guiding principles, such as for instance the *cosmological principle, and some established physical laws, for instance *general relativity. It is unlikely however that these considerations will define a unique model which makes a single well-defined prediction for observational data. Much more likely is that the model will leave

some questions unanswered, which can only be decided by looking at observational data. An example might be the relative amounts of the different constituents of the Universe, such as *baryons, *dark matter, and *dark energy.

Put another way, our model describes a set of possible Universes, all internally self-consistent, corresponding to different values of the cosmological parameters. But the real Universe is just one of that set, and we need to use observations to discover which it is. Or we may find that our model is not capable of explaining the data for any values of the parameters, in which case we need to refine or extend our model to include new physical effects.

Occasionally, the input of a model might not be a simple parameter, but might be an unknown function. An example would be the potential energy of a *scalar field driving a period of *cosmological inflation. It is not possible to determine a function directly from observational data; first it must be **parametrized** by approximating the function in a way that can be described by parameters, for instance by writing it as a Taylor series, truncating at some order, and then attempting to fit for the values of the coefficients. If successful, an approximation to the original function can be reconstructed.

Usual practice in cosmology has been for researchers to decide for themselves which set of parameters they will consider in fitting to data. An alternative approach, known as *model selection, is to allow the data to decide which set of parameters should be used, as well as finding their values.

Determining cosmological parameters

In order to use cosmological data to determine the values of parameters, one must be able to predict the values of observables, such as the *cosmic microwave background anisotropies or galaxy *power spectrum, as a function of those parameters. One then seeks to vary these parameters until the predictions are in good agreement with all the observations, and then map out the ranges of the parameters that can be regarded as consistent with the data.

Carrying out this procedure in practice is rather challenging, and requires access to considerable supercomputing resources. This is because the calculations of observable quantities even for a single choice of parameter values are time consuming, perhaps one minute of processor time, and because the number of parameters that can be varied is large, typically at least seven or eight and often more.

Early cosmological studies would attempt to search for the best models by evaluating the goodness-of-fit (formally, the *likelihood) by breaking up the parameter space into a grid from which each model would be tested. Accurate sampling however might require say 10 grid points in each parameter direction, so with 8 parameters a total of 10^8 calculations would be necessary to compile the grid, which was and is way beyond feasibility [10^8 minutes of processor time is equal to about 200 years on a computer with one processor, or one year of dedicated time on 200 processors]. Fortunately, such methods have been superseded by *Monte Carlo methods which explore the parameter space in a probabilistic way, in which the algorithm is guided to the high-likelihood regions where good sampling is needed. This yields superior results with typical runs of only 10^5 calculations or so, still enough to require supercomputer access for anyone but the most patient, but within the resources of many cosmology researchers. (For current favoured values of parameters, *see* STANDARD COSMOLOGICAL MODEL.)

cosmological perturbation theory

Cosmological perturbation theory is the study of the development of irregularities in the distribution of matter in the *Universe as it evolves. The term 'perturbation theory' indicates that the irregularities are assumed to be small deviations imprinted on an otherwise *homogeneous Universe; mathematically this leads to the important simplification that the equations describing the evolution can be **linearized**, a process in which one neglects any terms which are the product of two or more quantities on the grounds that if each quantity is small then the product of two of them must be so small as to be worth ignoring. For this reason, the subject is also known as 'linear perturbation theory'.

The underlying physical laws of the Universe are usually assumed to be Albert *Einstein's theory of *general relativity, describing *gravity, plus whatever laws describe the behaviour of matter and its interactions. Most of the complexity is due to general relativity, leading to yet another terminology of 'relativistic perturbation theory'.

Remarkably, the theory of linear cosmological perturbations was derived in essentially complete form as long ago as 1946, in an amazing paper by Soviet physicist Evgeny Lifshitz. This was long before any real understanding of the nature of our Universe had been developed. Since those early days, it has seen considerable development both in terms of building an underlying physical picture and in the inclusion of many kinds of particles and interactions. The ever-increasing precision with which observations are being made places great demands on the ability to carry out accurate perturbation theory calculations.

Although the mathematical equations of cosmological perturbation theory are uncontroversial, their physical interpretation remains challenging. This is because in general relativity we are allowed to make different choices of coordinates to describe physical processes, and there is no single well-motivated choice (different options are technically described as **gauges**) for perturbations on scales large enough that the expansion of the Universe must be taken into account. This 'gauge dependence' disappears only for perturbations on length scales much less than the *Hubble length, when all reasonable gauge choices then reduce to Newton's theory of gravity.

One aims to describe perturbations in both the matter and in the *metric of space-time, which are tied together by *Einstein's equation, but the coordinate freedom can be used to shuffle things around. For instance, one can choose coordinates which make the density of the Universe appear homogeneous (called the uniform-density gauge), in which case all of the 'perturbation' is encoded in the metric alone. Or alternatively the metric could be made homogeneous (the spatially flat gauge), or perhaps neither. The best choice depends on the physical situation, and perhaps also on the ease of implementing computer calculations.

An attempt to obviate this problem was devised by James Bardeen in 1980, known as the **gauge-invariant** formalism, which sought to deal only with quantities that were independent of the choice of coordinates. While the principle of this is attractive, in practice it becomes mathematically very complicated and is nowadays seldom used.

The most important recent landmark in cosmological perturbation theory was the release of a computer program called *cmbfast in 1996 by Uroš Seljak and Matias Zaldarriaga, which contained an efficient numerical calculation of the evolution of linear perturbations. This program tracks all the necessary types of material (*baryons, *dark matter, *photons, *neutrinos, and *dark energy) and computes their evolution from some initial conditions (e.g. such as those motivated by the *inflationary cosmology). This is sufficient to compute the *cosmic microwave background (CMB) anisotropies right up to the present, as they remain small, though it cannot follow the matter distribution right to the present as the process of gravitational collapse takes us beyond the regime of small linear perturbations.

Cmbfast and its successors have become an essential part of the infrastructure of working cosmologists, enabling them to make predictions from their models without necessarily having to understand the full apparatus of cosmological perturbation theory. For instance, without such programs the interpretation of CMB anisotropies measured by experiments such as the *WMAP satellite, in terms of *cosmological models, would not be possible.

Once the magnitude of perturbations is such that the linear approximation is no longer valid, they become known as **nonlinear perturbations**. For these, more sophisticated computer modelling is needed, with *N-body simulations used to evolve the dark matter, and *hydrodynamical simulations used to follow the baryons. These are essential to follow the evolution of the Universe once *structure formation is properly underway.

Another aspect of cosmological perturbation theory is the choice of the initial conditions for the perturbations, which are then to be evolved according to physical laws. For example, the simplest models of inflation yield *adiabatic perturbations, which are the easiest to follow. However, the possibility remains open that the initial perturbations might in part be *isocurvature perturbations.

cosmological principle

The cosmological principle, known also as the Copernican principle, states that we do not live at a privileged location in the *Universe. It is one of the fundamental principles of modern *cosmological models.

Historically it took considerable time for this viewpoint to be established. Copernicus may have overturned the view that the Earth was at the centre of the Universe, but only in order to put the Sun there. Subsequent widely held beliefs included that the Sun was located at the cente of the *Milky Way, and that the Milky Way lay at the centre of the *Universe. Only in the middle of the 20th century did the principle become firmly established, when it became apparent that the Milky Way was a fairly typical large *galaxy with the Solar System lying well off-centre.

In its pure form, the cosmological principle enforces the assumptions of *homogeneity and *isotropy, leading to the Universe being described by the *Robertson–Walker metric and the *Friedmann equation. Such cosmological models are capable of giving a good description of the large-scale properties of the Universe.

The cosmological principle does not however apply precisely to our Universe, as there are differences for instance between dense *clusters of galaxies and *voids. Nevertheless, it can still apply in a statistical sense. Indeed, cosmology as a subject can pretty much be defined as those questions whose answers do not depend on location, i.e. questions that would have the same answer whichever galaxy in the Universe our planet happened to be in. The question 'what is the typical separation between galaxies?' is a cosmological one, but the question 'how close is the nearest galaxy?' is not.

Sometimes one has to be careful to allow for statistical variations between locations in answering cosmological questions. For instance, when measuring the *power spectrum of galaxy clustering or of *CMB anisotropies one has to allow for this effect, known as *cosmic variance.

Although the cosmological principle does not hold in detail, it does become increasingly more accurate if one averages over larger and larger scales. The most powerful evidence for this comes from the near homogeneity of the *cosmic microwave background, and it is now also indicated in large-scale *galaxy surveys such as the *two degree field galaxy redshift survey and the *Sloan Digital Sky Survey.

The cosmological principle has proven quite effective when applied to different locations in space, so it is tempting to extend it to apply in time too. This was the assumption of the now-defunct *steady-state cosmology. However, in the big bang cosmology different epochs are clearly distinguishable as the Universe expands and cools and then structure forms. That we live at this particular epoch is usually seen as arising just from a timing argument that there needed to be enough time for galaxies to form, for *supernovae to generate heavy elements, and for life to evolve.

Perhaps the most puzzling violation of a temporal cosmological principle is the *coincidence problem, which points out that we happen to be discovering about the Universe exactly during a rather brief (in cosmological terms) epoch where the *dark energy density is comparable to the density of other material.

cosmological tests

The purpose of cosmological tests is to derive constraints on the *cosmological parameters from observations such as counts and angular sizes of distant objects such as *galaxies and *quasars. Since these classical tests were first proposed after the discovery of the *Hubble expansion in the 1930s, *anisotropies in diffuse radiation sources, in particular the *cosmic microwave background (CMB), have come to provide even more important constraints on the cosmological parameters.

In a paper published in 1961, Allan Sandage reviewed four classical cosmological tests that could be addressed using the 200-inch telescope on Mount Palomar. These were:

- The *redshift–*magnitude relation.
- Galaxy number–magnitude counts.
- The angular diameter–redshift relation.
- The timescale (comparison of the age of the oldest known objects with the theoretical *age of the Universe).

Redshift–magnitude relation—In principle, the redshift–magnitude relation (which is equivalent to a redshift–distance relation if one knows the intrinsic luminosity of the sources being observed) provides a stringent constraint on the *deceleration parameter q_0. The big problem in practice is identifying a *standard candle whose luminosity is known. For galaxies, luminosity evolution has a far bigger effect than *cosmology at moderate redshifts, $z \lesssim 1$, and so one needs to understand galaxy evolution far better than we do at

present to use galaxies in this test. However, it is thought that type Ia *supernovae are a suitable standard candle, and indeed it was the discovery in the late 1990s that high-redshift type Ia supernovae are fainter than expected that led to the currently favoured accelerating cosmology.

A closely related test is the redshift–volume relation, where one counts sources as a function of redshift. To perform this test, one needs to understand the *selection function of the survey being analysed, in addition to the luminosity and density evolution of the sources.

Galaxy counts—Sandage concluded that galaxy number–magnitude counts (that is the observed number of galaxies as a function of apparent magnitude) are insensitive to the *cosmological model. Magnitude errors of order $\Delta m \approx 0.3$ give rise to an error on the derived deceleration parameter $\Delta q_0 \sim 1$. Whilst nowadays one can perform significantly more accurate photometry, this test is still plagued by uncertainties in *galaxy evolution models.

Angular diameter–redshift relation—The angular diameter subtended by a physical distance can be predicted as a function of redshift z, and so in principal, one can use the angular sizes of galaxies and *clusters of galaxies as a cosmological test. In addition to uncertainties in the size evolution of galaxies and clusters, an equally serious problem here is the fact that the *surface brightnesses, and hence isophotal diameters of galaxies, vary strongly with redshift.

Nowadays, the *acoustic peaks in the CMB *power spectrum and *baryon oscillations in the galaxy power spectrum, provide modern implementations of this test, which do not suffer from the problems mentioned above.

Timescales—The age of the Universe in the *standard cosmological model, $t_0 = 13.7 \pm 0.2$ Gyr, is consistent with the ages of the oldest known *globular clusters (13.5 ± 2 Gyr).

cosmology

Cosmology, derived from the Greek *cosmologia*, meaning 'world discourse', is the study of the entire *Universe. Although modern scientific cosmology is a recent field of study, not really starting until the early 20th century, the philosophical, religious and esoteric study of the Universe has a long and ancient history.

Cosmology may be divided into four disciplines: physical cosmology, metaphysical cosmology, religious cosmology, and esoteric cosmology.

Physical cosmology

Physical cosmology, the topic of this book, is the study of the Universe through scientific calculation, observation, and experiment. (For an overview of our current understanding of physical cosmology, *see* STANDARD COSMOLOGICAL MODEL.)

Metaphysical cosmology

Metaphysical cosmology describes the philosophical study of cosmology. It addresses questions about the Universe which are beyond the scope of science, and makes conjectures which are not (yet) observationally testable. For example, what caused the *big bang? Is there a creator? What, if any, is the 'purpose' of the Universe? Is our Universe just one of a large number, making up a *multiverse?

As our scientific understanding of the Universe increases, some metaphysical topics may enter the realm of physical cosmology. As an example, prior to the 20th century, the existence of separate 'island universes' was a matter of conjecture. We now have sound scientific evidence for the existence of other *galaxies. We may, in the future, gain a scientific understanding as to the origin of the big bang.

Religious cosmology

Religious cosmology is the earliest form of cosmology. Many religions have beliefs about the creation of the world. In most religious cosmologies the Universe was created by a god or gods. Judaeo-Christian religions describe creation by a single God according to the Book of Genesis. According to Buddhist and Hindu religions, the Universe undergoes an infinite cycle of creation and destruction.

Esoteric cosmology

Esoteric cosmologies share many concerns with religious and metaphysical cosmologies, but they are often more sophisticated than religious cosmologies, relying on intellectual understanding rather than faith, and place a stronger emphasis on states of existence and consciousness than most metaphysical cosmologies. Examples of esoteric cosmologies are found in Kabbalah and Sufism.

counts in cells

Counts in cells provides a convenient way of quantifying the *clustering of objects such as *galaxies and *quasars. One places cells of specified identical size and shape (normally cubic or spherical when analysing a *redshift survey, square or circular when analysing a sky-projected catalogue) either at random or on a regular grid, and counts the number of objects that fall in each cell. The second moment, or variance, of the cell counts then yields a quantity equivalent to the integral of the two-point *correlation function over the volume of the cell. The third moment yields the integral of the three-point function and so on. It is a particularly fast technique for measuring clustering, since the time required is trivial when compared with a direct estimate of the two-point correlation function. The time saving is even more significant when measuring third and higher-order statistics.

critical density

The critical density is a particular value for the *density of the *Universe which has the special property that it corresponds to a flat spatial *geometry. According to the *Friedmann equation, it is related to the expansion rate of the Universe. It is not necessarily the true density of the Universe, but provides a useful yardstick against which to measure the density of the Universe.

In the *standard cosmological model, the numerical value of the present critical density is

$$\rho_{\text{critical}} = 1.0 \times 10^{-26} \, \text{kg/m}^3 \, .$$

This can be computed from the Friedmann equation using the value of the *Hubble constant.

The concept of critical density was originally applied to cosmological models without a *cosmological constant, and referred to the total density of matter. In that case there is a simple correspondence whereby densities greater than critical give a *closed Universe with a spherical geometry and finite volume, and those with less give an *open Universe with a hyperbolic geometry and infinite volume. Modern cosmologies however feature a cosmological constant. Provided its contribution to the total density is included, then the above correspondence of density and geometry remains true, but one can no longer conclude that a closed Universe will necessarily recollapse and open ones expand forever. Some astronomers however reserve the phrase 'critical-density Universe' for the case where the cosmological constant is zero and the total density equal to the critical density.

It is often convenient to measure densities with respect to the critical density, which is always indicated by the Greek letter Ω (capital omega) and known as the *density parameter. This can be done both for the total density, with the Universe having a critical density if $\Omega = 1$, and separately for different components of matter such as *baryons, *dark matter, and *dark energy. The type of material considered is usually indicated by a subscript, such as Ω_b for baryons, and the subscript '0' usually indicates the present value.

curvaton

The curvaton is a hypothetical particle, which is able to generate primordial *density perturbations in the *Universe. It is a spin-zero particle, being an example of a *scalar field. Its name arises because such density perturbations are most simply described through their effect on the *curvature of space, known as the *curvature perturbation, and through the habit of particle physicists to end the names of particles with '-on' (e.g. *proton, *photon, *neutron, and *inflaton). It was named by David Lyth and David Wands in one of the first papers to describe the general idea.

The curvaton mechanism for producing density perturbations is a variant on the way in which the *inflationary cosmology is able to produce perturbations, known as *inflationary perturbations. The material responsible for driving the inflationary expansion—the inflaton—develops irregularities due to quantum processes as inflation proceeds, and in the standard theory these irregularities are ultimately responsible for *structure formation.

In the curvaton theory, the inflaton perturbations are instead too small to be important, but perturbations in a second material, known as the curvaton, are also created and these become dominant sometime after inflation comes to an end. At the time when these become dominant, the curvature perturbation in the Universe is generated. In the standard version of the scenario, the curvaton particles then subsequently decay into conventional material, which ensures that the density perturbations generated are of *adiabatic form as favoured by observations.

The curvaton theory is thus somewhat different from the usual inflationary scenario, but still requires inflation to have taken place.

It is not presently known whether there really is such a particle as the curvaton. Such theories are significantly more complicated than the simplest inflation ones, and as the latter continue to give a good fit to observations there is no particular motivation for the curvaton scenario. One characteristic of the curvaton idea is that it is quite easy to generate primordial *non-Gaussianity, so if this is ever seen in future observations it will provide a boost to the curvaton idea.

If the curvaton scenario is ever verified, then it will become important to try to identify where the curvaton field fits into the picture of fundamental physics, for example *superstring theory or *M-theory. The curvaton should be a particular particle predicted by those theories.

curvature

Curvature is a property of surfaces and volumes. For instance, a flat piece of paper has no curvature, but the surface of a sphere has. Curvature is a geometrical property that can be measured without ever leaving the surface; for instance parallel lines cross or diverge on curved surfaces, rather than remaining a fixed distance apart as on a flat surface. While curvature is most easily envisaged in terms of two-dimensional surfaces as described above (illustrated in Fig. 1), three-dimensional volumes, or even higher-dimensional objects, can also possess curvature.

Two types of curvature commonly arise in discussions of cosmology, curvature of space and curvature of *space-time. We'll start with the latter.

Albert *Einstein's theories of *relativity unite space and time into a single concept, four-dimensional space-time. According to his *general relativity theory, this space-time can be curved, and that curvature is responsible for the gravitational force. This is difficult to visualize; in fact it is fair to say that four-dimensional space-time is a hard enough challenge to picture on its own, even without the possibility of it being curved. Indeed, the mathematics needed to describe such curvature, known as differential geometry, is quite advanced. According to *Einstein's equation, the curvature of space-time is caused by matter; for instance in a cosmological context the *Friedmann equation is just such an equation

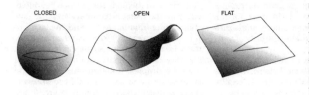

Fig. 1. Three possible curvatures for two-dimensional surfaces. In the flat case lines maintain fixed angles, in the spherical case lines converge, and in the final case, known as hyperbolic, the divergence of lines increases. The same three possibilities apply to the curvature of the three-dimensional space of our Universe.

relating the expansion rate of the *Universe to the density of matter within it. Calculations in general relativity, for example of *black hole properties, are all about trying to compute the curvature of space-time, and then determining how that curvature affects the way things, such as light rays, move.

A distinct concept is the curvature of space alone. Cosmological models are based on the *cosmological principle, which states that all observers (at a given time) are equivalent—there is no special location within the Universe. This greatly restricts the possible *geometries of space-time, which must be given by the *Robertson–Walker metric, but still permits space to be curved (the surface of a perfect ball is an example of a two-dimensional curved surface where each point is equivalent to any other; for the Universe we need analogous three-dimensional geometries).

The curvature of space appears in the Friedmann equation determining how fast the Universe expands. Model universes in which it has a positive or negative value are known as *open and *closed Universe models. In the former case, the Universe can in its late stages become dominated by the effect of curvature, and enter what is called a free expansion.

Measuring the curvature of space has been a goal of cosmology since Alexander *Friedmann first wrote down his cosmological models. It remained quite unconstrained until the late 1990s, when first the *Boomerang experiment and then the *Wilkinson Microwave Anisotropy Probe satellite used *cosmic microwave background anisotropies to show that the Universe was consistent with having no curvature, and that significant amounts of curvature were ruled out. This is in agreement with models of *cosmological inflation, which were designed

to explain why the Universe appeared to have the flat geometry.

As well as describing the overall curvature of the global Universe, it is possible to define a local curvature of space at each point. It turns out that this gives a particularly useful and elegant way of describing the *density perturbations in the Universe that lead to structure formation; rather than describing the variations in density from point to point, Einstein's equation is used to calculate the corresponding variations in the curvature of space at a given time and those are specified instead. This has the advantage that the curvature is well defined throughout the history of the Universe, whereas particular materials that dominate at some stage of the Universe may later decay into other materials (a classic example being the material driving inflation decaying to produce normal matter). This description is known as the *curvature perturbation.

curvature perturbation

The curvature perturbation tells us how the curvature of the *Universe is affected by the distribution of material within it. The Universe possesses *density perturbations, being variations in the density from point to point, and this induces perturbations in the space-time via the equations of *general relativity.

Rather than defining the curvature of the entire space-time, a more useful quantity arises from breaking up the space-time into slices of constant time (*see* SPACE-TIME, especially Fig. 1). These slices then are three-dimensional spaces, and one can consider the curvature of these slices, known as the **spatial curvature**. This spatial curvature perturbation, usually called the curvature perturbation for short and commonly indicated by the symbol $\mathcal{R}$, can then be evolved in time by following its progress from slice to slice.

A complexity arises in this procedure because in general relativity there is no unique choice of time coordinate, with people able to use whatever definition they like. Alternative definitions will slice up space-time in different ways, leading to different expressions for the curvature perturbation, and care must be taken to specify which slicing is being used. For the favoured choice of *adiabatic perturbations there is a particularly simple choice, which corresponds to slicing such that the density is constant on each slice. Normally the curvature perturbation is specified on these 'uniform-density' slices.

The curvature perturbation is a powerful quantity to analyse, because for adiabatic perturbations its value on large length scales (much bigger than the *Hubble length at a particular epoch) remains constant. The perturbations we observe are believed to have been created during *cosmological inflation and stretched to such large scales, subsequently spending most of the history of the Universe in this regime. Having identified the curvature perturbation created by inflation, it will then retain the same form regardless of the (possibly unknown) physical processes happening within the Universe until the epoch of *horizon entry.

cyclic universes

There is a long history to the idea that the *Universe might experience repeated cycles of expansion and contraction, sometimes known as a bouncing Universe, with the *big crunch leading to a new *big bang. The cycles might either be essentially identical, or there might be evolution whereby, for instance, the Universe bounces to a bigger size with each cycle. The sequence of bounces may be infinite into the past (thus sidestepping the issue of how the Universe began in the first place), and may also be infinite into the future.

Alternatively, if the bounces increase in size with each cycle there may eventually be a final bounce which is sufficient to send the Universe into eternal expansion. It has been conjectured that the idea of increasing bounce size, which may for instance be associated with the creation of entropy at each bounce, might explain why our Universe is so large (the *flatness problem)— only once the bounces have become large enough do the cycles last long enough for stars and planets to form and intelligent life to develop.

Early work on cyclic models was stymied by the inability of known physical laws, particularly *general relativity, to describe what happened at the instant of the *singularity that transformed the collapsing phase into a new expansion phase. Researchers were forced to make unjustified assumptions about how the transition might take place, preventing the idea from being taken very seriously.

The cyclic Universe idea has attracted new interest lately in work by Paul Steinhardt and Neil Turok, who created a cosmological model based on the *braneworld scenario. This idea proposes that space has more spatial dimensions than the three we are familiar with, and that our Universe corresponds to a three-dimensional surface, called a *brane, running through this higher-dimensional space. There may be several branes, and Steinhardt and Turok postulate that there might be two branes which repeatedly collide, passing through each other and then being drawn back by gravitational attraction. As seen from the viewpoint of one of the branes, these collisions correspond to a big crunch/big bang, followed by a sequence of expansion and collapse into a new big crunch/big bang pair. However as seen from the higher-dimensional point-of-view, there is no singularity and the evolution can be followed from one phase into the next. At the time of writing the fundamental basis of this theory is highly controversial, although observationally it is not in conflict with what we know about our Universe.

damped Lyman alpha systems

Damped *Lyman alpha systems are dense clouds of neutral hydrogen gas that absorb all Lyman alpha (Lyα) photons incident upon them. They are thus an extreme form of *quasar absorption system.

Most hydrogen clouds are diffuse enough that at least some fraction of Lyα photons can pass through them. These will contribute a 'U' or 'V' shaped absorption line to the *Lyman alpha forest (*see* LYMAN ALPHA FOREST, Fig. 1). Damped Lyman alpha systems instead produce a broad trough in absorption. The absorption occurs not just at the wavelength of the Lyα transition, but over a range of wavelengths, with a width and intensity (line shape) determined in part by the lifetime of the excited $n = 2$ state of hydrogen and also by the *velocity dispersion of the cloud. Damped Lyman alpha systems are believed to be associated with galaxies that are just starting to form. By looking for these systems along the line of sight to distant *quasars, one is thus able to map the locations of proto-galaxies at high *redshift.

http://astro.berkeley.edu/~jcohn/lya.html

dark energy

According to the *standard cosmological model, the most abundant form of material in the present *Universe is dark energy. Unfortunately, dark energy is also the least understood component. Indeed, it would be fair to say that the term 'dark energy' is covering the fact that although it is quite well established that the expansion of the Universe is accelerating, the cause of this is presently unknown. The label 'dark energy' for the substance or phenomenon responsible for the *acceleration of the Universe may be convenient, but does not by any means indicate that the mechanism is understood.

Cosmic acceleration

The first accelerating Universe model was *de Sitter space, discovered soon after *Einstein published his theory of *general relativity.

This model describes a Universe dominated by a *cosmological constant, which to this day remains a viable possibility for dark energy. The first indication that such an acceleration might have cosmological benefits came with the discovery of *cosmological inflation by Alan Guth in 1981, who realized that an epoch of acceleration in the *early Universe could explain some otherwise puzzling features of the global Universe. At around the same time, the *cold dark matter model was becoming favoured as the explanation for *structure formation, and in 1984 James Peebles of Princeton University noted that a version including a cosmological constant gave a good account of the then-available data. Despite massive improvements in data quality in the intervening years, the model as he proposed it was essentially the standard cosmological model in its present form, and featured a present-day acceleration of the Universe.

Despite that work, the cosmological constant model achieved only moderate popularity for many years, perhaps mainly due to the *coincidence problem (*see* COSMOLOGICAL CONSTANT). Typically, the model would be discussed as one possibility amongst several alternatives. However, in 1998 evidence from the luminosity–redshift relation of distant type Ia *supernovae convinced the cosmology community that the Universe is presently undergoing acceleration.

The acceleration of the Universe follows from the *Friedmann equation and *fluid equation, which together imply

$$\frac{1}{a}\frac{d^2a}{dt^2} = -\frac{4\pi G}{3}\left(\rho + 3\frac{p}{c^2}\right),$$

which is known as the acceleration equation. Here a is the *scale factor measuring the size of the Universe (the second derivative d^2a/dt^2 being the acceleration), ρ is the *density, and p is the *pressure. Positive acceleration therefore requires

$$\rho c^2 + 3p < 0,$$

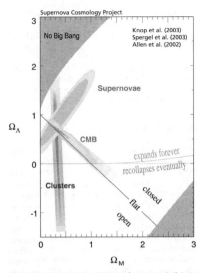

Fig. 1. Constraints on the densities of matter and of dark energy, shown in terms of their density parameters. This plot assumes that the dark energy takes the form of a cosmological constant. The shaded areas marked 'supernovae' are the 68% and 95% confidence limits from the Supernova Cosmology Project. CMB shows constraints from WMAP, and 'clusters' those from the number density of galaxy clusters. The three sets of constraints meet at the location of the standard cosmological model, with $\Omega_M \approx 0.3$ and $\Omega_\Lambda \approx 0.7$.

and given the usual assumption that the energy density ρc^2 is always positive this implies that the total pressure must be negative.

Evidence for dark energy

There are several separate strands of evidence pointing towards the existence of dark energy. Most of these are inferences from combining independent types of observational data, but the first item in the list gives direct evidence on its own.

Type Ia supernovae—Type Ia supernovae are believed to take place when a white dwarf star accretes material which takes it over the Chandrasekhar limit, which is the maximum mass for stable white dwarf stars. This accretion renders it unstable to collapse into a neutron star, releasing considerable energy. As the Chandrasekhar limit is a universal constant (apart from a weak depen-

dence on elemental abundances), such supernovae have very similar brightnesses, and hence can be used as *standard candles to map out the expansion history of the Universe. In the 1990s it became possible to systematically search for supernovae in distant galaxies, and large observational programmes carrying out such *supernova surveys were undertaken by two independent collaborations, the Supernova Cosmology Project and the High-redshift Supernova Search Team.

They uncovered an expansion law that was incompatible with known cosmological models, unless those models incorporated dark energy. A sample plot is shown in Fig. 1. This shows three different types of data. The supernova data on their own are already incompatible with no dark energy ($\Omega_\Lambda = 0$), and in combination with other data types indicate favoured values for the matter and dark energy densities of 0.3 and 0.7 respectively, in units of the *critical density.

While there had been previous observational indications that dark energy might be necessary, they had always relied on compilations of diverse types of data and were not seen as compelling. The publication of the supernova results in 1998 was the step which convinced the majority of the cosmology community that dark energy was inevitable.

Combining CMB and large-scale structure—One of the most direct inferences from study of *cosmic microwave background (CMB) anisotropies is that the *geometry of the Universe is close to spatial flatness. This on its own does not indicate the existence of dark energy, but one can also bring in information from *cosmic structure, such as the numbers of galaxy *clusters or measures of galaxy clustering, which indicate that the total density of matter is much less than the critical density. As the spatial flatness condition requires that the densities add to the critical density, this means that there must be extra material in the Universe which does not undergo *gravitational instability. Only dark energy fits this bill. In Fig. 1, we see constraints from the *WMAP satellite measurements of CMB anisotropies and from an analysis of galaxy clusters by Steve Allen and collaborators. These two measures are only consistent if the matter and dark energy densities are around 0.3 and 0.7 respectively. This provides evidence for dark energy independent of the supernova measurements, and which is clearly seen to give a result in good agreement with them.

Combining flatness and baryon fraction— Alternatively, we can combine the spatial flatness inferred from the CMB with the measured *cluster baryon fraction, i.e. the relative amounts of baryonic and dark matter in galaxy clusters, which is believed to be representative of the Universe as a whole. The cluster baryon fraction is observed to be about 15%. From cosmic *nucleosynthesis the baryon density is measured to be about 4% of the critical density, implying a dark matter density around 30% of critical. Again, this leaves the other 70%, needed for spatial flatness, to be explained, and as the material must avoid falling into galaxy clusters dark energy is once again indicated.

Combining Hubble constant plus age— Further evidence comes from comparing measurements of the *Hubble constant, giving the present expansion rate, with measures of the *age of the Universe. Objects are seen with ages of at least 12 billion years, and the Hubble constant measured to be approximately $h = 0.7$. However in a critical-density Universe without dark energy the age is given by $t_0 = 6.5h^{-1}$ billion years, which is no more than about 10 billion years, incompatible with the observed object ages. The predicted age is increased by lowering the matter density, and more so if the matter density is replaced by dark energy density (the corresponding acceleration means that the Universe was expanding more slowly earlier on, and hence took longer to reach the present observed size). The standard cosmological model predicts an age of about 14 billion years, comfortably able to accommodate the ages of objects within the Universe.

Etymology

The term 'dark energy' was coined by Chicago astrophysicist Michael Turner in 1998, and has come to refer to any material capable of explaining the observed acceleration. While undoubtedly a catchy title, it is technically rather inaccurate, as the distinguishing feature of a dark energy candidate is not its energy, but that it contributes a negative

pressure. However it is hardly surprising that the more accurate but clumsy 'cosmic anti-pressure' has failed to catch on. Nevertheless, given the lesson from *relativity that matter and energy are interchangeable concepts connected by $E = mc^2$, one might be misled into thinking that dark matter and dark energy must be one and the same, whereas they are entirely distinct topics.

Models for dark energy

The simplest dark energy model is the cosmological constant, which maintains a fixed density as the Universe expands. Equivalently, it can be described as a fluid whose pressure equals minus its energy density. Thus far, the cosmological constant has proven capable of explaining all relevant observational data, and thus is the chosen ingredient of the standard cosmological model.

Nevertheless, cosmologists have considered several other possibilities, the most powerful motivation being to try to overcome the coincidence problem. In order to do so, it is necessary to allow the dark energy density to change with time, though it must do so slowly to stay consistent with the ability to drive an accelerated expansion. Forthcoming dark energy projects mainly focus on methods for discovering whether the dark energy density does indeed evolve.

The most popular idea is *quintessence, which mimics the mechanism of early Universe inflation by assuming that the dark energy is due to a *scalar field with a potential energy. Alternatives come under such names as **k-essence**, **tachyon dark energy**, and the **Chaplygin gas**.

An alternative to dark energy is to postulate that the law of *gravity may be modified from the usual assumption of *general relativity, the main recent examples being related to the *braneworld idea. So far, this idea has proven less popular than the others listed above, but it remains a logical possibility.

Dark energy equation of state

The *equation of state of a material is the relation between its pressure and its density. Dark energy models are often discussed in terms of the dark energy equation of state w, defined by

$$w = \frac{p}{\rho c^2},$$

where p and ρ are the pressure and density of the dark energy alone. This gives a general description of the dark energy which is not restricted to some specific physical model such as quintessence. In order to give acceleration, the condition $w < -1/3$ is necessary, while the special case of a cosmological constant corresponds to $w = -1$, which gives a constant energy density. One goal of upcoming projects to study dark energy is to investigate whether w can be shown to be different from -1, which would mean that the dark energy density evolves. Such measurements would hopefully shed light on the physical nature of dark energy.

The simplest models of dark energy would assume that w was a constant. However it turns out that none of the physical models under consideration, such as quintessence, are well approximated by a constant equation of state. Instead, the equation of state w should evolve with epoch. It is hoped that upcoming experiments might be able to measure evolution of the equation of state parameter as well as its present value.

If the equation of state is less than -1, this means that the dark energy density is increasing with time. This is sometimes called *phantom dark energy, which could lead to the Universe ending in a *big rip. It turns out to be quite hard to obtain physical models which give phantom dark energy, but at present such models are consistent with observational data (and will remain so as long as the cosmological constant model is viable, since there exist phantom models which give predictions arbitrarily close to those of the cosmological constant model).

Observational consequences of dark energy

The principal reason for introducing dark energy is the need to explain the observed acceleration of the Universe, as indicated for instance by observations of the luminosity of distant supernovae. As explained above, however, there are independent strands of evidence requiring its existence, including the need to reconcile the observed spatial flatness of the Universe with the sub-critical matter density. There are a number of other observational implications.

Once dark energy begins to dominate, the *gravitational potential begins to evolve and this modifies the pattern of cosmic microwave anisotropies via the integrated

*Sachs–Wolfe effect. This modifies the CMB anisotropies on large angular scales, though this signature has not been detected due to *cosmic variance. However there is also a predicted correlation between the CMB anisotropies and the matter distribution which generates the gravitational potential, and this correlation has been seen at the expected level and with reasonable statistical *significance. This is fairly direct evidence that dark energy does indeed exist.

The existence of dark energy modifies the rate at which cosmic structure develops, as it provides an extra repulsion that must be overcome by gravitational instability. Many future probes of dark energy will focus on this, by analysing the development of structure across a wide range of *redshifts, either by direct measurements of galaxy clustering or via *gravitational lensing. These complement geometrical measurements such as the supernova luminosity–redshift relation and measurements of *baryon oscillations in the matter power spectrum at different redshifts.

dark matter

It is almost universally accepted by astronomers that the *Universe contains large quantities of a material known as dark matter. This material is responsible for providing the gravitational attraction that enables objects such as *galaxies to form, and holds them together once they have formed. Apart from this gravitational influence, however, its nature is unknown. The most common assumption is that the dark matter is made up from fundamental particles of Nature, whose properties are such that they do not interact significantly with normal matter such as *baryons and *photons.

History

Dark matter was first suggested by Fritz *Zwicky in 1933. He studied a nearby rich *cluster of galaxies, the *Coma cluster, and compared the motions of galaxies in the outskirts to the total mass of the galaxies. He found a discrepancy of nearly 1 000 between these two numbers, indicating the presence of gravitational attraction 1 000 times greater than could be explained by the galaxies. He concluded that there was some other form of matter within the galaxy cluster, though he had no idea what it might be.

Dark matter did not however become a widely accepted phenomenon until much later. Its acceptance owes much to the work of American astronomer Vera Rubin and collaborators from the 1970s onwards, who studied *galaxy rotation curves. These curves measure the rotational velocity of stars and gas in spiral galaxies, as a function of distance from the centre. If the light profile of the galaxies followed the mass, the rotation curves should fall at large distances meaning that stars orbit more slowly (this is the Galactic version of Kepler's Law, which in the Solar System ensures that the planets closest to the Sun move the fastest). In fact, the rotation curves are found to be very flat even at large distances, indicating that the stars are orbiting much faster than they should. Faster orbits imply they are feeling a stronger gravitational force, which implies that there is more matter than is visible, typically by a factor of 10 or more in the galaxy outskirts. This had already been noticed soon after Zwicky's galaxy cluster work by Horace Babcock and Jan Oort, but it took until the 1970s for the evidence to become compelling.

In the 1980s, models of the formation of galaxies and other structures began to be properly formulated. It rapidly became apparent that they too required dark matter, in order to provide sufficient gravitational attraction to allow structures to form given the known level of regularity of the *cosmic microwave background (the cosmic microwave background anisotropies would not be discovered until the following decade). Indeed, it became commonplace to carry out numerical calculations of the matter distribution that *only* included dark matter, known as *N*-body simulations, with the galaxy locations to be later inferred from the locations of high-density dark matter regions (an identification which however was plagued by the phenomenon of *biased galaxy formation).

Dark matter in cosmology

The key role of dark matter in *cosmology is to provide the gravitational attraction that allows the observed structures to form, given the constraint that the *density perturbations at the time of *decoupling must match the observed *cosmic microwave background anisotropies. Detailed calculations show that a baryon-only Universe cannot achieve this; there simply isn't time since decoupling for the perturbations to grow.

Dark matter gives an extra boost to structure formation, because being non-interacting it has no pressure and can begin gravitational collapse earlier than the baryons which will ultimately make up the visible galaxy, which are prevented from collapsing until after decoupling. By the time of decoupling, density perturbations in the dark matter on galactic scales are much more prominent than those in the baryons. Once decoupling has taken place, the baryons are able to fall into the *gravitational potential wells created by the dark matter. This extra kick is just what is needed to speed up the *galaxy formation process in order to comply with the presently observed Universe.

In addition to the cosmological effect of structure formation, the gravitational impact of dark matter makes itself felt in a number of arenas. The *bulk motions of galaxies can only be explained if the gravitational force between them is much larger than visible matter alone allows. Within galaxy clusters, where Zwicky saw the first evidence of dark matter, it remains true to this day that a substantial dark matter component is needed to prevent the cluster from flying apart (though modern measurements suggest that the dark matter dominates the galaxies by a factor of somewhat less than a 100, and the total baryons by a factor of around 8, rather less than the numbers that Zwicky obtained). And the evidence from galaxy rotation curves remains as strong as it has ever been.

The evidence for the gravitational effect of dark matter is therefore extremely strong, and modern cosmological models will not work without it. The question, then, is what is this dark matter?

Baryonic dark matter

Some of the dark matter is normal baryonic matter, which has not been incorporated into stars and hence does not shine brightly for us to see. Observations of primordial element abundances (*nucleosynthesis) and of *cosmic microwave background anisotropies give compelling evidence that the density of baryonic matter in the Universe is around 4 or 5% of the *critical density. A direct counting of the visible material gives a much smaller answer, perhaps 0.5% of the critical density. The conclusion is that only around one-tenth of baryonic material is in stars, the rest remaining in diffuse gas.

This conclusion is directly affirmed by studies of galaxy clusters, because in the process of their formation the ambient gas gets heated up to such temperatures (typically tens of millions of degrees) that it emits brightly in *X-rays, and can easily be detected by satellites such as *XMM–Newton and *Chandra. Indeed, in a galaxy cluster there is roughly ten times more material in hot gas than in galaxies. Other regions of the Universe must surely have the same sort of ratio, even though the gas is too cool to be seen there, since many of those regions will one day themselves collapse into galaxy clusters and ought to have the same composition as those clusters which have already formed.

Baryonic matter cannot, however, be all the dark matter in the Universe, both because the total amount is inadequate, and because baryons alone are not able to give the boost to structure formation that is required. The majority of the dark matter is believed to be **non-baryonic dark matter**. There are a range of possibilities.

Cold dark matter

Cold dark matter refers to dark matter candidates which exhibit negligible velocity. Their present location in the Universe is then attributable entirely to the gravitational forces they have been subjected to due to density perturbations in the Universe, and is independent of their mass. Cold dark matter is the standard hypothesis cosmologists now make, and it is usually assumed to be comprised of individual fundamental particles such as those predicted by *supersymmetry.
See COLD DARK MATTER.

Hot dark matter

Hot dark matter refers to dark matter candidates whose velocity plays a significant role in determining their eventual location. In practice this means that they must be moving with *relativistic speeds for a significant portion of the Universe's evolution, though they need to have slowed down by the present if they are to be incorporated into galaxies and provide the gravitational attraction characteristic of dark matter.

The classic hot dark matter candidate is the *neutrino, a type of elementary particle produced in radioactive decays and believed also to have been copiously produced in the early Universe. Neutrinos are extraordinarily weakly interacting particles, meaning that

their fundamental properties are very difficult to measure. They come in three types, each believed to have a characteristic mass, but it is quite uncertain what those masses are. The neutrino mass determines both their characteristic velocity and their contribution to the total density of the Universe. For light neutrinos, there is a range of masses where they can contribute a significant density of the Universe and act as hot dark matter.

Hot dark matter cosmologies hit a peak of popularity in the 1980s, when experimental evidence suggested that the electron neutrino (the type associated with nuclear decay) might indeed have a mass lying in the crucial range that predicted a substantial cosmic abundance. For a time, hot dark matter and cold dark matter models competed as the best description of galaxy formation and clustering; cold dark matter models were known as 'bottom-up' since the smallest galaxies formed first and were accumulated into ever-larger objects, while in the 'top-down' hot dark matter model large objects formed first and then fragmented into small galaxies. However, the experimental measurements of the electron neutrino mass were subsequently shown to be incorrect, with present upper limits lying well below the values claimed then.

The possibility that all the dark matter in the Universe is hot has long been ruled out—the dark matter motions oppose gravitational collapse and prevent galaxies from forming. A mixture of cold and hot dark matter remains possible, and is the expected situation provided neutrinos do indeed have a mass. However it is quite likely that the neutrino mass is small enough that neutrino hot dark matter is not a major influence on the evolution of the Universe, and hot dark matter is commonly ignored in attempts to model the evolution of the Universe.

Exotic dark matter

A variety of more exotic proposals have been made for the nature of dark matter. In some cases there is motivation from some observations, but there is no compelling evidence in favour of any of the following alternatives.

Warm dark matter—If it can be cold or hot, it can also be warm. Warm dark matter would exhibit some level of *free streaming, but if the particles are massive enough this might not have a disastrous effect on

Fig. 1. A large-scale numerical simulation of the dark matter distribution in the standard cosmological model. The location of galaxies will closely, but not exactly, correspond to the locations of high-density dark matter clumps. The image shows a slice through the Millennium Simulation.

galaxy formation. It might have observable effects such as modifying the central cores of galaxies, where there has been some dispute over whether the cold dark matter model really matches the observed distributions.

Interacting dark matter—By its nature, dark matter is supposed to interact only weakly with other things, otherwise its presence would have become obvious to us long ago. However, it is possible that it could interact more strongly with itself, so that interactions between dark matter particles are no longer negligible. This may change the dark matter distribution, particularly in the regions of highest density, and this may or may not be desirable depending on how some observations of galaxy central cores are interpreted.

Annihilating dark matter—If it can interact, why not annihilate completely? Again, this can modify the distribution in high-density regions. However, the annihilation energy has to go somewhere, and it can't go into standard particles or they would be spotted easily. This scenario also has to overcome the objection that the highest dark matter densities in the present Universe were routinely exceeded in the young Universe.

Condensated dark matter—If the dark matter particles are extremely light, their quantum properties can be important and they no longer behave as classical particles. A large number of them may, for instance, assemble into a giant version of a *boson star, with the visible galaxy embedded within this huge diffuse structure. How such a structure would form is, however, completely unclear.

Primordial black holes (PBHs)—These are miniature black holes which might form in the early Universe. Provided they are massive enough (more than 10^{15} grammes) that Hawking evaporation is negligible, then they behave to all intents and purposes like very large cold dark matter particles. For a wide range of masses, it would indeed be very difficult to prove or disprove the hypothesis that the dark matter is PBHs. On the plus side, the hypothesis does away with the need to invent a new dark matter particle (though candidates abound in particle theories anyway). On the minus side, no one knows of a plausible mechanism to generate the right number of black holes, rather than far too many (ruled out) or far too few (allowed but uninteresting). The PBHs would have to have formed before nucleosynthesis in order to allow their density to exceed the baryon density predicted by that theory.

Out-of-this-Universe—According to the *braneworld scenario of modern fundamental physics, there may be a number of Universes lying parallel to one another in an extra dimensional direction. *Gravity is the only means of communication between these branes. It might be that the dark matter resides on a different brane to ours, in which case any search for non-gravitational evidence for it is doomed to failure.

There isn't any—Much of the quantitative evidence for dark matter is obtained under the assumption that gravity is described by Einstein's *general relativity. Particularly for galactic dark matter, it has been hypothesized that the data could instead be explained by a modification to the law of gravity in the regime where it becomes very weak. The most popular such theory is known as *modified Newtonian dynamics (MOND), invented by Israeli physicist Mordehai Milgrom in 1983 and extended to a relativistic formulation by Jacob Bekenstein in 2004. Such a theory does appear quite successful in avoiding the need for galactic dark matter, but whether it can also remove the need for dark matter in the other contexts described above remains to be seen.

Large-scale dark matter distribution

The present dark matter distribution in the Universe cannot be observed directly, but it is crucial in determining the distribution of galaxies and galaxy clusters, and will soon be probed in detail via *gravitational lensing. Accordingly, considerable effort has gone into carrying out computer simulations, known as N-body simulations, to predict the dark matter distribution. Following the dark matter is much simpler than the gas of baryons, as only gravitational forces are important, and so the largest simulations to date have been of N-body type. Fig. 1 shows a slice of the dark matter distribution from the largest ever (as of 2007) simulation, known as the Millennium Simulation, which follows the position of over 10 billion individual N-body particles (each representing a large clump of dark matter particles).

The cold dark matter model is a *hierarchical structure formation model, where the smallest structures (dwarf galaxies) form first, and are then subsequently assembled into large and larger structures as the continued gravitational attraction pulls them together. In the present Universe, this process has reached the stage of assembling large galaxy clusters of total mass around 10^{15} solar masses, corresponding to roughly 1 000 times the mass of the *Milky Way. Such large objects are rare presently, but will become commonplace in the future as structure formation continues.

Predictions for the dark matter distribution are highly accurate, so it is rather a shame that it cannot be observed directly. Ultimately gravitational lensing experiments will be the best tool for exploring its distribution. Predicting the corresponding galaxy positions is much trickier, and can be done either by including gas in the simulations, or by inventing algorithms (known as *semianalytic galaxy formation) which try to associate galaxies to dark matter clumps.

Small-scale dark matter distribution

At the level of individual galaxies and clusters, the dark matter is highly clumped, and the dark matter enshrouding a galaxy is usually known as its *galactic halo. Because the dark matter has no dissipative forces, it is believed to end up in roughly spheroidal configurations reminiscent of elliptical galaxies, even if the galaxy embedded within is a spiral disk galaxy.

The precise profile of these halos is a matter of intense discussion, particularly in the core regions where there have been some indications of a discrepancy between the predictions of cold dark matter models and actual observations, though this remains controversial. On average, both theoretically modelled and observed halos appear well described by an NFW profile, named after Julio Navarro, Carlos Frenk, and Simon White who first found the universal form in numerical simulations. It is also believed that dark matter halos contain substantial substructure, being the surviving remnants of the smaller objects from which they have been built up.

Direct detection of dark matter

Many experiments are being undertaken which seek to directly detect dark matter, meaning to find evidence of its presence other than from its gravitational influence. The cold dark matter model predicts that the Universe is full of dark matter, and for a candidate such as a *WIMP large numbers pass through your body every second without you ever realising it, simply because they are so weakly interacting.

A dark matter detection experiment hopes that although the interactions are indeed very weak, they are not non-existent altogether. If so, then from time to time there will be an interaction between dark matter and normal matter which reveals the presence of the dark matter. The main predicted reaction is the dark matter particles bouncing off protons or neutrons, leading to a measurable nuclear recoil. In order to isolate the experiments from less subtle effects than dark matter, they are typically sited in heavily shielded laboratories deep underground, to minimize *cosmic rays and radioactivity. Superficially, dark matter detection and neutrino detection experiments are quite similar, particularly in their location.

Such experiments are very subtle, but can rely on one important piece of assistance, which is that any signal should be annually modulated as the Earth circles the Sun, sometimes moving generally with the local dark matter flow and sometimes against it. As the motion of the Earth around the Sun is around one-tenth the speed of the Earth–Sun system through the galaxy, this modulation should be approximately 10%. Such a signal has been found by an Italian-led experiment DAMA (DArk MAtter), but has not been verified by other experiments and is presently highly controversial. So far, all other dark matter experiments in operation around the world, of which there are at least twenty, have only been able to set upper limits on dark matter properties such as the interaction rate.

Eventual discovery of particle dark matter, if it does happen, would be a landmark achievement for astrophysics in stimulating research in fundamental particle physics.

Millennium Simulation home page (for movies and visualizations): http://www.mpa-garching.mpg.de/galform/millennium/

dark matter halo
Estimates of the total mass of *galaxies from observed dynamics or from *gravitational lensing show that the visible parts of galaxies account for less than 10% of their total mass. To explain these observations, it is assumed that galaxies sit within a roughly spherical distribution of *dark matter, known as the dark matter halo. The form that this dark matter takes is presently unknown, but the majority is thought to be in the form of slowly moving particles described as *cold dark matter.

*Numerical simulations of *structure formation model these dark matter halos using typically hundreds or thousands of fictitious 'particles' which are significantly more massive than the actual dark matter particles.

DASI
See DEGREE ANGULAR SCALE INTERFEROMETER.

de Sitter space
One of the first cosmological models to be created from Albert *Einstein's theory of *general relativity was de Sitter space, proposed in 1917 by Dutch astronomer Willem de Sitter (not to be confused with the *Einstein–de Sitter Universe). It corresponds to an expansion rate proportional to the exponential of time, and hence corresponds to *acceleration of the Universe. It has the special property that it is not just *homogeneous in space, as are other cosmological models, but it is unchanging in time as well with the *density remaining constant.

In modern times, it is understood that this type of expansion arises if the *Universe is completely dominated by a *cosmological constant, or by the *potential energy of a *scalar field. The latter case is the standard implementation of *cosmological inflation; in Alan Guth's original inflation model the expansion was precisely of de Sitter type, while in other models where the scalar field evolves the expansion may only be approximately of de Sitter type. In either case, this early de Sitter phase must give way to *radiation-dominated and *matter-dominated epochs to give a viable cosmology.

There is currently convincing evidence that the Universe contains *dark energy, and the simplest viable model of the dark energy is a cosmological constant. If this is true, then the Universe is presently evolving towards a de Sitter state, which may be its ultimate destiny.

De Sitter space has a curious mathematical property in that one can describe it using different coordinates in which the *space-time appears to be static rather than expanding, instead resembling the space-time of a *black hole. Any observational property is however independent of the choice of coordinates used, as according to the rules of general relativity.

deceleration parameter
There is no reason to believe that the current expansion rate of the *Universe was the same in the past and will be the same in the future. Since the discovery of the expansion of the Universe in the 1920s until roughly the turn

of the 21st century, it was widely thought that the gravitational attraction of matter in the Universe would slow down the expansion rate as time went on, so that the expansion would be decelerating. A dimensionless deceleration parameter, $q(t)$, was thus introduced to parameterize this deceleration as a function of time since the *big bang.

Its present value, denoted q_0, is expressed in terms of the *scale factor $a(t)$ and the *Hubble constant H_0 as

$$q_0 = -\frac{\ddot{a}(t_0)}{a(t_0)}\frac{1}{H_0^2} = -\frac{a(t_0)\ddot{a}(t_0)}{\dot{a}^2(t_0)}.$$

Here the time t_0 denotes the present day and an overdot denotes a derivative with respect to time [the Hubble constant H_0 is defined as $H_0 = \dot{a}(t_0)/a(t_0)$]. Thus $\dot{a}(t_0)$ is the current expansion rate of the Universe (the rate of change of the scale factor) and $\ddot{a}(t_0)$ is the acceleration of the scale factor. The deceleration parameter q_0 is defined so that it is dimensionless and so that positive values imply deceleration, as was thought to be the case when the parameter was introduced.

In a matter-dominated Universe (one in which radiation provides a negligible contribution to overall density compared with matter) with no *cosmological constant, one may show that the deceleration parameter and the *density parameter Ω_0 are simply related by $q_0 = \Omega_0/2$, and so determination of the deceleration parameter would tell us the density parameter.

However, in the late 1990s, observations of distant *supernovae suggested that the Universe is actually *accelerating today: the distant supernovae are more distant than one would expect from their *redshifts. This result is independently supported by observations of *cosmic microwave background anisotropies, and it now seems hard to escape the conclusion that today, not only is the Universe expanding, but this expansion is also accelerating.

The cause of this acceleration is as yet unknown. Two possible causes are a cosmological constant or *dark energy.

decoupling

Decoupling refers to the epoch where a given type of particle ceases to interact with other particles as the *Universe expands. This happens either because the density of particles becomes so low that the chances of two meeting becomes negligible, or because the parti-

cle energy becomes too small to interact. By far the most common use of the term refers to *photons of light, which will ultimately form the *cosmic microwave background, though the term is also applied to other particles.

Photon decoupling

Photon decoupling is the last stage of the sequence that leads to the production of the cosmic microwave background, when the Universe was around one-thousandth of its present size. The sequence begins with *recombination, when *electrons first join with atomic nuclei to form proper atoms. Once electrons reach their ground state, there is a minimum energy required for photons to be able to interact with them, and due to ongoing *redshifting by the expansion of the Universe the photons find themselves below this energy. Recombination is delayed somewhat because the number of photons is much greater than electrons, and rare high-energy photons are able to keep the electrons ionized so that the lower-energy photons can continue to interact with them. Eventually, however, with the plasma at around 3 000K, even these photons become too feeble to prevent atoms forming.

With no free electrons left, photons have nothing to interact with and travel freely through the Universe—they are said to have decoupled from the other matter. This decoupling only takes place after most of the electrons have formed atoms, because scattering from free electrons is efficient enough that even a small population of ionized electrons is enough for all the photons to interact with. Decoupling therefore happens after recombination. [There is actually a small level of *residual ionization left after recombination, but it is too low to have a significant effect.]

A technical definition of decoupling is the epoch at which the time between scatterings first exceeds the age of the Universe. As the scattering efficiency continues to sharply reduce as the Universe expands, this is more or less equivalent to the epoch after which a typical photon will not scatter again. The location where the photons we presently observe last underwent scattering is known as the *last-scattering surface, and corresponds to a giant sphere centred on our location.

A useful concept is the *visibility function, which gives the probability that a photon rescattered at a given epoch. It is sharply peaked at the time when decoupling takes place, but has some spread as some photons by chance have their final scattering earlier or later than others. The visibility function defines how wide the last-scattering surface is, confirming it as a thin shell. Its thickness affects the calculation of *cosmic microwave background anisotropies on small angular scales.

At the time when the photons decoupled, they were still fairly energetic, mainly occupying the near *infrared part of the *electromagnetic spectrum. However the continued expansion of the Universe causes them to redshift, and by the time they are eventually detected by us they are in the *microwave part of the spectrum and are known as the cosmic microwave background.

All of the above refers to processes taking place when the Universe was around one-thousandth of its present age. The Universe underwent *reionization at a much later epoch, when the first generation of stars and *quasars generated enough energy to reionize all the atoms. This creates a new population of free electrons from which the microwave photons can scatter at relatively recent epochs, but such scattering is usually considered separately and, since almost all photons scatter at most a single time, does not correspond to the recreation of a thermal state between photons and electrons.

Decoupling of other species

Several other species of particles have experienced a decoupling, though not with such a visual result as the microwave background. *Neutrinos experience negligible interactions in the present Universe due to its low density, but during its early stages were able to interact frequently (primarily with electrons) and maintain a state of thermal equilibrium with the other particles. This era came to an end when the Universe was about one second old (more or less simultaneously with the *nucleosynthesis process where atomic nuclei first formed), with neutrinos subsequently travelling essentially uninterrupted. This travel is often referred to as neutrino *free streaming, and is particularly relevant if neutrinos have a cosmologically significant mass as neutrino free streaming then opposes the gravitational collapse of density perturbations.

Dark matter also does not interact directly with other particles in the present Universe. Whether it did so in the early Universe depends on its (currently unknown) properties, but if it did then it too will have undergone a decoupling. Under the hypothesis that it is a weakly interacting massive particle (*WIMP), it will have decoupled quite early in the history of the Universe. Indeed, decoupling of particles which interact with the characteristic strength of weak nuclear interactions, and also have the characteristic mass–energy of the electro-weak scale, leads to an abundance of dark matter consistent with that observed in the Universe (at least in an order-of-magnitude sense). Accordingly, WIMPs are widely regarded as the most plausible dark matter candidate particle. A further curiosity of the decoupling calculation is that the more weakly such a particle interacts, the higher its final abundance is predicted to be, since it will then decouple at an earlier epoch.

degeneracy

In *cosmology, one is frequently trying to fit a multi-parameter model to a set of data. For instance, one may be attempting to estimate the values of the *cosmological parameters given observations of the *cosmic microwave background (CMB) anisotropies. One often finds that an equally good fit may be obtained by simultaneously varying the values of two or more parameters: there is no unique set of 'best' parameters. In this case, the parameters are said to be degenerate.

There are two ways of overcoming degeneracy. The ideal solution is to obtain additional data that provide complementary constraints on the parameters one is trying to constrain. For example, when one is working with CMB anisotropy data alone, there is a strong degeneracy between the matter density parameter Ω_m and the vacuum density Ω_Λ. In contrast, observations of type Ia *supernovae result in a degeneracy in the same parameter space almost perpendicular to that from CMB anisotropies. By combining the two independent data sets, vastly improved constraints on both Ω_m and Ω_Λ may be obtained (see DARK ENERGY, Fig. 1).

In other cases, one has to make what are hopefully reasonable assumptions about the values of some parameters in order to estimate values of others. Such assumptions about the values of certain parameters are known in the language of *Bayesian inference as **priors**. For example, a value for the *Hubble constant of $H_0 = 72$ km/s/Mpc, or a range around that, is frequently assumed.

Fig. 1. An image of the DASI experiment on site at Antarctica, photographed against the Sun.

Degree Angular Scale Interferometer (DASI)

The Degree Angular Scale Interferometer was an experiment to study the *cosmic microwave background (CMB) anisotropies. Its most important achievement was to make the first measurements of *polarization of the CMB, which were announced at a conference in Chicago in 2002 by project scientist John Carlstrom.

The DASI experiment operated at the Amundsen–Scott South Pole Station in Antarctica between 2000 and 2004, and was led by scientists from the University of Chicago. The South Pole is a favoured site for CMB experiments, due to the low water vapour content of the atmosphere avoiding what would otherwise be an important source of contamination. Fig. 1 shows the experiment on site. It consisted of 13 separate detector elements, operating at the low-frequency end of the *microwave spectrum, the signal being obtained by allowing the information from different detectors to 'interfere'. As the name suggests, it had an angular resolution around the one-degree scale. The challenging observing schedule required scientists to stay on site to operate the instrument throughout the southern hemisphere winter.

DASI measured CMB temperature anisotropies, which had previously been studied on various angular scales by many instruments. These results were useful in further constraining *cosmological models, but the ground-breaking result was the first measurement of CMB polarization, the existence of which had long been predicted. DASI was able to measure a type of polarization known as E-mode polarization (*see* CMB ANISOTROPIES), and also a correlation between the temperature anisotropies and the polarization. Both signals were consistent with the level expected from theoretical models. While the DASI polarization observations are not in themselves powerful for constraining models, the discovery of CMB polarization opened a new field of CMB study which ultimately is expected to provide cosmological information which extends and complements the CMB temperature anisotropies.

density

The **mass density** of a material is simply the amount of mass within a standard volume (for instance a cubic metre), and is normally indicated by the Greek symbol ρ (pronounced *rho*). In *cosmology we are usually interested in the densities of the various materials making up the content of the *Universe. Often they are expressed as fractions of the *critical density, known as the *density parameter.

Directly related to the mass density, and often used interchangeably thanks to Einstein's famous $E = mc^2$ equation, is the **energy density**, given by ρc^2. Use of *particle physics units entails setting the speed of light equal to one, and then the mass and energy densities become identical. An important subtlety is that the mass density is meant in the *relativistic sense, and includes all contributions to the mass–energy of a system, so that for instance the kinetic energies of particles is included in the calculation of their density.

The density is one of the quantities which determines the expansion rate of the Universe, via the *Friedmann equation. Indeed, in a spatially flat Universe the total density is the only quantity needed to compute the expansion rate.

density parameter

The density parameter, Ω (Greek capital omega), is the ratio of the actual mean *density of the *Universe to the *critical density, that density required for a Universe to have zero *curvature (a flat spatial *geometry.)

The *Friedmann equation relates the *Hubble parameter (the expansion rate of the Universe) to its total density and curvature. For a given value of the Hubble parameter H, there is then a particular value of the density which is consistent with zero curvature. This value is known as the *critical density, $\rho_c(t)$, and may be expressed as

$$\rho_c(t) = \frac{3H^2}{8\pi G},$$

where G is Newton's gravitational constant. Note that the critical density changes with time, since it depends on the Hubble parameter, which is also time-varying.

In models with no *cosmological constant, if the total average matter density is ρ, then the density parameter is defined as $\Omega(t) = \rho/\rho_c$ and its present value written Ω_0. In this case a density parameter greater than one gives a *closed Universe with spherical geometry and finite volume, and a density parameter less than one gives an *open Universe with hyperbolic geometry and infinite volume. A density parameter of exactly one corresponds to a flat Universe with Euclidean geometry.

However, since the discovery of an *accelerating expansion in the late 1990s, the cosmological constant has been included in cosmological models. If its contribution to total density is included, then the above correspondence between density and geometry still holds, but it is no longer necessarily true that closed Universes will recollapse and open Universes expand forever.

One can separate the contributions to the density parameter into components due to

matter, Ω_m, due to a cosmological constant, Ω_Λ, and due to curvature, Ω_k. In this case the Friedmann equation may be rewritten

$$\Omega_m + \Omega_\Lambda + \Omega_k = 1.$$

Current observations, particularly those from the *WMAP satellite, are consistent with zero curvature ($\Omega_k = 0$) and with matter providing about 25% of the critical density of the Universe ($\Omega_m \approx 0.25$), with the remaining 75% coming from a cosmological constant or *dark energy ($\Omega_\Lambda \approx 0.75$).

density perturbations

One of the most fundamental properties of the *Universe is the *density of material at different locations. Since this varies from point to point, the density is said to have *perturbations. During the Universe's early stages it is believed to have been approximately uniform, and the perturbations are then referred to as **linear perturbations**. Following the evolution of such density perturbations throughout the Universe's evolution is one of the most important parts of modelling the Universe, and is known as *cosmological perturbation theory. This theory aims to follow the complete evolution of the density from the early linear epoch through to the formation of *galaxies in the present Universe.

Types of perturbation

If the Universe contained only a single type of material, the density perturbations would be completely described by specifying how the density varies with position. In reality, however, the Universe contains a variety of different materials. The present constituents are believed to be *dark energy, *dark matter, *baryons, *photons, and *neutrinos. In its early history, the Universe may have been dominated by one or more *scalar fields. Usually, it is necessary to follow the perturbations in each constituent separately.

The simplest type of perturbation is an *adiabatic perturbation, in which the perturbation to the total density is shared out amongst the different components, so that they share the same profile; e.g. in locations where the dark matter density has a maximum so do the densities of all the other materials. There is a precise rule about how this sharing takes place, known as the **generalized adiabatic condition**, but it is too technical to include here. An adiabatic perturbation

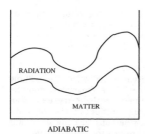

ADIABATIC

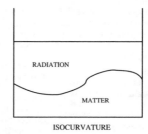

ISOCURVATURE

Fig. 1. An illustration of adiabatic and isocurvature perturbations for a two-component fluid of matter and radiation. In the adiabatic case, wherever there is more matter there is also more radiation, so that the total density varies from place to place. In the isocurvature case, wherever there is more matter this is compensated by there being less radiation, so that the total density is constant.

therefore corresponds to a perturbation in the total density.

An alternative type of perturbation is where two materials have equal and opposite perturbations. In this case the total density is initially unperturbed, and we have what is known as an *isocurvature perturbation (see Fig. 1). In general one may have several types of isocurvature perturbations, as the cancellation could be between any combination of the materials present.

An arbitrary perturbation can always be written as a sum of an adiabatic perturbation plus a set of isocurvature perturbations. If the Universe contains M different types of materials, there will be one adiabatic perturbation and $M - 1$ independent isocurvature perturbations. However as the Universe evolves these perturbations may interact with one another.

If at any time the Universe contains only a single type of material, then its perturbations must be adiabatic and this property will then be conserved on large scales (bigger than the *Hubble length) by the future evolution. A classic example of this is the simplest models of *cosmological inflation, during which the Universe is dominated by a single *scalar field which later decays to create the material filling the present Universe. Another situation where perturbations are guaranteed to be adiabatic is if the Universe enters *thermal equilibrium, as then energy will be exchanged between the different materials until the adiabatic condition becomes satisfied.

Density and curvature
The presence of a density perturbation in the Universe implies that there will be a perturbation in the *curvature of *space-time, because Einstein's theory of *general relativity tells us that space-time curves in response to the

presence of matter. In fact, it tends to be convenient to describe the density perturbation in terms of its effect on the curvature, known as the *curvature perturbation.

The curvature perturbation is a particularly useful quantity for describing adiabatic perturbations, because in that case it is possible to define a time coordinate so that the density of *every* fluid is constant on a slice of constant time, meaning that the curvature perturbation is the only perturbation. The curvature perturbation is normally analysed using such a space-time slicing, known as the uniform-density slicing.

Statistical description
It is neither possible nor desirable to attempt to predict the full three-dimensional features of the density of the Universe. Rather, the aim is to predict statistical properties, such as the typical magnitude of the perturbations as a function of scale. This is described by the most important such statistic, the *power spectrum. Comparison with observation then relies on assuming that, statistically speaking, we occupy a typical location in the Universe. The sorts of questions we wish to ask are those that would have the same answer regardless of which galaxy in the Universe we happened to be living in; an unsuitable question would be 'how close is the nearest galaxy to ours?', whose answer clearly depends on which one we are in, while an example statistical question that everyone can agree on the answer to would be 'what is the typical separation between galaxies?'.

If the perturbations obey a property known as *Gaussianity, then the power spectrum provides a complete description of them and the value of any observable quantity can be obtained using it. Thus far all observations have been consistent with the initial density perturbations being Gaussian. However, if instead the perturbations are *non-Gaussian then additional statistical properties can be derived.

Generation of perturbations

The initial generation of density perturbations is usually thought to have taken place during cosmological inflation, and is described fully later (*see* INFLATIONARY PERTURBATIONS). An alternative possibility is generation by *topological defects such as *cosmic strings. While observations show that they cannot be solely responsible for the perturbations (as they predict *cosmic microwave background anisotropies in conflict with the observed form), it remains possible that they contribute a subdominant but potentially measurable part of the total perturbations.

Evolution of density perturbations

Density perturbations evolve in time according to the physical laws described by *cosmological perturbation theory. Ordinarily this evolution is so complex that it must be followed using large computer calculations, and several programs exist to do this, including *cmbfast. Density perturbations typically grow under the influence of *gravitational instability.

An exception where it is easy to follow the evolution is when perturbations are both adiabatic and have wavelengths much larger than the Hubble length. The simplest models of inflation produce adiabatic perturbations, and those which we observe today did spend most of their history on scales larger than the Hubble length, so this regime is quite widely applicable. In this situation, it is known that the perturbation in the curvature is conserved (i.e. is unchanging). This is a consequence of *causality. This means that the curvature perturbations predicted by these simplest models do not depend on the way the Universe evolved while they are greater than the Hubble length, which includes the most uncertain parts of cosmological evolution such as post-inflationary *reheating.

distance

The measurement of distance in *cosmology is complicated by the fact that the *Universe is expanding, and thus distances between objects are not constant.

Practical methods for actually measuring extra-Galactic distances are discussed at *distance ladder. Here we discuss different ways of defining distance in cosmology.

Proper distance D_P

This is the distance one would obtain if one could somehow 'freeze' the expansion of the Universe and measure the distance between two points using a series of measuring rods. Clearly the proper distance measured will increase if one freezes the Universe later in its expansion. Of course, this measurement cannot be carried out in practice, although practical distance measurements do approach the proper distance for small separations of less than 1 billion light years or so.

Line-of-sight comoving distance D_C

This distance is measured in *comoving coordinates, a coordinate system that expands along with the Universe. In this system, objects which have no *peculiar velocity relative to the *Hubble expansion will maintain a constant distance from each other. For two nearby objects at redshift z, the comoving distance between them δD_C is equal to the proper separation δD_P divided by the *scale factor of the Universe, i.e. $\delta D_C = (1 + z)\delta D_P$. The total line-of-sight comoving distance from us to a distant object is obtained by integrating the infinitesimal comoving separations between nearby points between $z = 0$ and the object, $D_C(z) = \int_0^z \delta D_C(z)$.

Transverse comoving distance D_M

The comoving separation between two objects at the same distance from us but separated by some angle $\delta\theta$ is $D_M\delta\theta$, where D_M is the transverse comoving distance. It is also the proper motion distance, the ratio of physical transverse velocity (distance per unit time) of an object to its proper motion (radians per unit time). For a Universe with zero *curvature, known as a flat Universe, the comoving distance is the same whether measured along the line-of-sight or transverse to the line-of-sight, i.e. $D_M = D_C$. This is not the case for a Universe with non-zero curvature, in which transverse and line-of-sight comoving distances are not identical.

Angular-diameter distance D_A

The *angular-diameter distance D_A is defined as the ratio of an object's physical transverse size to its angular size (measured in radians). It is related to transverse comoving distance by the simple formula

$$D_A = \frac{D_M}{1 + z}.$$

Angular-diameter distance does *not* therefore increase indefinitely with *redshift, but reaches a maximum at $z \sim 1$, beyond which more distant objects actually appear larger in angular size. The reason for this at-first surprising behaviour is as follows. Light from an object at redshift z was emitted when the Universe was a factor $(1 + z)$ smaller than its present size. Other galaxies were thus closer to us when their light was emitted, and so appear to be a factor $(1 + z)$ larger in angular size. Transverse comoving distance is not affected in this way, since, while a single object is assumed to be gravitationally bound, i.e. not participating in the Hubble expansion of the Universe, two nearby objects are assumed to be moving apart with the Hubble expansion.

If galaxies appear larger at higher redshift, for $z \gtrsim 1$, why then is the night sky not dominated by apparently giant galaxies? The reason is that the *surface brightness of galaxies decreases rapidly with redshift, scaling as $(1 + z)^{-4}$, and so these distant galaxies are very hard to detect above the sky brightness.

Luminosity distance D_L

The *luminosity distance D_L is defined by the relationship between the intrinsic luminosity of a source and its observed flux. If a source is at a distance D_L and assuming that it is radiating isotropically, i.e. the same in all directions, then the flux F we measure is given by the luminosity L of the source divided by the surface area $A = 4\pi D_L^2$ of a sphere of radius D_L. Thus $F = L/(4\pi D_L^2)$. Rearranging this equation, we arrive at the definition of luminosity distance:

$$D_L = \sqrt{\frac{L}{4\pi F}}.$$

Several assumptions must be made when applying this formula as stated:

- We know the intrinsic bolometric luminosity L of the source.
- We measure the bolometric flux F.
- There is no extinction of the light, e.g. due to *dust between the source and the observer.

In practice, both luminosity and flux will be measured through some passband, and a *K correction must be made to allow for the fact that the object is emitting radiation in a different band to that which is observed. An additional correction is made for the estimated extinction along the line of sight.

Luminosity distance is related to transverse comoving distance and angular diameter distance by

$$D_L = (1 + z)D_M = (1 + z)^2 D_A.$$

distance estimates
See DISTANCE LADDER.

distance ladder

Determining distances to extra-Galactic objects is a multi-step process, likened to climbing the steps on a ladder, hence the term 'distance ladder'. Here we describe some of the more important distance estimates, starting on small scales, with the Earth–Sun distance and the distance to the nearest stars, progressing out to more distant stars and then beyond the *Milky Way to extra-Galactic systems.

Trigonometric parallax

The first rung on this ladder is provided by trigonometric parallax, the apparent displacement of the positions of nearby stars as the Earth orbits the Sun. Knowing the Earth–Sun distance, which is most accurately measured via the distance to Venus determined from laser-ranging (see Fig. 1), one astronomical unit ≈ 150 million km, then simple trigonometry allows us to find the distance to a star whose position appears to change by a measured angle over a six-month period. A star is defined to be at a distance of one parsec, short for parallax-arcsecond, and further abbreviated 1 pc, if it has a parallax of one second of arc, see Fig. 2. (In analogy with measurement of time, one arcsecond is one-sixtieth of one arcminute, which in turn is one-sixtieth of one degree.)

The accuracy of measuring trigonometric parallax from the ground is limited by the blurring effect of the Earth's atmosphere, known as 'seeing'. Only at the world's best observing sites does the seeing regularly drop below one arcsecond. One can do much better by employing an **adaptive optics** system to correct for the rapidly changing atmospheric conditions, or by going above

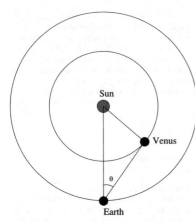

Fig. 1. Measuring the angle θ between Venus and the Sun and the distance between Earth and Venus enables the Earth–Sun distance to be calculated from simple trigonometry.

the atmosphere. The HIPPARCOS satellite (HIgh Precision PARallax COllecting Satellite, named after the ancient Greek astronomer Hipparchus), which operated from 1989–1993, measured parallaxes for more than 2.5 million stars within 150 pc and for 120 000 brighter stars to 1 kpc.

Standard candles

Most remaining rungs on the distance ladder rely on having a *standard candle, an object whose intrinsic luminosity is known. The observed flux of the object is diminished by a factor proportional to the square of the distance to the object, the inverse square law. Thus by comparing observed flux to intrinsic luminosity, one may estimate the distance. The two main problems with this technique are:

• The assumption of known luminosity.

• Applying an appropriate correction for absorption of light by intervening dust, which would cause the flux to diminish more rapidly than given by the inverse square law.

If a standard candle has a distance determined from parallax, then the intrinsic luminosity of the candle can be estimated from its measured flux. Ideally one would like many examples of a particular standard candle to be calibrated in this way, so that one may ascertain the intrinsic scatter in luminosity of that candle.

Examples of standard candles include stars of particular spectral type (see next subtopic), *Cepheid variables, RR Lyrae stars, and *supernovae.

Spectroscopic parallax

This method entails estimating the intrinsic luminosity of a star by observing its spectrum. The particular features present in the spectrum determine the **spectral type** of the star. The spectral type is tightly correlated with the surface temperature of the star. One can also estimate the size of the star by measuring the width of its spectral features. Compact stars have a higher surface gravity than diffuse stars, and thus exhibit wider spectral features. Having estimated the surface temperature T and radius r of the star, one can then further estimate its intrinsic luminosity L from Stefan's law:

$$L = 4\pi r^2 \sigma T^4.$$

Here $\sigma = 5.67 \times 10^{-8}$ W m^{-2} K^{-4} is a constant of Nature known as the Stefan–Boltzmann constant.

Spectroscopic parallax works well for main-sequence stars (a typical 10% uncertainty in luminosity results in a 5% uncertainty in distance), but for giant stars luminosity can only be estimated to $\pm \sim 0.5$ mag and hence distance to $\sim 25\%$.

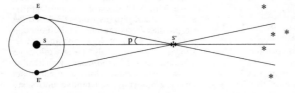

Fig. 2. Illustration of trigonometric parallax. As the Earth orbits the Sun (S) from point E to point E', the position of a nearby star S' appears to change by an angle $2p$ relative to much more distant stars. The parallactic angle p is the angle subtended by the Earth–Sun separation at the distance of the star.

Photometric parallax

Obtaining a high-quality spectrum for a star is quite observationally expensive and so the method of photometric parallax uses observations of the colour of a star to estimate its luminosity. This is much less accurate than the luminosity obtained from spectroscopic parallax, but one can observe the colours of many thousands of stars in a star cluster using just two or three observations and obtain a reliable estimate of the distance to the cluster.

Note that the word 'parallax' is loosely applied in both spectroscopic and photometric parallax since it is a geometric term, and is used to imply the measure of distance.

Surface-brightness fluctuations

This is the first of three methods that utilize physical properties of *galaxies as distance indicators.

The method of surface-brightness fluctuations applies only to spheroidal stellar distributions, i.e. elliptical and lenticular galaxies and the bulges of spiral galaxies. The idea is to measure the fluctuations in brightness between the pixels of a charge coupled device (CCD) image arising from variations in star counts between the pixels. One is thus measuring the 'graininess' of the galaxy's image.

Consider two identical galaxies at different distances: it is not possible to distinguish them by mean flux per pixel, since *surface brightness is a distance-independent quantity if expansion of the Universe is neglected. This is because the number of stars per pixel increases with the square of distance, which is compensated by the fact that the flux per star decreases with the square of distance. Total flux per pixel is therefore independent of distance.

We *can* however distinguish the galaxies by the *variance* in flux per pixel. If the mean number of stars per pixel is $\bar{N}$ and the mean flux per star is $\bar{f}$, then the mean flux per pixel $= \bar{N}\bar{f}$. If the stars are randomly distributed within the galaxy, then they obey Poisson statistics, where the variance in star counts per pixel is equal to the mean, $\langle (N - \bar{N})^2 \rangle = \bar{N}$. Thus the variance in flux is related to distance d by

$$\langle (N\bar{f} - \bar{N}\bar{f})^2 \rangle = \bar{N}\bar{f}^2 \propto d^2(d^{-2})^2 \propto d^{-2},$$

that is the variance is inversely proportional to the square of distance. Or in other words, a

galaxy at twice the distance will appear twice as smooth.

This method assumes that there is no inherent pattern in the stellar distribution within galaxies. This will not be true if the galaxy contains any traces of spiral arms or shell-like structure. Another problem with the method is that there are no nearby elliptical or lenticular galaxies to use as calibrators. However, the method has successfully been applied to estimate the distance to the *Coma cluster.

The 'fundamental plane'

This technique is applicable only to elliptical galaxies, and is an empirical finding that the three quantities luminosity L, surface brightness Σ_e measured within the effective radius R_e, and *velocity dispersion σ occupy a fairly narrow plane within this three-dimensional space. This means that if any two of L, Σ_e and σ are known, the third parameter can be inferred from the fundamental plane relation:

$$L \propto \sigma^{8/3} \Sigma_e^{-3/5}.$$

Note that σ and Σ_e can be measured without knowing the distance to the galaxy, and thus the intrinsic luminosity L and hence the distance can be inferred once this relation has been calibrated in a given passband.

Tully–Fisher relation

The *Tully–Fisher relation correlates luminosity with the rotation speed of spiral galaxies. By assuming that the density of a spiral galaxy drops inversely with the square of radius from its centre, and that galaxies have constant central surface brightness and constant mass-to-light ratio, one can show that luminosity is proportional to the fourth power of rotation speed, $L \propto v^4$. One can measure the rotation velocity v from the width of the hydrogen 21cm line using a single-dish radio telescope if the inclination of the galaxy's disk to the line-of-sight is known. Drawbacks of the method include difficulty of calibration, internal extinction within the (dusty) spiral arms of the galaxy, influences of the galaxy's environment and a varying mass-to-light ratio.

Supernovae

Type Ia supernovae provide estimates of distance to much higher redshift than any of the above methods, and currently form the final rung on the distance ladder and provide our only distance estimates beyond redshift $z \sim$ 0.1 (*see* SUPERNOVAE). In brief, one compares the observed peak brightness of the supernova with that predicted from the decay time of its light curve, and so estimates the distance to the supernova. Supernovae provided the first evidence that the expansion of the universe is actually accelerating today, contrary to previous expectations.

distance modulus

Since the observed brightness of an object depends on its *distance, one may express this distance as the difference between the apparent (observed, m) and absolute (intrinsic, M) *magnitude of the object, a quantity known as the distance modulus, DM = $m - M$. From the definition of apparent and absolute magnitude, and the fact that the perceived brightness of an object decreases with the square of the distance, one can easily show that distance modulus and physical distance are related by

$$\text{DM} = m - M = 5\log_{10}(d/10\text{pc}).$$

Here the distance d is measured in units of 10 parsecs, since by definition, absolute and apparent magnitude are identical for an object at this distance.

In *cosmology, it is easier to quote distances in units of Mpc (one million parsecs), in which case the distance modulus becomes

$$\text{DM} = m - M = 5\log_{10}(d/\text{Mpc}) + 25.$$

One also needs to apply two corrections to allow for the fact that the *Universe has expanded by a factor of $(1 + z)$ in the time it takes light from an object of *redshift z to reach us.

The first correction multiplies the *comoving distance to the object by the expansion factor $(1 + z)$; this is a geometric correction to allow for the fact that light has travelled further than the comoving distance of the object.

The second correction, known as the *K correction, corrects for the fact that light observed with a wavelength λ was emitted at a rest wavelength $\lambda_0 = \lambda/(1 + z)$. Thus measuring the apparent magnitude in the optical part of the spectrum may actually correspond to the *ultra-violet region in the restframe of the object. To make a consistent comparison between the luminosities of objects at different redshifts, one must estimate the magnitude difference (the K-correction) between the restframe and observed passbands. This correction depends on three factors:

- The response function (sensitivity as a function of wavelength) of the observed passband, λ.
- The *spectral energy distribution (SED) of the object, S.
- The redshift of the object, z.

With both of these corrections, the **cosmological distance modulus** becomes

$$\text{DM} = m - M = 5\log_{10}(d(1 + z)/\text{Mpc})$$
$$+ K_{\lambda, S}(z) + 25.$$

Given an observed apparent magnitude m and redshift z of an object, this formula allows one to determine its absolute magnitude M.

Doppler effect

The Doppler effect, named after Christian Doppler, is the name given to the change in wavelength of a wave-propagated signal, such as a soundwave or lightwave, when the source of the signal and the observer are moving relative to each other.

It can be understood as follows. Imagine that somebody in a boat close to a lakeshore is slapping the water with an oar at regular intervals, say once per second. This causes waves to propagate out in all directions at a speed of order 1 metre per second. The wave then has a wavelength (distance between adjacent peaks) equal to the distance travelled in 1 second, i.e. 1 metre. Now suppose the boat starts moving away from the shore at a speed of 0.5 m/s. In the time between one oar-slapping and the next, the first wave will have travelled a distance of 1 metre toward the shore while the boat will have travelled 0.5 m away from the shore. The wave peaks reaching the shore will thus now be separated by a distance of 1.5 m, i.e. the wavelength of the signal is stretched to 1.5 m. Note that the frequency of waves reaching the shore is decreased by the same factor of 1.5.

The same effect can occur with soundwaves, resulting in an increased siren pitch (shorter wavelength soundwave) for an ambulance travelling towards you, and a decreased pitch (longer wavelength soundwave) for one moving away. Similarly with lightwaves, light appears bluer (of shorter wavelength) for a source moving toward the observer and redder (of longer

wavelength) for a receding source. Since the
*Universe is expanding, nearly every extra-
Galactic source exhibits a *redshift rather
than a blueshift.

If the wavelength of the emitted radiation
λ_0 and the wave speed c are known, then the
recession velocity v of the source can be calculated from the observed wavelength λ from
the formula

$$\frac{v}{c} = \frac{\lambda - \lambda_0}{\lambda_0}.$$

Thus by measuring the observed wavelength
λ of a spectral feature of known rest wavelength λ_0 in a distant object's *spectrum, the
redshift of that object can be determined.

dust

Dust is the name given to small particles,
or 'grains', of solid matter, which range in
size from molecular (a few nanometres) to
around a tenth of a millimetre in size. Cosmic
dust is often named after its location: **zodiacal dust** is concentrated in the plane of the
Solar System, **circumstellar dust** lies in a shell
around the star that produced it, **interstellar
dust** lies between the stars in *galaxies and
intergalactic dust lies between galaxies. Most
dust grains are based on silicates or carbon,
formed in the atmospheres of cool stars.

Dust grains absorb or scatter light whose
wavelength is comparable to or smaller than
the size of the grain. Thus the effects of dust
between a source of light and an observer
are both to diminish the light in intensity,
and to make it appear redder: it is dust in
the Earth's atmosphere that makes the Sun
appear redder around sunset. It is thus important to correct the *magnitudes and colours of
extra-Galactic objects for the effects of dust:
primarily interstellar dust in our own Galaxy.
This may be done using a dust reddening map derived from COBE-DIRBE observations of the *infrared background (see Fig. 1).
This map is commonly referred to as the
'SFD map' after the initials of its authors'
names: David Schlegel, Douglas Finkbeiner,
and Marc Davis. In comparison with interstellar space, intergalactic space is a far 'cleaner'
environment, and intergalactic dust reddening is generally neglected. However, interstellar dust reddening *within* a source being
observed can be significant and should not be
neglected. This is particularly true for spiral
galaxies observed edge on: dust in the disk of
the galaxy absorbs optical light from stars in
the far regions of the disk.

Star formation in recently formed galaxies is frequently shrouded by clouds of dust.
*Ultra-violet radiation from the hot young
stars is absorbed by the dust grains, which are

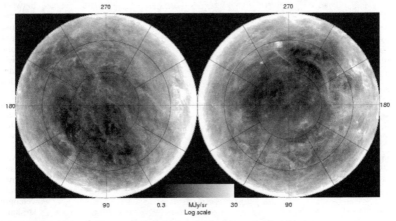

Fig. 1. Maps of the dust distribution in our Galaxy, as measured by the intensity of its infrared
radiation, for the northern (left) and southern (right) Galactic hemispheres. The Galactic poles
are at the centre of each plot, the Galactic plane lies around the edge of each.

heated, and re-radiate the energy absorbed in the mid–far infrared part of the spectrum.

dwarf galaxies

Despite their name, dwarf *galaxies are defined to be of low luminosity, rather than of small size, although in practice they do tend to be smaller, and also of lower surface brightness, than their giant counterparts. There is no sharp division between dwarf and giant galaxies, although historically, galaxies with a blue absolute magnitude fainter than about $M_B \approx -17$, corresponding to roughly one billion Solar luminosities, have been classified as dwarfs.

Since dwarf galaxies are intrinsically faint, they can only be seen nearby, and so most *galaxy surveys include very few dwarfs. However, they are in fact more numerous than giant galaxies: the *Local Group con-

tains only four giant galaxies and at least 40 dwarfs, most of which are *satellite galaxies of the *Milky Way or M31. Extrapolating beyond the local group, it is likely that dwarfs are at least a factor of ten times more numerous than giants. While they dominate in numbers, dwarfs make only a minor contribution to the *luminosity density of the Universe due to their extreme faintness.

Like giant galaxies, dwarfs may be classified by morphological appearance. The most common classifications are dwarf elliptical (dE), dwarf spheroidal (dSph, low-luminosity dwarf elliptical satellites), dwarf spiral (dSA), dwarf barred spiral (dSB) and dwarf irregular (dIrr).

Dwarf galaxies often show signs of rapid star formation. In general, the star-formation rate per unit mass is higher in dwarf galaxies than in giant galaxies.

e-foldings

The number of e-foldings, usually denoted N, is a way of quantifying the amount of expansion that the *Universe experienced during *cosmological inflation. One simply compares the final value of the *scale factor, measuring the size of the Universe, relative to the initial one. However this number is typically so vast that it is convenient to take its logarithm to get a manageable number. Conventionally it is the natural logarithm that is used. If the amount of inflation is for instance $N = 100$, that means that the Universe has expanded by a factor of e 100 times (giving a total expansion by a factor of about 10^{43}), where $e = 2.718\cdots$ is Euler's number, one of the key mathematical constants.

The term e-foldings is sometimes used in other contexts where processes take place according to an exponential law. For instance, in radioactive decay, the decay timescale is defined as the time it takes for the amount of material to have reduced by a factor of $1/e$.

early Universe

'Early Universe' is a generic term for the branch of *cosmology which studies the earliest instants of the *Universe's evolution. While there is no precise definition, it is usually taken to refer to epochs when the Universe was so hot that the relevant physical laws are uncertain, being out of reach of terrestrial experiments. By this definition, the early Universe refers to the first 10^{-10} seconds of the Universe's existence. This research area is also known as *particle cosmology, and sometimes, less accurately, as *particle astrophysics.

The main aims of the early Universe researcher are quite different from those of other astronomers, who are normally trying to apply well-established physical laws in complicated physical circumstances. Early Universe cosmologists are instead trying to infer the nature of physical laws that might

have applied during the early epochs. This sounds a fairly hopeless task, but luckily the young Universe is predicted to have been very close to *homogeneity, i.e. a uniform distribution of matter, meaning that the physical situation was as simple as it could possibly be. That makes it practical to investigate a wide range of hypotheses about how the young Universe might have behaved.

Although the physical laws are uncertain, that is not to say that there is no guidance at all. While established high-energy physics is limited to the *standard model of particle physics, enhanced by the inclusion of *neutrino masses, particle physicists have offered many ideas for how that model might be extended to high energies. These include *supersymmetry, *superstrings, and *M-theory. Not only do these ideas indicate to cosmologists the type of processes that might take place; it might well be that cosmology is actually the only practical way to test them.

Of course, we cannot directly observe the first 10^{-10} seconds of the Universe's existence. However, we can learn about what happened through relics that might have been created during that era and which survive to the present. There are actually many of these; according to current thinking, it is during that first fraction of a second that the present abundances of *baryons and *dark matter were laid down, and when the *density perturbations which lead to *structure formation were created. Additionally, there may be relics of early cosmic *phase transitions, such as *topological defects, still present in our Universe, and perhaps even *primordial black holes.

The most important current idea in early Universe studies is *cosmological inflation, which is the leading paradigm for the origin of density perturbations, and which is also able to explain the large-scale properties of the Universe, including its spatial flatness. Other key areas of early Universe research include cosmic phase transitions, *baryogenesis, and the impact of new ideas from fundamental physics such as the *braneworld.

Kolb, E. W. and Turner, M. S. *The Early Universe*, Addison–Wesley, 1990 [Technical].

Eddington, Arthur (1882–1944)

Sir Arthur Eddington (Fig. 1) was one of the most important astrophysicists of the early 20th century. Amongst his wide-ranging achievements, he was the first astronomer to build detailed working models of the internal structure of stars, able to reproduce their main observed properties such as temperature and size. In *cosmology, his main fame derives from his contributions to the development of *general relativity, and in particular his leadership of an expedition to use observations during a Solar eclipse to overturn Newton's theory of *gravity in favour of *Einstein's.

Eddington was born in 1882 in England, and was quickly recognized as a mathematical prodigy. He studied natural sciences at Trinity College Cambridge, and then took up a position at the Royal Observatory in Greenwich. In 1913, at an age of just 30, he was awarded the Plumian Professorship at Cambridge, to this day seen as the most senior astrophysics professorship in the UK, which he held until his death in 1944.

Fig. 1. Sir Arthur Eddington.

As remarked above, Eddington's astrophysical achievements were wide-ranging, and the most notable was his theory of stellar structure, culminating in his seminal textbook *The Internal Constitution of the Stars* published in 1926. His success is usually attributed to his willingness to use physical insight to make approximations that turned out to be accurate. He is also well known for an acrimonious dispute with a young Indian physicist Subrahmanyan Chandrasekhar over the nature of white dwarf stars and whether they had a maximum possible mass (now known as the Chandrasekhar limit), with history showing in this case that Eddington was on the wrong side. However the rest of this article will concentrate on his contributions to cosmology.

Einstein's publication of his work on general relativity in 1915 coincided with the grim heights of the First World War. Eddington, as a pacifist and internationalist who had had strong pre-war links with German colleagues, was uniquely placed amongst British scientists to appreciate the significance of this development, and one of the very few able to grasp its mathematical basis. He became a tireless advocate for the theory, perhaps driven by his eagerness to show that science transcends national boundaries.

One of the predictions of general relativity was for the bending of light around massive objects. Within the Solar System only the Sun was massive enough for a detectable effect, but measurements were only possible during a Solar eclipse by the Moon, which blocks the light from the Sun enabling the light from stars almost behind it to be seen. After the war ended, a Solar eclipse was due in May 1919, but visible only from Brazil and Western Africa. Eddington took a team to the island of Principe off Africa, and despite many trials experienced during the trip they were able to make measurements sufficient to support Einstein's theory. In particular they were able to rule out Newton's theory of gravity, which did predict a bending, but only of half the magnitude.

Another important result bearing his name is the Eddington limit or Eddington luminosity, which is the maximum luminosity that can be emitted through a spherically symmetric layer of gas such as the surface of a star. If the limit is exceeded then the gas layer is expelled by the radiation pressure. The Eddington limit is important in a range of systems including *active galactic nuclei, though it can be exceeded if spherical symmetry is not present or if radical structural changes are taking place as, for instance, in *supernovae.

Some of his later work led to notoriety, as he became increasingly obsessed with numerology, for instance arguing first that the fine-structure constant (a fundamental constant of nature) should be precisely 1/136, and then as new measurements came in that it should be precisely 1/137 (neither is true). These arguments led him to make an absurdly precise, though doubtless wrong, prediction that the number of protons in the Universe was 15 747 724 136 275 002 577 605 653 961 181 555 468 044 717 914 527 116 709 366 233 425 076 185 631 031 296 (i.e. $1.57 \cdots \times 10^{78}$), sometimes known as the Eddington number.

Eddington won a number of awards during his life, including the Gold Medal of the Royal Astronomical Society, a knighthood, and the UK Order of Merit. Both a lunar crater and an asteroid are named after him, and the Royal Astronomical Society awards an Eddington Medal every three years to scientists making 'investigations of outstanding merit in theoretical astrophysics'. He has also been attributed as the originator of the cliche that a team of monkeys with typewriters would, given infinite time, type the complete works of Shakespeare, which may have evolved from his 1929 comment, 'If an army of monkeys were strumming on typewriters, they might write all the books in the British Museum'.

Douglas, A. V. *The Life of Arthur Stanley Eddington*, Nelson, 1954.

Stanley, M. *Physics World, 18, 9*, pp. 33–8, 2005.

Einstein, Albert (1879–1955)

Albert Einstein (Figs. 1 and 2) is the most iconic of scientists, his name synonymous with genius. His most lasting contributions to physics are the special and general theories of *relativity, which encode principles that must be obeyed by all physical laws and thus transcend the normal division of physics into distinct subject areas such as atomic physics or thermodynamics. Other key contributions include the explanation of the photoelectric effect, one of the discoveries which led to the creation of *quantum mechanics, and the Bose–Einstein distribution in statistical physics which describes fundamental particles with integer spin (*bosons). Curiously, he also invented a type of refrigerator.

Fig. 1. Albert Einstein photographed around 1905, his annus mirabilis.

Einstein was born in March 1879 in Ulm in southern Germany. His family lived in several European countries while he was growing up, and he eventually studied at the ETH (Federal Institute of Technology) in Zurich, Switzerland, receiving a teaching diploma in 1900. The following year he became a Swiss citizen.

Famously, Einstein's early work was carried out not in a university department but in a Swiss patent office where he was employed as a technical examiner (a rather unlikely claim is that his most significant achievement in this post was awarding a patent for Toblerone chocolate bars). This is because his record was not strong enough to secure him a university position. During this time, he worked towards a doctorate, which he would receive in 1905.

1905 was also the year in which he single-handedly revolutionized physics, in what is often called his annus mirabilis. He wrote three papers in unrelated areas of physics, each of which broke new ground. His study of the photoelectric effect, attributing it to the

Fig. 2. Einstein in 1933, in characteristically dishevelled style.

quantization of the energy of light into what are now known as *photons, ultimately led to the development of the quantum theory. His work on Brownian motion indicated the reality of atoms. And with the special theory of relativity, he rewrote the rules of space and time, uniting them into a single concept of *space-time. A fourth 1905 paper developed this idea further to yield the famous equivalence of mass and energy, $E = mc^2$.

Ironically his doctorate did not include any of this work, which he viewed as too challenging for his examiners. He instead gave them a relatively pedestrian theory of the measurement of molecular dimensions.

With his growing reputation, in 1908 Einstein was able to obtain a teaching role at the University of Bern. In quick succession, he moved to Zurich, Prague, back to Zurich again, and then to Berlin in 1914 where he would remain until 1933.

During the period 1905 to 1915, Einstein worked on the theory of *general relativity, a theory of the gravitational force which overturned Newton's law of *gravity. In it, gravity is interpreted as due to a *curvature of space-time, that curvature being caused by the presence of massive objects. This theory is widely recognized as his greatest achievement; while arguably other physicists might have obtained his earlier results soon after, had he not, general relativity could easily have gone undiscovered for decades. Its construction

required a unique combination of prowess in the advanced mathematics of differential geometry and crystal-clear physical insight.

In 1919, an expedition led by Arthur *Eddington to observe a solar eclipse dramatically confirmed Einstein's theory, which predicted that the path of light rays would bend to follow the curvature of space-time as they passed near the Sun. Einstein was catapulted into the role of media star, the public proving to have considerable appetite to hear of, though perhaps not to understand, his theories.

Einstein also used his general theory of relativity to make the first models of the entire *Universe, around 1917. Unfortunately, he was hampered by a belief that the Universe was static (it would be more than a decade until Edwin *Hubble discovered the expansion of the Universe), causing him to modify his theory by introducing a quantity known as a *cosmological constant. Einstein later realized that the cosmological constant was unnecessary and dismissed its inclusion as a blunder. However, in the modern *standard cosmological model, the dominant material in the Universe is in fact a cosmological constant!

By 1921 it was clear that Einstein's many achievements made him the leading candidate for a *Nobel Prize. It is commonly stated that the award was solely for the photoelectric effect, the relativity theory being too controversial, though the actual citation, 'For his services to Theoretical Physics, and especially for his discovery of the law of the photoelectric effect', leaves some room for ambiguity.

During the 1920s, the development of quantum theory was in full flow, and Einstein found himself in the unfamiliar role of resistant scientific conservative. In particular, he was repelled by Neils Bohr's view of the quantum world as inherently unpredictable, famously (mis)quoted as saying 'God does not play dice with the Universe'. [The actual quote, in a letter to Max Born, is that quantum theory '... does not really bring us any closer to the secret of the Old One. I, at any rate, am convinced that He does not throw dice.'] He spent quite a lot of effort seeking to undermine the probabilistic interpretation of quantum mechanics, preferring to believe that it might have an underlying deterministic layer. It was however later shown by John Bell that such a layer would violate the laws of relativity.

In 1933, Hitler came to power in Germany and all Jewish professors, including Einstein, were forced from their jobs. He moved to Princeton in the US, where he had already established links, to join the Institute for Advanced Studies. He would remain there until his death in 1955.

During this late phase of his scientific work, he focused mainly on attempts to unify the *fundamental forces, a topic which still pre-occupies theoretical physicists to this day. He was not helped in this by the nuclear forces being unknown, and this work ultimately ended in failure.

Einstein's main political stance was pacifism, and he used his fame to campaign for civil rights and against tyranny around the world. Despite that, he was in part responsible, through a letter sent to President Roosevelt (it is thought that the letter was actually written by his friend Leó Slizárd, but signed and sent by Einstein), for the setting up of the Manhattan Project which created the first nuclear weapons. He did so through fears that Nazi Germany might be the first to develop such weapons. He worked tirelessly in his later years to campaign against nuclear weapons, which were ultimately a consequence of the theories he had developed.

Einstein was married twice, first to mathematician Milena Marić in 1903 (with whom he had three children, the first born before their marriage and believed to have been put up for adoption) and then, following a divorce, to his cousin Elsa Löwenthal in 1919.

Einstein is commemorated in numerous ways, including the Albert Einstein Peace Prize and the chemical element Einsteinium. His image rights are owned by the Hebrew University in Jerusalem and generate significant income. The centennial of his annus mirabilis was celebrated in UNESCO's 2005 World Year of Physics, also known as Einstein Year. In 1999 *Time* magazine declared him the 'Person of the Century'.

Pais, A. *Subtle Is the Lord: The Science and the Life of Albert Einstein*, Oxford University Press, 2005 (first published 1982).

White, M. and Gribbin, J. *Einstein: A Life in Science*, Free Press (2005).

Einstein–de Sitter Universe

The Einstein–de Sitter cosmological model assumes a homogeneous, isotropic *Universe with zero *curvature, *cosmological constant, and *pressure. In this case, the *Hubble parameter H, which gives the expansion rate of the Universe, is given by

$$H^2 = \left(\frac{\dot{a}}{a}\right)^2 = \frac{8}{3}\pi G\rho,$$

where G is Newton's gravitational constant, a is the *scale factor and ρ is the mean matter-density.

One can show that in such a model, the Hubble parameter decreases with time as

$$H = \frac{\dot{a}}{a} = \frac{2}{3t}.$$

Note that the expansion rate tends slowly to zero as the age of the Universe tends to infinity. Inverting this equation, and substituting in the current value of the Hubble parameter H_0, one can express the *age of the Universe as

$$t = \frac{2}{3H_0}.$$

This model grew out of correspondence between *Einstein and the Dutch astronomer Willem de Sitter in 1916–18, *before* *Hubble's discovery of the expanding Universe. Until Hubble's discovery, Einstein was generally unhappy with non-static solutions to the *Friedmann equation, and he introduced his cosmological constant as a way of avoiding expansion. Once he was convinced of the validity of Hubble's discovery, Einstein abandoned the cosmological constant, which he later allegedly called the biggest blunder of his life (an anecdote related by *Gamow). The recent discovery that the expansion rate of the Universe is accelerating is inconsistent with the Einstein–de Sitter model, and so Einstein's cosmological constant has made a comeback.

Einstein's equation

*Einstein is responsible for the world's most famous equation, $E = mc^2$, relating energy mass, and the speed of light. However it is not this equation which bears his name. Einstein's equation refers to the central equation of his theory of *general relativity, which supplanted Newton's theory as the modern description of *gravity.

According to general relativity, gravity is an apparent rather than actual force. Its origin is that the presence of matter curves *spacetime. Objects in motion then follow the curvature of space-time, and in doing so act pretty much as if they are responding to a gravitational force.

The mathematics of Einstein's equation, known as tensor calculus, are well beyond the scope of this book, but for completeness we quote the equation. It can be written

$$R_{\mu\nu} - \frac{1}{2}g_{\mu\nu}R = \frac{8\pi G}{c^4}\,T_{\mu\nu}.$$

In the highly-condensed notation of tensor calculus, this expression actually corresponds to ten separate equations, essentially of the same form but referring to the ten different ways of combining the three space coordinates and one time coordinate into pairs (allowing for duplication).

The details need concern us no further, except to note that the expression on the left-hand side represents the curvature of space-time, while $T_{\mu\nu}$ on the right-hand side is the energy–momentum tensor representing the properties of matter. Schematically, then, the equation can be written

Curvature = matter.

In words, due to the great American relativist John Wheeler, 'Matter tells space how to curve, space tells matter how to move'. [A pedant (this note apparently putting us in that category) would point out that he should say 'space-time', not 'space', but one shouldn't argue with one of the most renowned popularizers of relativity theory.]

The right-hand side includes as constant of proportionality Newton's gravitational constant G, now reinterpreted as giving the strength of relativistic gravity. Its presence indicates that we are indeed talking about gravity; if G were set equal to zero then matter would no longer be responsible for causing space-time curvature. The precise combination of constants on the right-hand side is to ensure that general relativity reproduces Newton's theory of gravity in situations where the gravitational force is very weak.

An essential part of this equation is that it remains non-trivial even at locations where there is no matter, so that the right-hand side vanishes. Even when it does, there can still be curvature of space-time; the vanishing of the left-hand side of Einstein's equation is not enough to remove all curvature. That's just as well, as we know the effects of gravity can still be present in the vacuum around objects, for instance guiding the Earth around the Sun. An extreme example of such a situation is the gravitational field outside a *black hole.

ekpyrotic Universe

The ekpyrotic Universe is a bold proposal to explain the *big bang phenomenon. It was introduced in 2001 by Justin Khoury, Burt Ovrut, Paul Steinhardt, and Neil Turok. The term originates from an ancient cosmological idea known as *ekpyrosis*, whereby the *Universe is consumed in flames and then reconstituted. It is an attempt to explain the large-scale properties of the Universe without the need for *cosmological inflation.

The ekpyrotic Universe is grounded in the advanced modern physics ideas of *M-theory and *braneworlds. In brief, these ideas suggest that our visible Universe may be embedded in a higher-dimensional space— a convenient mental picture is to think of a piece of paper, on which are drawn all the stars and *galaxies. The terminology is that our visible Universe lies on a *brane, the word being short for membrane. Our own brane may only be one of many scattered throughout the full higher-dimensional Universe.

The idea behind the ekpyrotic scenario is that these branes may move around, and in particular may collide, just as you could collide two pieces of paper. For example, we might begin in a situation where a second brane (i.e. a completely separate entire Universe!) lies parallel to our own but some distance away. Due to mutual attraction the branes are drawn together, and ultimately collide. Obviously you don't really want to be around when entire Universes are colliding, and indeed from the perspective of some hypothetical person living in the Universe when this happens, things would look very much like a big bang and subsequent expansion. Insofar as you are willing to believe the initial condition that the branes are perfectly parallel, this even explains why the big bang appears to have happened at the same time everywhere in the Universe (*see* HORIZON PROBLEM for discussion of why this is mysterious).

The scenario even claims a mechanism for generating the initial *density perturbations thought to be responsible for structure formation. Due to quantum effects, the branes will not remain perfectly parallel as they approach, but rather will develop small ripples which spoil the simultaneity of the collision. With the big bang now happening at slightly different times in different regions of our brane, small differences in density can be created that are later amplified by *gravitational instability to form galaxies.

Surprisingly, despite its highly speculative nature the ekpyrotic scenario does appear to make some clear cut predictions which permit it to be falsified by future observations. Perhaps the most important is that it predicts the absence of large-scale *gravitational waves in the Universe. If such waves are ever detected, the ekpyrotic scenario must be wrong.

The ekpyrotic scenario is ambitious and controversial, and has not received widespread acceptance in the cosmology community. Experts in fundamental physics theories have argued that, while motivated by it, the ekpyrotic theory does not properly incorporate the physics of M-theory. Whether or not the density perturbations produced are in accord with observations is also in dispute; so far it has not proven possible to model the moment of impact itself, and that leads to ambiguities with different research groups claiming different forms for the perturbations produced. In any event, ekpyrosis has not displaced inflation as the standard explanation of the large-scale Universe and its density perturbations; indeed it draws heavily on the ideas and mechanisms of inflation and it is really a matter of taste as to whether it is thought of as distinct from inflation or instead a particularly exotic implementation of the inflationary idea.

The ekpyrotic Universe scenario features a single collision of branes. It inspired a later model known as the *cyclic Universe where there would be a repeating, perhaps endless, cycle of brane collisions and re-expansions. While equally controversial, the cyclic Universe model is usually taken to have superseded the original ekpyrotic idea, being a somewhat simpler construction.

electromagnetic radiation

Electromagnetic radiation is energy that travels as a wave, and that is transmitted as oscillations in electric and magnetic fields. These fields can exist in vacuum, and so radiation can travel through empty space.

All electromagnetic radiation travels at the same speed, known as the *speed of light, with value $c \approx 3 \times 10^8$ metres per second.

The type of radiation is determined by its wavelength λ: the distance between successive peaks in the electric or magnetic fields. In the visible part of the spectrum, short wavelengths are perceived as blue, longer wavelengths perceived as red. One may also define frequency for electromagnetic radiation, the rate at which peaks in the wave pass a given point along the direction of travel. This frequency is normally denoted by the Greek letter ν and is related to wavelength by $\nu = c/\lambda$. The quantum unit of electromagnetic radiation is called the *photon. This possesses properties of both a wave and a particle, and has an energy given by $E = h\nu$, where h is the Planck constant (see PLANCK SCALE).

As well as its wavelength, one is also able to measure the intensity of radiation (how bright it appears), and also its *polarization. Polarized light has an electric field oscillating in a preferred direction perpendicular to the line of sight; the degree of polarization provides useful information about the astrophysical processes generating the radiation.

electromagnetic spectrum

The electromagnetic spectrum is the name given to all different types of *electromagnetic radiation in the Universe. The electromagnetic spectrum is commonly broken down into several regimes. These regimes are described in order from long-wavelength, low-frequency and low-energy, to short-wavelength, high-frequency and high-energy (see Fig. 1).

Radio—Radio waves have wavelengths longer than about 10 cm. They are used on Earth to transmit radio and television signals. They are also emitted by many astrophysical objects, such as *active galactic nuclei and by clouds of neutral hydrogen gas. The 21cm line of hydrogen (due to a flip in the spin of the *electron orbiting a *proton) has proved extremely valuable in mapping *galaxy rotation curves and in finding giant hydrogen clouds surrounding some *dwarf galaxies.

Microwave—Microwaves are more accurately referred to as millimetre waves, since their wavelengths cover the approximate range 100 to 0.1 mm. Microwaves at the long-wavelength end of the range are used to heat food and are also used in radar systems. The most important

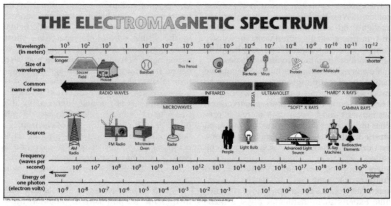

Fig. 1. Graphical representation of the electromagnetic spectrum.

astrophysical source of microwaves is the *cosmic microwave background (CMB), a direct relic of the *big bang. Studying the CMB can tell us a great deal about the *early Universe.

Infrared—*Infrared radiation covers the approximate wavelength range 1 mm to 1 micron. It is familiar to us on Earth as the heat felt from the Sun or from a toaster. The radiation from evolved stars peaks in the near-infrared, around 2 microns. At longer wavelengths, infrared light is a valuable tracer of *dust that has been heated by starlight.

Visible—Our eyes are sensitive between roughly 400 and 750 nm (1 nanometre = 10^{-9} m). Astrophysical sources of visible light include stars and collisions of some fast-moving particles.

Ultra-violet—Radiation of shorter wavelength than can be perceived by the human eye, down to about 10 nm in wavelength. It is emitted by hot stars (including the Sun; UV light is responsible for sunburn) and other hot objects in the Universe. It is particularly useful as a tracer of young, recently formed *galaxies which contain a large fraction of hot, massive stars.

X-ray—*X-ray radiation is often subdivided into 'soft' X-rays (around 10^{-8} to 10^{-10} m in wavelength), and 'hard' X-rays (wavelength shorter than 10^{-10} m). X-rays are used for medical imaging (flesh is transpar-

ent to them, revealing the bone structure underneath) and are produced by high-temperature gases, such as are found in *clusters of galaxies. Arguably the best way of finding galaxy clusters is to search in the X-ray part of the spectrum.

Gamma-rays—Gamma-rays overlap in wavelength with hard X-rays, being shorter in wavelength than about 10^{-11}m, but generally produced by nuclear decay processes or extremely high-energy particle collisions rather than thermal processes.

Note that some types of 'radiation' emitted in radioactive decay are not electromagnetic waves. Alpha particles are fast-moving helium nuclei (two protons and two *neutrons bound together) and beta particles are fast-moving electrons.

electrons

Electrons are fundamental particles. According to the *standard model of particle physics, they are indivisible particles, belonging in a class known as **leptons** which also contains *neutrinos. They carry an electric charge of -1.

The electron has a partner particle known as the electron neutrino, and there are two heavier particles with similar properties to the electron, known as the muon and the tau particle. Those are however unstable, and exist only fleetingly in the Universe.

The Universe is believed to be charge neutral on average (otherwise electromagnetic forces would dominate gravitational ones). This means that the number of electrons should equal the number of *protons, which carry an electric charge plus one. Because of this, electrons are usually not listed separately in the inventory of the Universe's contents, but instead are included under the title of *baryon even though electrons are definitely not baryonic particles. As an electron has a mass only about one two-thousandth that of a proton, their contribution to the *density of the Universe is small.

Electrons can either be bound into atoms, where they are localized near a positively charged atomic nucleus, or they can exist freely. Within atoms electrons occupy discrete energy levels, and can move between them by emission or absorption of a *photon of light. In hot gas (at temperatures above 20 000 kelvin or so for hydrogen), the ambient energy is sufficient to prevent electrons combining with atoms and we have a plasma state. The process of removing electrons from atomic nuclei is known as ionization, and the *ionization history describes how the fractional ionization of the Universe changes with time.

The main electron properties of interest astrophysically are their interaction with photons of light. Electrons in atoms can absorb and emit photons at characteristic frequencies depending on the energy level structure of the particular atom in question. The study of these energy levels is known as *spectroscopy, one of the main tools of the observational astrophysicist. Free electrons can scatter photons of light, known as **Thomson scattering** for the usual case where the electron is not moving relativistically (in the centre of mass frame), and **Compton scattering** when it is. Thomson scattering changes the direction that the photon is travelling in but not its energy, and this gives rise to the signature of *reionization in *cosmic microwave background anisotropies.

Electrons can also emit radiation if they undergo acceleration. Electrons following circular trajectories (caused for instance by the magnetic fields in a *galaxy) emit low-energy radiation known as **synchrotron radiation**. Electrons in a hot ionized gas, which are accelerated when passing close to a positively charged proton, emit high-energy radiation known as **bremsstrahlung radiation**

(translated as 'braking radiation'). This is also sometimes known as free–free radiation. Bremsstrahlung is the main mechanism by which *clusters of galaxies emit *X-rays.

electro-weak phase transition

The electro-weak phase transition is an epoch of the *Universe's evolution where its properties are believed to have changed significantly. According to modern particle physics, Nature presently has four *fundamental forces, being the electromagnetic force, the weak and strong nuclear forces, and *gravity. However at high energies, such as would have existed in the *early Universe, it is believed that some of those forces can be unified.

When the Universe was less than 10^{-12} seconds old, it was hot enough that the electromagnetic and weak nuclear forces were united into a single force known as the electro-weak force. As the Universe cooled, those two forces obtained their separate identities, the transition being caused by a particle known as the Higgs particle. At the same time, many particles changed from being massless to being massive.

Curiously, this phase transition does not lead to the formation of any *topological defects, even though almost all phase transitions do. It might however be of relevance to the process of *baryogenesis.

The energy scale of the electro-weak transition is just at the limit of what particle physics can probe, at giant accelerators such as CERN. So far the existence of the Higgs particle has not been confirmed, and its discovery is a key goal of the Large Hadron Collider (LHC) facility due to open at CERN in 2007.

elliptical galaxies
See GALAXY CLASSIFICATION.

energy density
See DENSITY.

equation of state

The equation of state is a property of materials which fall into a class known as *perfect fluids. The properties of such a fluid are completely specified by giving its *density and its *pressure, and the equation of state gives the relation between density and pressure. It is commonly indicated by the symbol w, defined by $w = p/\rho c^2$ where p is the pressure and ρ the density (ρc^2 then being the energy density, as according to Einstein's famous $E = mc^2$ formula). The equation of state is crucial in understanding how a material will behave as the Universe expands, via the *fluid equation, and also governs the way in which the Universe will expand if dominated by a fluid of that equation of state. Fluids where w is a constant are sometimes referred to as barotropic fluids.

The simplest case is $w = 0$, corresponding to no pressure. This refers to material where no forces are acting other than gravity and motion is non-relativistic. It might refer to fundamental particles, for instance *cold dark matter, or the 'fluid' in question could be made up of individual stars or even *galaxies, neither of which have significant non-gravitational interactions. As the Universe expands, the density of material simply reduces in proportion to the volume, as it becomes spread more thinly.

Another important case is $w = 1/3$, which represents radiation, or more generally the fluid made from any relativistic particles. The most common examples are normal radiation, made from *photons, and *neutrinos. In this case the positive pressure means that the fluid loses energy more quickly; this can be viewed either macroscopically as work done by expansion of a fluid with pressure, or microscopically as the *redshifting of the individual particles. A Universe dominated by radiation decelerates more strongly as it expands.

Not all types of material possess a unique equation of state, which requires there to be a unique pressure associated with any particular density. A *scalar field is an example of a material where density and pressure can be specified independently. Scalar fields crop up in various cosmological contexts, particularly as the type of material believed responsible for *cosmological inflation in the early Universe, and as a candidate to explain the *dark energy in the present Universe. Nevertheless, though there is no unique equation of state for a scalar field, it is possible to define an effective equation of state simply by dividing the pressure by the density. Unlike the case of a perfect fluid, such an effective equation of state would be expected to vary in time.

One of the most common modern usages of the term equation of state refers to dark energy, and indeed often the symbol w is used specifically to mean the (effective) equation of state of dark energy. The simplest model for dark energy, a *cosmological constant, corresponds to a fluid with a negative

pressure equal in magnitude to the energy density, $p = -\rho c^2$ and hence $w = -1$. This is in good agreement with all reliable observations. This is quite a special case: a negative pressure fluid actually gains energy as the Universe expands, and with $w = -1$ this gain is exactly sufficient to cancel out the reduction in energy density due to the expansion, so that the fluid keeps a constant density as the Universe expands. One of the keys to understanding dark energy is to determine whether its density does change as the Universe expands, which is quantified by measuring w. A key goal of upcoming observational programmes is to search for evidence that w is different from -1, which would rule out the cosmological constant model.

event horizon

An event horizon is a property of *spacetime, normally described by Albert *Einstein's theory of *general relativity. It separates the regions of space-time that we will be able to see (i.e. receive radiation from) from those that we never will. Event horizons arise in two main contexts, *black holes and accelerating *cosmological models.

The black hole event horizon marks the edge of the region which, if entered, can never be left again. Its existence first became apparent in the static black hole space-times discovered by Karl Schwarzschild in 1916, but it exists also in more complicated black holes with charge or rotation. In fact, it took around fifty years for the Schwarzschild space-time to be fully understood, through use of coordinate systems that simplified the description of the space-time.

In *cosmology, event horizons may or may not exist depending on the details of the model. None of the simple *Friedmann cosmologies without a *cosmological constant has an event horizon; even in the recollapsing case the entire *Universe has become visible by the time of the *big crunch. However, if the Universe is accelerating, objects far enough away from us will be receding so swiftly that their light can never reach us, even if the Universe becomes infinitely old. This is sometimes known as the de Sitter horizon, as its simplest example is *de Sitter space.

A distinction between the black hole and cosmological settings is that for a black hole all observers outside the black hole infer the event horizon to be in the same place, whereas in cosmology each observer has their own distinct event horizon centred on their location.

While in the classical theory of general relativity the event horizons are absolute, Stephen Hawking showed in 1975 that once quantum effects are included black holes are able to radiate energy and shrink. The precise physical picture of what goes on is, as usual with *quantum mechanics, rather unclear, but should perhaps be thought of as negative energy particles (in the sense of their negative gravitational potential energy exceeding their positive mass–energy) entering the event horizon, rather than anything leaving it. Related to this is the **black hole information paradox**, whereby information entering a black hole appears to be lost even though quantum mechanics insists that information is preserved. Recently, *superstring theory has led to some progress in understanding this issue.

The quantum creation of *inflationary perturbations can be regarded as analogous to black hole evaporation, with the cosmological event horizon taking the role of the black hole event horizon.

The existence of cosmological event horizons is sometimes said to cause problems for *superstring theory, which presently is only formulated properly if interacting strings can end up infinitely separated. Whether this is a genuine problem or a shortcoming of present theoretical understanding remains to be seen.

The event horizon is sometimes confused with a related concept, the *particle horizon. The particle horizon is the limit to the region of the Universe that we can see at the present time, whereas the event horizon limits the regions we will be able to see at any point in our future, even if that future is infinitely long.

expanding universe
See HUBBLE EXPANSION.

extinction
Extinction is the name given to the reduction in apparent brightness of an astronomical source by absorption or scattering of radiation as it travels towards us. Attenuation would be a more technically correct name, as only in the most extreme cases is the light totally extinguished.

The effects of *redshift dimming and the *K-correction are *not* counted as extinction, but as separate cosmological effects. When using *standard candles as distance estimators, it is important to make allowances for extinction, as otherwise the distance to the source will be overestimated.

There are several sources of extinction. Here some of the most important are listed in order as we trace back the path of a *photon from observer to source.

Atmospheric extinction
*Dust in the Earth's atmosphere scatters optical light, as is apparent in the dimming and reddening of the Sun as it is seen through a greater depth of the atmosphere ('airmass') around sunrise and sunset. Airmass is proportional to the reciprocal of the cosine of the angular distance of the source from the zenith. Atmospheric extinction, which typically amounts to a few tenths of a *magnitude, can vary during the course of a night, as well as from night to night. When making photometric observations, atmospheric extinction is determined by observing a number of standard stars, whose brightness is known, throughout the course of each night. Published data are always corrected for atmospheric extinction.

Galactic extinction
The space between the stars in our Galaxy is filled with gas and dust, the **interstellar medium** (ISM). About 1% of the mass of the ISM consists of small dust particles, mainly silicates and carbon, with diameters 0.01–1 μm. Radiation with wavelengths less than this will be scattered and absorbed by the dust grains, which re-emit the energy in the *infrared. Galactic dust along the line of sight to a source will make it appear both dimmer and redder, since blue light is more strongly scattered and absorbed than red light. Radiation of wavelength λ is attenuated by a factor $e^{-\tau(\lambda)}$ where the *optical depth is $\tau(\lambda) \propto 1/\lambda$.

The distribution of dust in the Galaxy has been mapped from its infrared emission by the Diffuse InfraRed Background Experiment (DIRBE) on the COBE satellite (*see* COSMIC BACKGROUND EXPLORER SATELLITE, Fig. 4). Combining this infrared emission with galaxy counts in the *APM Galaxy Survey, a Galactic extinction and reddening map has been published, known as the SFD map, after its authors David Schlegel, Douglas Finkbeiner, and Marc Davis. Published photometry generally does *not* include a correction for Galactic extinction, since improved extinction maps may be available in the near future. It is straightforward for the user to apply their preferred correction (the SFD map is now routinely used) since Galactic extinction depends only on the Galactic coordinates of the object and on the waveband of observation. In particular, unlike atmospheric extinction, it is independent of the observational conditions at the time of observation.

Extinction by the intergalactic and intra-cluster media

Dust extinction in the media between galaxies and within clusters of galaxies is thought to be negligible at optical wavelengths. However, neutral hydrogen gas, ubiquitous throughout the low-redshift Universe, strongly absorbs photons of *ultra-violet light with sufficient energy to ionize hydrogen, namely a wavelength $\lambda < 912$Å.

Intrinsic extinction

Many sources will contain their own absorbing dust. This is true of all spiral galaxies, and recently formed galaxies of any type, as well as *active galactic nuclei, including *quasars. Since the amount of dust extinction is very uncertain, it is rarely corrected for. When a correction is made, it is based on a simple model, and so the uncertainty in the correction is large.

extra dimensions

It is pretty self-evident to us that the world around us has three spatial dimensions, for instance left–right, forwards–backwards, and up–down. In more mathematical terms, three coordinates are required to completely and uniquely specify the location of any object. Add time to the mix, and we have the four-dimensional *space-time of *relativity, where the three spatial coordinates and one time coordinate indicate each possible event occurring within the *Universe. Despite this perception, which has been verified by all experiments carried out to date, modern ideas in fundamental physics theory indicate that there may in fact be extra spatial dimensions, hidden from us but there nonetheless.

The earliest extradimensional theory was due to separate work by Theodor Kaluza and by Oskar Klein in the 1920s. Not long after *general relativity was devised, Kaluza sought to unify it with *electromagnetism (the other fundamental forces then being unknown) and discovered that this could be done by postulating a fourth spatial dimension. In order that this dimension is not readily accessible (no one has ever avoided a mugger by sneaking off in an entirely new direction), it must be extremely small, even by sub-atomic standards. The standard explanation is *compactification, introduced in 1926 by Klein and described under that topic heading, and the consequent theories are known as Kaluza–Klein theories.

The Kaluza–Klein mechanism is still invoked today, in the context of modern theories of fundamental physics. *Supergravity theory was the first to make widespread use of it; while supergravity theories can be formulated in the normal four space-time dimensions, they work more naturally with higher dimensions, which then however have to be hidden from view. Supergravity theory led naturally on to *superstring theory and later *M-theory, where the situation became more focused—these theories can *only* be self-consistently formulated in space-times with ten and eleven dimensions respectively. In each case, six of the dimensions are hidden by being extremely small (the extra dimensions taking a special shape known technically as a Calabi–Yau manifold). In M-theory that still leaves one dimension too many, which is perhaps removed by a different mechanism we now discuss.

M-theory has led to the introduction of a new way of hiding extra dimensions, known as the *braneworld scenario. As discussed more fully under that topic, the braneworld scenario suggests that we live on a three-dimensional *brane stretching through the higher-dimensional space, with particles and interactions confined to the brane (with the exception of gravity, which is able to enter the higher dimension). In particular, the braneworld mechanism may be responsible for concealing the final dimension of the eleven-dimensional M-theory.

All of the above concerns extra spatial dimensions, which can be formulated mathematically as a straightforward extension of normal three-dimensional space. Another question entirely is whether there could be extra time dimensions. While it does appear possible that such theories can be

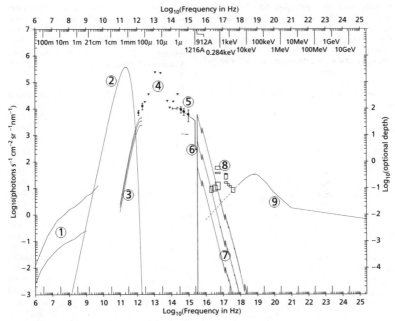

Fig. 1. Spectrum of the extra-Galactic background light from (1) radio to (2) microwave, (3–4) infrared, (5) optical, (6–7) ultra-violet, (8) X-ray and (9) gamma-ray regions.

constructed mathematically, it is very hard to visualize what it might be like to live in one, though it would be nice to think that one could avoid an unpleasant but inevitable event by making a sharp turn into a completely new time direction.

extra-Galactic background light (EBL)
The term extra-Galactic background light (EBL) refers to all radiation (not necessarily just in the visible part of the spectrum) that originates outside of the *Milky Way and which cannot be attributed to any individual source. A spectrum of the EBL is given in Fig. 1.

At optical, *ultra-violet, and near-*infrared wavelengths, the EBL is thought to consist mainly of starlight from unresolved *galaxies at a range of *redshifts, with possible additional contributions from stars or gas in intergalactic space, and from decaying elementary particles. In the mid- and far-

infrared, the main contribution is thought to be redshifted emission from *dust particles, heated by starlight in galaxies. Studies of the EBL spectrum can serve as important tracers of the history of the formation of stars and galaxies.

The majority of the radiation in the *microwave region, the *cosmic microwave background, is a relic of the hot *big bang fireball. Its study is a direct probe of the physical conditions in the *early Universe.

Most of the *X-ray background comes from hot gas in *clusters of galaxies and from the accretion of material onto *black holes in *active galactic nuclei (AGN). AGN are also the most likely source for the gamma-ray background.

Measurement of the EBL is extremely difficult, as one has to subtract off contributions from terrestrial and atmospheric sources (if using a ground-based telescope) as well as the zodiacal light from dust grains in the Solar

System scattering sunlight, and light from stars in the Milky Way Galaxy. In most parts of the spectrum, EBL is much fainter than these foreground sources, and so many null detections provide only *upper limits* to the intensity of the EBL. One can do better using space-based telescopes. Indeed this is essential when probing those parts of the electromagnetic spectrum to which the Earth's atmosphere is opaque.

Overduin, J. M. and Wesson, P. S. *Dark Sky, Dark Matter*, IoP Publishing, 2003.

Faber–Jackson relation

The Faber–Jackson relation relates the *velocity dispersion of the stars within an *elliptical galaxy to the luminosity of the galaxy. Since velocity dispersion can be determined purely from *spectroscopic observations, this relation provides an important way of estimating *distances to elliptical galaxies.

The relation was discovered empirically in 1976 by Sandra Faber and Robert Jackson. They obtained spectra of 25 galaxies and a number of standard stars. Since the stars are effectively point sources of radiation, the widths of absorption features in the spectrum are determined primarily by the spectroscopic resolution of their instrument. Galaxy light comes from the integrated flux of millions of stars, each moving relative to the centre of the galaxy. The relative motion of the stars leads to *Doppler broadening of absorption features. By comparing the galaxy and standard star spectra, Faber and Jackson were able to estimate the velocity dispersion of the stars in each galaxy. This was done both by eye, by broadening the stellar spectra by various amounts and comparing with the galaxy spectra, and by using Fourier techniques to mathematically deter-

mine the velocity dispersions. When comparing the measured velocity dispersions v with the already-known luminosities L for these galaxies (Fig. 1), Faber and Jackson found that the data were well fitted by the relation

$$L \propto v^4,$$

where the constant of proportionality is determined by the passband in which the luminosity is measured. This is known as the Faber–Jackson relation and enables one to estimate distances to galaxies independent of their *redshift. One measures the velocity dispersion as above, and from it determines the luminosity using the Faber–Jackson relation. The apparent magnitude is determined, using, for example, CCD photometry. The distance to the galaxy may then be determined using the *distance modulus.

The accuracy of the Faber–Jackson relation is limited by an intrinsic scatter in galaxy properties, rather than measurement errors. One can improve on the accuracy of the Faber–Jackson relation by introducing an additional parameter which is also (approximately) independent of distance: the *surface brightness I. Observations show that the three quantities luminosity L, velocity dispersion v, and surface brightness I for elliptical galaxies are constrained to lie in a thin plane (known as the 'fundamental plane') in their three-dimensional parameter space, with equation

$$L \propto I^{-2/3} v^{5/2}.$$

The Faber–Jackson relation is the projection of this equation along the surface brightness direction.

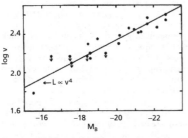

Fig. 1. Line-of-sight velocity dispersion v versus absolute magnitude ($M_B = \text{const} - 2.5 \log L$) for the 25 galaxies observed by Faber and Jackson. The straight line shows the relation $L \propto v^4$.

feedback

In a cosmological setting, feedback refers to the way in which the consequences of structure formation might affect the later development of structures. When the first structures form, the initial generation of massive stars is very short-lived, outputting both strong radiation and stellar winds, and ends in violent *supernovae. Additionally, accretion of material onto *supermassive black holes in *quasars and *active galactic nuclei can release further energy. That energy can emerge from the regions where it was generated and affect distant structures.

Feedback takes the form both of energy injection (typically via radiation) and injection of material, especially heavy elements created in stars and then released during supernovae.

A simple example of feedback is *reionization of the *Universe, where the energy released is responsible for ionizing the *inter-galactic medium. This is known to happen soon after the first structures form. The reionization heats the gas throughout the Universe, and this makes subsequent gravitational collapse harder. This is therefore a negative feedback where early structure formation impedes later structure formation. More generally, feedback processes are expected to affect the observable properties of *galaxies, and modify the properties of the *intra-cluster medium in *clusters of galaxies.

Modelling feedback has become an important part of carrying out *numerical simulations of the formation and evolution of structures. Unfortunately computer power does not allow simulations to model individual stars, only large collections of a million or a even a billion stars at a time. Feedback therefore has to be modelled in an approximate way which tries to capture the essence of effects happening on scales far too small for the simulation to resolve. *Semi-analytic galaxy formation models also need to include the effects of feedback.

fermions

Fundamental particles can be divided into two classes, known as *bosons and fermions, according to their spins. All particles possess a spin, which in the case of a boson is given by a positive integer, and for a fermion is given by a half-integer $1/2$, $3/2$, etc. *Quarks and *leptons are examples of fermions, and indeed we ourselves are primarily made of fermions.

Whether particles are bosons or fermions determines how they behave *quantum mechanically. Bosons behave as indistinguishable particles, meaning that if the locations, velocities etc. of two particles are swapped, the quantum state is unchanged.

Fermions, by contrast, are distinguishable particles: if two particles are swapped the quantum state becomes the negative of what it was before. This is known as Fermi–Dirac statistics, from which the name fermions derives. An immediate consequence of this is the Pauli exclusion principle, which states that the maximum occupancy of 'fermionic' quantum states is one: if two fermions were in the same state then they could be exchanged, implying that the negative of the quantum state was the same thing as the state itself, i.e. that the state is zero. This is why, for instance, all the *electrons in an atom can't fall into the lowest orbital energy state.

flatness problem

The flatness problem, along with the *horizon problem and *monopole problem, was one of the motivations leading to the development of the *inflationary cosmology. While observations indicate that the *Universe has a *geometry which is either flat or very close to flat, the standard hot *big bang cosmology (without inflation; see 'Overview') indicates that the flat geometry is unstable and that only very special choices of initial conditions for the Universe can lead to a present Universe matching our observations.

The problem

*Cosmological models are described on average by the *Robertson–Walker metric, which features a constant k indicating the spatial *curvature of the Universe. Once specified, this constant is unchanged during the evolution of the Universe, with zero value corresponding to the flat geometry. Whether or not the curvature is observable depends on whether it is an important contributor to the expansion rate of the Universe.

It has long been clear that the total *density of the Universe is close, say within an order of magnitude, to the *critical density which would imply a flat Universe. Recent observations of the *cosmic microwave background anisotropies by the *WMAP satellite have refined this, placing the density within a percent or so of the critical value.

While the flat geometry is the simplest possibility, and attractive for that reason, the big bang cosmology unfortunately predicts that the Universe can only be so close to flat today if the conditions in the young Universe are very special—the density must be extraordinarily close to the critical density at that time (perhaps within one part in 10^{30}). The reason is that the flat geometry is unstable; as the Universe evolves the geometry is predicted to deviate more and more from the flat case. Arranging the Universe to have negligible curvature by the present epoch is rather like trying to balance a pencil on its tip with such accuracy that after ten billion years it has still not fallen over.

In terms of the *Friedmann equation, the flatness problem can be understood from the evolution of the different terms. As a function of the *scale factor $a(t)$, the density of matter would evolve in proportion to $1/a^3$, and radiation as $1/a^4$. Each of these reduces more swiftly with expansion than the curvature term, which falls proportional to $1/a^2$. Accordingly, if the curvature is not to be totally dominant over matter and radiation today, it has to be tiny in comparison to them in the young Universe.

The flatness problem is, on its own, not particularly compelling however, as one could always argue that there is some as-yet-undiscovered principle which means that the Universe must be created with precisely the flat geometry, which would then persevere throughout its existence.

Solution by inflation

In 1981 Alan Guth proposed that the flatness problem could be solved by postulating that the early Universe underwent a period of accelerated expansion, which he named inflation. During such an epoch, the geometry of the Universe is driven towards the spatially flat case, which has become a stable rather than unstable situation. The inflationary period must last for long enough that the geometry finishes extremely close to flatness, since once inflation ends flatness will become unstable again and the geometry will begin to deviate. But provided enough inflation occurs, then even the long epoch between the end of inflation and the present is not enough for the Universe to move significantly away from the nearly flat geometry established by the inflationary epoch.

A more intuitive way to understand the inflationary solution may be to realize that the curvature of the Universe is a characteristic length scale which expands with the Universe. The purpose of the inflationary expansion is to drive the curvature scale to be much larger than the size of our present *observable Universe, and hence the portion of the Universe

we can see appears to be flat, just as the surface of an ocean appears on average to be flat provided you can only see a small region of it.

To solve the flatness problem, the inflationary expansion would have to increase the size of the Universe by a factor of at least 10^{27}. Such an expansion would simultaneously solve the horizon and monopole problems.

fluid equation

The fluid equation, also known as the energy conservation equation or the continuity equation, governs how the *density of material in the *Universe changes as it expands. Partnered with the *Friedmann equation, which determines the expansion rate of the Universe, it can be used to construct cosmological models.

In its simplest form, the fluid equation can be written as

$$\frac{d\rho}{dt} = -3H\left(\rho + \frac{p}{c^2}\right) ,$$

where ρ is the density of the material in the Universe, p its *pressure (with c the *speed of light), and H the *Hubble parameter measuring the expansion rate of the Universe. The term $d\rho/dt$ is a derivative—the rate of change of density with time.

The expansion rate H is to be found by solving this equation simultaneously with the Friedmann equation. To do so, however, we need to specify the pressure p to close the system of equations. The pressure will depend on the type of material filling the Universe; different materials will therefore have different time evolutions, and hence give different expansion histories for the Universe. The simplest hypothesis is that the Universe contains a *perfect fluid, which is a material whose pressure is directly related to its density. Examples are non-interacting, non-relativistic matter, which is pressureless, and radiation which has pressure $p = \rho c^2/3$ (i.e. the pressure is one-third of the energy density). The simplest *cosmological models are based on those assumptions.

Having only one type of material in the Universe is oversimplistic. The real Universe is believed to contain several differents kinds of material: *baryons, *photons, *cold dark matter, *neutrinos, and *dark energy. Each of these materials obeys its own fluid equation containing the appropriate expression for its pressure.

Fourier transform

The Fourier transform is a mathematical transformation, named after French mathematician and physicist Joseph Fourier, that expresses any given function as a sum of sinusoidal functions. Its principal use in cosmology is in calculating *power spectra and in transforming between power spectra and *correlation functions.

In the general case of anisotropic clustering, the power spectrum $P(\mathbf{k})$ and the correlation function $\xi(\mathbf{r})$ depend on the orientation of the wave-vector $\mathbf{k}$ and the separation vector $\mathbf{r}$. In this case, they are related by

$$P(\mathbf{k}) = \int \xi(\mathbf{r})e^{i\mathbf{k}\cdot\mathbf{r}}d\mathbf{r}.$$

If we assume that clustering is isotropic, which is expected to be true on large scales, the vector form of the Fourier transform reduces to the scalar form

$$P(k) = 4\pi \int_0^\infty \xi(r)\frac{\sin kr}{kr}r^2 dr.$$

Algorithms exist to compute Fourier transforms very efficiently, using a technique known as the Fast Fourier Transform (FFT). In particular, the power spectrum of a data cube from a *numerical simulation may be very simply calculated by calculating the FFT of data points within the cube.

free streaming

Free streaming refers to the motion of non-interacting particles across the *Universe due to them possessing an initial velocity. The most common example is *neutrinos, which cease interacting with all other particles very early in the Universe's history, at neutrino *decoupling.

Free streaming has the effect of erasing any irregularities in the material on scales smaller than the free-streaming length. This is because the overdense regions contain more particles, and so the net flux of particles away from them is higher than that coming in from the lower density regions nearby. A related but distinct phenomenon is *Silk damping, which describes dissipation when interactions are present.

Free streaming is particularly damaging if the *dark matter particles exhibit it, as this can erase the initial *density perturbations which are supposed to induce *galaxy formation. Indeed, neutrinos are excluded as a dark matter candidate for precisely this

reason. Such scenarios, known as hot *dark matter, ceased to be viable once *cosmic microwave background anisotropy measurements pinned down the primordial amplitude of density perturbations.

Friedmann, Alexander (1888–1925)

Alexander Friedmann (sometimes translated as 'Friedman'; Fig. 1) was a Russian mathematician, born in St Petersburg in June 1888. He was educated and spent most of his life in that city. [During his time there, St Petersburg was renamed first to Petrograd and then to Leningrad. Nowadays it is known as St Petersburg again.] He discovered the first realistic *cosmological models to feature an expanding *Universe, by solving the equations of Albert *Einstein's *general relativity. His ideas were vindicated by Edwin *Hubble's observations of the expanding Universe in the 1930s, though only after his early death from typhoid in 1925. Today cosmological models are often referred to as Friedmann models or Friedmann–Robertson–Walker models, and the key equation governing the Universe's expansion is the *Friedmann equation.

Friedmann's contributions to *cosmology came late in life, when relative peace came to the Soviet Union around 1920 and Einstein's work began to be disseminated there. His ground-breaking work 'On the curvature of space' was completed in 1922 and

Fig. 1. Alexander Friedmann.

submitted to leading German physics journal *Zeitschrift für Physik*. The expert referee was none other than Albert Einstein himself, who believed it was in error and rejected it. Only one year later did Einstein admit that his own calculations were in error, and that '...Mr Friedmann's results are correct and shed new light'. By the time Friedmann published a second paper, he had covered the three classic cosmological cases of an *open Universe, a *closed Universe, and a *critical-density Universe.

In addition to his cosmological work, Friedmann was active politically as a student and played a role in the First World War as a pilot. In July 1925 he made a record-breaking balloon ascent (over 7000 metres) on a meteorological expedition. It was soon after that flight that he contracted the typhoid which led to his death, aged 37.

Friedmann equation

The Friedmann equation determines the expansion rate of the *Universe, based on the *density of material within it and the *curvature of space. It is a consequence of Albert *Einstein's theory of *general relativity. It was first derived by the Russian scientist Alexander *Friedmann in the 1920s, who used it to build the first *cosmological models of the expanding Universe. These models are still relevant today, and are often called Friedmann, or Friedmann–Robertson–Walker, cosmologies.

For completeness, we write the Friedmann equation in its usual form:

$$H^2 = \frac{8\pi G}{3}\rho - \frac{k}{a^2}.$$

Here H is the *Hubble parameter, measuring the expansion rate of the Universe as a function of time, ρ is the density of material within the Universe, G is Newton's gravitational constant, k a constant measuring the curvature of space, and a the *scale factor describing the size of the Universe. Schematically, this could be written

expansion rate (squared) = density − curvature.

Construction of a cosmological model requires solution of the Friedmann equation, which gives the expansion rate as a function of time, and hence the size of the Universe. However in order to solve it, one must know how the density ρ changes. That requires a separate equation, known as the *fluid equation. Those two equations must therefore be solved simultaneously. The Friedmann equation tells us how the contents of the Universe affect the expansion rate, and the fluid equation tells us how that expansion rate affects the properties of those contents.

fundamental forces

The standard view of physics is that Nature has four fundamental forces, namely electromagnetism, *gravity, and the strong and weak nuclear forces. These forces, also known as **interactions**, govern the way in which particles respond to each other's presence. While only the first two of these impact directly on our day-to-day existence, the strong force is responsible for holding atomic nuclei together, and the weak force for nuclear reactions including those that power the Sun. All four forces are therefore vital to our existence.

Three of these forces are described by a single theory, the *standard model of particle physics, while the fourth, gravity, stands separate and is usually described by Albert *Einstein's *general relativity theory. One of the goals of theoretical physics is to establish a 'theory of everything' which unites gravity with the others, and presently *superstring theory and *M-theory are the leading candidates to achieve this.

While in the present *Universe the three forces of the standard model each have a separate identity, the theory predicts that in hot enough conditions electromagnetism and the weak interaction become different aspects of a single force, the electro-weak force. This is expected to have held in the early Universe, with the Universe undergoing a *phase transition known as the *electro-weak phase transition when cooling below the critical temperature. The corresponding energy is accessible by the most powerful *particle accelerators, which have provided strong evidence that this will indeed have taken place.

In the standard model the strong force remains separate from the other two forces it describes, but according to *grand unification theories it too becomes united at sufficiently high energies. Those energies are however way out of reach of conceivable technology, and at present it is unknown whether the very young Universe might have undergone a transition corresponding to the breakdown of this unification.

Occasionally physicists discuss the possibility of one or more extra forces of Nature, sometimes referred to as a 'fifth force'. This might refer to a force on a terrestrial scale, though actual existence of such a force has never been convincingly demonstrated, or on a cosmological scale, where again nothing has been observed. However theories featuring very light *scalar fields do predict forces associated with them which would be of fifth-force type, and the lack of observation of such forces does constrain such theories.

future of the Universe

Quantum physics pioneer Neils Bohr wryly stated that 'prediction is very difficult, especially about the future'. This quotation reflects the fact that physicists are often struggling to explain phenomena which have already happened, and often under carefully constructed laboratory conditions. It shouldn't therefore come as a surprise to learn that predicting the future of the *Universe is fraught with difficulties.

The classic early *cosmological models of Alexander *Friedmann and Georges Lemaître gave a straightforward view of the Universe's destiny. If the *density of the Universe exceeded the *critical density, the Universe would be *closed and destined to recollapse and end in a so-called *big crunch. If less, we would have an *open Universe which would expand forever. Dividing these cases, a critical-density Universe would also last forever, but expanding ever more slowly. Indeed, much of the early work in observational cosmology sought to identify which of those possibilities represented our actual Universe. The three models are shown in Fig. 1.

This simple set of options has been overturned by the recent discovery that our Universe is presently experiencing *accelerated expansion, presumed to be caused by *dark energy. Once dark energy is included in cosmological models, the simple correspondence of closed Universes to finite age and open Universes to infinite age is broken, to be replaced by a more complicated classification based on the relative amounts of normal matter and dark energy. This classification includes both closed Universe models which expand forever, and open Universe models which collapse. The latter are disfavoured by current observations, while the former remain a possibility.

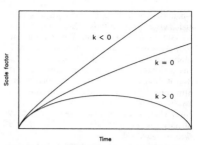

Fig. 1. The three standard types of cosmological model due to Friedmann. The curves marked $k > 0$, $k = 0$, and $k < 0$ correspond to closed, flat, and open Universes respectively.

The simplest model for dark energy is Albert *Einstein's *cosmological constant, and this is the basis for the *standard cosmological model. This model has a flat spatial *geometry, meaning that the combined densities of normal matter and the cosmological constant add to give the critical density. Such a cosmology expands forever; indeed the cosmological constant becomes more and more dominant as the Universe evolves, with the *Hubble parameter approaching a constant value, a situation known as *de Sitter space.

However even that need not be the whole story, as the dark energy density might evolve with time, altering the balance between it and normal matter. In the most extreme cases, the density of dark energy might even evolve to become negative. If so, normal matter will then exceed the critical density (so that it, plus the negative density of the dark matter, add up to the critical density), and that will lead to recollapse. At present we have no way of knowing whether such evolution is possible, though it has been shown that the above sequence of events is a plausible outcome of current ideas in fundamental particle physics. Unless and until the properties of dark energy are definitively known, only speculation is possible. Until then, the unknown future evolution of dark energy always has the possibility of overturning our expectation about the future of the Universe.

A big crunch is not the only way that the Universe can come to a sudden end. It might be that the dark energy density increases with time, rather than remaining constant or decreasing. This possibility is known as *phantom dark energy, and it has the particularly drastic effect that the Universe expands so rapidly the separation between objects becomes infinite in a finite time, known as the *big rip. During the late stages, the repulsive force of the dark energy becomes so great that first *galaxies, then solar systems, and ultimately even individual atoms are ripped apart by it. This alarming possibility is quite compatible with observations of the effects of dark energy, and indeed even somewhat favoured by them. On the plus side, if a big rip does take place, it will not be for at least 20 billion years.

As well as considering the future of the global Universe, one can speculate about physical processes that might occur within it. While remembering the caveats above, let's assume that the standard cosmological model applies so that the future of the Universe is to become dominated by a cosmological constant, but not to experience a big rip. This domination will not have serious effects within structures which have already formed, such as the *Milky Way Galaxy, as they are no longer expanding with the Universe and so within their environs the cosmological constant will remain subdominant in density as compared to matter (unlike in the big-rip scenario). However the rapid expansion induced by the cosmological constant will prevent further structure formation in the Universe, so the largest objects that there will ever be, the great *clusters of galaxies, are already forming. Ultimately, those objects not gravitationally bound to each other will be pulled apart by the rapid expansion, and indeed eventually pulled so far away that they are no longer visible. In the distant future the view from our Galaxy will be rather boring, with perhaps just a few *Local Group galaxies visible and everything else beyond the edge of the *observable Universe.

Even if the Milky Way survives being torn apart by dark energy, that is not to say it will last for ever. Firstly, it is scheduled for a major collision with the neighbouring Andromeda galaxy in about three billion years, which is within the lifetime of the Sun. However, galaxies are extremely diffuse objects, and even when galaxies merge stellar collisions are extremely rare. The Sun will therefore survive the merger; whether the Solar System does depends how near Andromeda's stars get.

Still, having survived that, individual stars have only a finite supply of nuclear energy, and when it comes to an end they evolve to become white dwarfs, neutron stars, or *black holes. And even the fundamental constituents of matter may not be stable; *grand unification theories predict that eventually even *protons and *neutrons will spontaneously decay. For protons the expected lifetime is colossal; current measurements indicate that the half-life of protons is at least 10^{35} years (this is not of course achieved by watching a single proton to see if it decays on that timescale, but rather by assembling a large enough group of protons so as to have some chance of seeing a decay within a reasonable time). Still, the Universe may well survive that long, and so eventually proton decay may become an important phenomenon.

All the above should be taken with a pinch of salt; predicting the future evolution presumes our current understanding of physical laws is correct. Perhaps more significantly, it is non-scientific in that we have no way of confirming observationally that our predictions are correct. So really there shouldn't be a 'Future of the Universe' topic in this book at all. Or, to quote footballer Paul Gascoigne, 'I never make predictions, and I never will'.

galactic halo

A galactic halo is a roughly spherical distribution of matter surrounding and extending beyond the visible parts of a *galaxy. It may comprise old populations of stars, gas, and *dark matter but contains no young, recently formed stars or *dust.

The best-studied galactic halo is that of our own *Milky Way Galaxy. The stars in the halo tend to be much older, and with a smaller fraction of heavy elements, than those in the bulge and disk of our Galaxy, and are known as Population II stars. The Galactic halo is also home to *globular clusters: old, compact, and nearly-spherical groups of hundreds of thousands to millions of stars. About 140 globular clusters are now known in our Galaxy. The globular clusters are thought to be distributed isotropically about the Galactic centre, and have been used to estimate our distance from the centre of the Galaxy. This was first done by Harlow Shapley in 1915–19, who measured distances to the clusters based upon the brightness of individual stars in each cluster. Shapley determined the clusters to be centred about a point 15 kpc away from the Sun towards the constellation of Sagittarius. More recent investigations with better observational data find a distance to the Galactic centre of about 8 kpc.

Globular clusters may also be resolved in images of other nearby galaxies. Studying their dynamics enables the total mass of their host galaxies to be estimated.

The majority of the mass of a galactic halo is thought to comprise dark matter. Evidence for the existence of dark matter in galaxy halos comes from **dynamical** observations. These entail study of the motions of galactic systems which are determined by the total mass of the system through *gravity. All such observations indicate that visible matter contributes only a few percent to the total mass of most galaxies.

Candidates for dark matter in galactic halos include gas clouds (although these are only 'dark' in visible light, and may be detected through radio observations), *MACHOs such as small *black holes, brown dwarfs, and Jupiter-sized planets, and *WIMPs (weakly interacting massive particles), a type of *cold dark matter.

*Numerical simulations of *cosmic structure find that the distribution of dark matter particles in a galaxy halo follows a characteristic distribution, known as a Navarro, Frenk, and White (NFW) profile.

galaxies

A galaxy is a gravitationally bound system, containing from millions ($\sim 10^8$), up to thousands of billions ($\sim 10^{13}$) of stars. As well as stars, galaxies may contain gas and *dust, collectively known as the interstellar medium. The masses of galaxies are dominated by another invisible component: *dark matter. The individual constituents of a galaxy orbit about the centre of the system under the gravitational attraction of the rest of the galaxy. In some galaxies, new generations of stars are forming from condensing gas clouds; other stars return material to the interstellar medium of the galaxy when they reach the end of their lives. Galaxies are thus dynamic, evolving systems, not inert lumps of matter.

The light we observe from a galaxy comes from two sources. First we see the light of its constituent stars. Because galaxies are so distant, in most cases we cannot resolve individual stars, but see diffuse light from billions of faint stars. Second, we may see radiation emitted by hot gas which has been ionized by young, luminous stars. These luminous gas clouds are particularly prominent in the arms of spiral galaxies.

Galaxies are vast systems; it takes light about 100 000 years to cross the *Milky Way, a fairly typical galaxy. Stars are thus thinly spread within a galaxy: their typical separation is orders of magnitude larger than their size. This is why the *surface brightness of galaxies is so low, and why it is difficult to see any galaxies with the naked eye.

Although diffuse *nebulae had been observed since Charles Messier compiled the *Messier catalogue in the late 18th century, it was not realized that many of these nebulae were in fact galaxies separate from our own Milky Way. A debate took place between Harlow Shapley and Heber Curtis in 1920 about whether the nebulae were gas clouds within the Milky Way (as Shapley argued) or 'island universes' (independent systems outside of our own Galaxy, as Curtis thought, and as had originally been proposed by Immanuel Kant in 1755). Observations of *Cepheid variables in M31, the 31st object in Messier's catalogue, by Edwin *Hubble in 1923, showed that the distance to this nebula was at least 300 kpc, greater than Shapley's proposed extent of the Milky Way galaxy. Later measurement of *Doppler shifts showed that nearly all nebulae were moving away from us, thus confirming the existence of separate galaxies (and the expansion of the *Universe).

Galaxies exist in many shapes from essentially spherical to flattened disk systems, many of which are observed to be rapidly rotating. The classification of galaxies is discussed further under the topic *galaxy classification.

Unlike stars, most galaxies appear to be of comparable age, having been formed some ten billion years years ago. Galaxy ages are estimated by studying the oldest stellar populations within them, typically *globular clusters. (Some stars in the Galaxy are only a few million years old; others are essentially as old as the Galaxy.) Thus in order to study *galaxy evolution, we have to look at galaxies in the distant Universe, where the light has taken a significant fraction of the age of the Universe to reach us.

Galaxies themselves like to clump together into *galaxy groups, *clusters and *superclusters, tracing out the *cosmic structure of the Universe. It is believed that galaxies formed at the locations of very small overdensities in the matter distribution generated during *cosmological inflation. The *clustering of galaxies today thus reflects the physical conditions in the *early Universe, and so by studying the clustering properties of galaxies, one can obtain useful constraints on *cosmological parameters including the

*density parameter and the type of dark matter.

Hubble, E. *The Realm of the Nebulae*, Dover Publications, 1958.

galaxy bias
See BIAS PARAMETER; BIASED GALAXY FORMATION.

galaxy classification
Galaxies exist in many shapes from essentially spherical to flattened disk systems, many of which are observed to be rapidly rotating. The first attempt at galaxy classification was made by John Henry Reynolds in the early 1900s, before the extra-Galactic nature of the *nebulae was widely known. It was however Edwin *Hubble who devised a classification system that is still in use today, his famous 'tuning fork' sequence (Fig. 1). The handle of the tuning fork consists of elliptical galaxies, the two prongs consist respectively of barred and unbarred spiral galaxies. An intermediate class, the lenticular galaxies, sit at the intersection of the prongs and the handle. Yet another class of galaxies, the irregulars, do not quite fit on this diagram, and are often shown to the right of the spirals.

Hubble originally thought that this diagram described an evolutionary sequence for individual galaxies, progressing from elliptical through to spiral galaxies. For this reason, elliptical galaxies are also known as 'early type' galaxies, and spirals as 'late type'. It is now known that this evolutionary picture is incorrect (in fact, according to *hierarchical structure formation models, ellipticals may form from the merger of spiral or irregular galaxies), yet one still sees ellipticals described as 'early type' galaxies, and even 'early' versus 'late' type spirals.

Others have developed and built upon Hubble's classification system. Gerard de Vaucouleurs and Allan Sandage introduced Sd and Sm galaxies and created intermediate types such as Sab. De Vaucouleurs also introduced a numerical system known as the 'T-type' which runs from -5 to -1 for ellipticals to lenticulars, 0 to 9 for spirals and 10 for irregulars. Sidney van den Bergh classified spiral galaxies by luminosity, which he observed to correlate with spiral arm structure. Luminous spirals with well-defined spiral arms are type I; low-luminosity spirals with indistinct arms are type III.

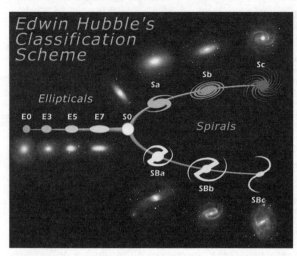

Fig. 1. Hubble 'tuning fork' diagram, showing the morphological classification of elliptical, lenticular and spiral galaxies. Ellipticals form the handle of the tuning fork, barred and unbarred spirals form the lower and upper prongs respectively.

Ellipticals

Elliptical galaxies comprise roughly 20% of the galaxy population, although this fraction varies strongly depending on environment, ellipticals being in the majority in the cores of rich *clusters of galaxies. They are characterized by a smooth, ellipsoidal light distribution, showing no evidence of spiral arms or disks (see Fig. 2 for an example). They are subdivided by their apparent (projected) ellipticity with the designation En, where the number n lies in the range 0 (for a perfectly circular galaxy image) to $n = 7$ for the most elongated ellipticals. Mathematically, n is given by $n = 10(1 - b/a)$, where a is the apparent major (long) axis of the projected ellipse, and b is the minor (short) axis. It is important to bear in mind that we see a two-dimensional projection of a three-dimensional system, and so an E0 galaxy is not necessarily spherical.

Fig. 2. The elliptical galaxy NGC4636 (centre frame). Note how the diffuse light from this galaxy extends over most of the image.

Fig. 3. The beautiful spiral galaxy M101 as imaged by the Hubble Space Telescope.

Elliptical galaxies consist mostly of old stars, and to that extent the label 'early type' is still appropriate, since the constituent stars of the galaxy at least formed fairly early on in the history of the *Universe. As already mentioned, ellipticals occur most commonly in high-density environments, being particularly common in rich clusters. This observation provides some circumstantial evidence that ellipticals form from the merger of smaller systems.

Lenticulars

Lenticular galaxies, also known by the designation S0, are flatter than ellipticals with intrinsic axis ratio $b/a < 0.3$. They are named for their lens-like (lenticular) profile. As well as displaying a prominent central bulge, lenticulars exhibit a flattened, disk-like stellar distribution. However, they show no sign of spiral arm structure within this disk; the zero in the designation S0 indicates this lack of spiral arms (cf. the spiral galaxies Sa, Sb etc., below). They may be sub-classified into $S0_0$, $S0_1$ and $S0_2$, progressing from bulge-dominated to disk-dominated systems. Like elliptical galaxies, they consist mostly of old stars.

Spirals

Spiral galaxies form some of the most spectacular and beautiful galactic systems, due to the presence of their eponymous spiral arms

(see for example Fig. 3). These galaxies have several distinct components.

- A central **bulge**, containing mostly old stars. The bulge is similar in its properties to a small, approximately spherical, elliptical galaxy.
- The **disk** of the galaxy consists of stars, gas (mostly neutral hydrogen), and *dust. Most obvious are the spiral arms, containing young stars and star-forming regions. In between the spiral arms live intermediate-age stars. It is important to note that the spiral arms do not rotate with the galaxy, but are thought to be generated by a **density wave**, a compression in the gas density that travels through the galaxy triggering the formation of new stars. The *Milky Way disk comprises three components: thin and thick stellar disks and a gas disk. It is likely that other spiral galaxies possess multi-component disks.
- A **halo** consisting of old stars, globular clusters of stars and *dark matter.

Spiral galaxies are sub-classified from Sa–Sd: Sa galaxies are bulge-dominated and have tightly wound, smooth spiral arms. Sd galaxies have a relatively small bulge, and loosely wound, clumpy spiral arms. Types Sb and Sc are intermediate to these.

Roughly half of spiral galaxies have an elongated bulge which has the appearance of a

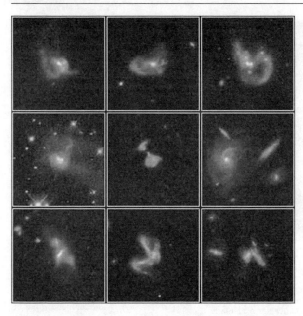

Fig. 4. Hubble Space Telescope images of nine merging galaxy systems, most of which have an irregular appearance.

bar, and so these are known as barred galaxies and designated SBa, for example.

Irregulars

Irregular galaxies are neither smooth, like ellipticals and lenticulars, nor do they show obvious spiral structure, but instead have an irregular surface brightness distribution. They comprise roughly 3% of the total galaxy population at low *redshift, although they are seen more commonly in deep images, such as the *Hubble Deep Field, which typically survey much younger galaxies. The closest examples of irregular galaxies to us are the Large and Small Magellanic Clouds, which are visible to the naked eye from the southern hemisphere. Many galaxies are thought to be irregular in shape due to merging events with other galaxies (see for example Fig. 4).

Problems with visual classification

Several possible problems with visual classification are worth bearing in mind.

- The relative dominance of features may depend on the passband in which the galaxy is imaged. For example, the bulges of spiral galaxies, which contain older, redder stars than the disk, appear more dominant in near-*infrared light. Spiral arms, containing young, blue stars are more prominent in the near-*ultraviolet. Thus the same galaxy might be classified as Sa in near-infrared light but as Sc in the near-ultra-violet.

- The appearance of a galaxy will depend on how faint the image reaches. In a very shallow image of a spiral galaxy, only the bulge may be visible.

- We observe a two-dimensional projection on the sky of an intrinsically three-dimensional system. It may be hard to recognize spiral arms in a galaxy that is edge on to us; the measured ellipticity of an elliptical galaxy may not reflect its true shape.

- Selection effects. Faint, low surface brightness galaxies are hard to see and may be missed completely. Conversely, very compact galaxies may be confused with stars.

Morphology–density relation

As noted above, ellipticals are found most frequently in clusters. This observation

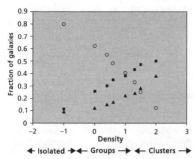

Fig. 5. A plot showing the relative fractions of spiral galaxies (circles), lenticular or S0 galaxies (squares) and elliptical galaxies (triangles) as a function of local density. Density is given by the number of galaxies per unit volume on a logarithmic scale relative to the mean, so that 0 is the mean density, 1 is ten times the mean, −1 is one-tenth of the mean, and so on.

is generalized and extended by the morphology–density relation discovered by Alan Dressler in 1980 and illustrated in Fig. 5. Note that while elliptical galaxies comprise less than 10% of the galaxy population in low-density environments, they dominate in cluster environments. This relation is believed to indicate that galaxy evolution is affected by environment. In particular, star formation appears to have been suppressed in high-density regions. An alternative explanation ('nature' rather than 'nurture') is that elliptical galaxies preferentially form in high-density regions.

Hubble, E., *The Realm of the Nebulae*, Dover Publications, 1958.

Sandage, A. *Annual Review of Astronomy and Astrophysics, 43, 581*, 2005.

galaxy clusters
See CLUSTERS OF GALAXIES.

galaxy evolution
Galaxy evolution is the name given to describe the changes in the properties of *galaxies as they age. Most nearby galaxies appear to be of comparable age, around ten billion years old, and any individual galaxy evolves too slowly to detect any changes in its observed properties. We are able to study galaxy evolution by observing distant galaxies, where the light has taken a significant fraction of the *age

of the Universe to reach us (*see* LOOK-BACK TIME).

Evolution can occur for different reasons, and is generally categorized as passive evolution, secular evolution, and environmental evolution.

Passive evolution—The stars of which galaxies are most conspicuously comprised go through cycles of birth, life, and death. Dying stars provide a source of heavy elements which will be incorporated into future stars, thus affecting their properties. Massive stars will end up as *supernovae which input energy as well as heavy elements into the galaxy. This change in the observational properties of galaxies as their constituents evolve is known as passive evolution, and results in a long-term dimming and reddening of galaxies with time as the material to form new stars gradually runs out.

Secular evolution—Secular evolution describes the slow rearrangement of energy and mass – dynamical evolution – that results from internal phenomena within a galaxy such as rotating bars, spiral structure, and non-circular disks. Such processes can lead to the formation of bulges within spiral galaxies.

Environmental evolution—Galaxies may evolve much more rapidly due to rapidly changing environmental influences, such as close encounters with other galaxies and merger events. These will disrupt the regular shape of the galaxies and may induce a renewed burst of star formation. Conversely, the stripping of a galaxy's gas supply due to infall into a *cluster of galaxies may quench star-formation in a spiral galaxy.

Galaxy evolution makes itself apparent observationally in several ways.

Morphological evolution
On the whole, young galaxies tend to be of irregular shape, due to the uneven distribution of star formation within the galaxy. This is seen in the large proportion of irregular galaxies in the *Hubble Deep Field. As the gas and *dust supply, from which stars form, is used up, galaxies tend to appear more regular. *Numerical simulations of two merging spiral galaxies have shown that the spiral arm structure is rapidly destroyed, resulting in an elliptical galaxy.

Colour evolution

As the constituent stars age, evolving from blue young stars, to old red stars, galaxies also redden in colour. Thus distant galaxies tend to be intrinsically bluer than nearby galaxies. This is not immediately apparent in images such as the Hubble Deep Field since distant objects *appear* redder due to their high *redshift. Only once the colours of distant galaxies are corrected for redshifting do their blue intrinsic colours become apparent.

Luminosity evolution

Again due to the passive evolution of their stellar populations, distant (and hence young) galaxies tend to be more luminous than their nearby (older) counterparts.

Number density evolution

According to *hierarchical structure formation, small objects form first and then merge together to form larger objects. This scenario is in principle observationally testable by looking for a decreasing comoving number density of galaxies with time. In practice, this test is complicated by the fact the sub-galactic clumps may not be readily observable: one needs to understand the *observational selection effects of a galaxy survey in great detail in order to detect the effects of galaxy mergers.

Clustering evolution

As well as evolution in the intrinsic properties of individual galaxies, the way that galaxies cluster together will evolve as the *Universe expands and *density enhancements grow due to *gravitational instability. Specific models of *clustering evolution depend on the assumed *cosmological parameters, including the spectrum of primordial *density perturbations, and the *semi-analytic recipe for forming galaxies within *dark matter halos.

The galaxy *correlation function $\xi(r, z)$, when measured in physical (as opposed to *comoving) coordinates r, is assumed to evolve with redshift z via the following simple parameterization: $\xi(r, z) = \xi(r, 0)(1 + z)^{\varepsilon - 3}$. The factor $(1 + z)^{-3}$ allows for the mean change in galaxy density due to expansion. Consequently, if galaxies maintain a fixed physical separation as the Universe expands, then $\varepsilon = 0$. If galaxies remain fixed in comoving coordinates then $\varepsilon = \gamma - 3$, where $\gamma \approx$ 1.8 is the slope of the correlation function power-law, $\xi(r) = (r/r_0)^{-\gamma}$. A positive value of ε would indicate that the galaxies are moving physically closer as the Universe expands, i.e. the clustering is getting stronger. Provisional results suggest that $\varepsilon \approx -0.8 \pm 0.2$ over the redshift range zero to 0.65, hinting that clustered galaxies are expanding only slightly more slowly than the *Hubble expansion.

Models of evolution

Models are available that describe the passive evolution of isolated galaxies with redshift. The most widely used are those produced by Gustavo Bruzual and Stéphane Charlot. These models are able to output the *spectral energy distribution of a galaxy as a function of time, given choices for such parameters as *initial mass function and *metallicity.

In order to model the dynamical evolution of galaxies, one must resort to numerical simulations, which allow one to follow environmental influences, such as near-interactions or mergers with other galaxies. A successful approach here has been to combine semi-analytic models with *N-body simulations.

Kormendy, J. and Kennicutt, Jr., R. C., *Annual Review of Astronomy and Astrophysics*, 42, 603, 2004.

galaxy formation

Galaxy formation, the creation of the *galaxies we see today from clouds of gas, is one of the least understood processes in astrophysics. The physics describing galaxy formation is complicated because as well as *gravity, it must deal with gas thermodynamics and energy production by stars and *supernovae. There are likely to be many mechanisms operating together and feeding into each other. For instance, stars form from cooling gas clouds which are subsequently reheated and dissipated by the newly formed stars.

The basic picture, however, is as follows. After *recombination, the *Universe was remarkably *homogeneous, as witnessed by the very small variations in temperature of the *cosmic microwave background, of around one part in one hundred thousand. It is mostly accepted that the *cosmic structure we see today formed from these tiny primordial fluctuations by the process of *gravitational instability: slightly overdense regions grew in *density as the Universe expanded by attracting material, both gas and *dark matter, from surrounding underdense regions.

Once these overdensities reach a critical mass known as the **Jeans' mass**, the gravitational binding energy exceeds the kinetic, or thermal, energy of the constituent particles, and the overdensity shrinks in volume. There will be density variations within the overdensity, with the denser sub-regions collapsing faster than the whole. These may then fragment into smaller and smaller sub-regions, in a process known as **successive fragmentation**. These fragments increase in density and temperature until hydrogen nuclei (*protons) are colliding with sufficient energy to overcome their electrostatic repulsion and are bound together by the strong nuclear force (*see* FUNDAMENTAL FORCES). Thus begins the process of nuclear fusion: the conversion of hydrogen into helium and the heavier elements, and the first stars are born.

Once the first stars appear, the surrounding gas will be heated by their *ultra-violet radiation, thus affecting subsequent star-formation. Massive stars will rapidly end their lives as *supernovae, enriching the surrounding gas with heavy elements which will also affect subsequent star-formation by making the stellar atmospheres less transparent to radiation. Due to the complex nature of these processes, most attempts at modelling galaxy formation to date have taken the *semi-analytic galaxy formation approach, in which *numerical simulations are combined with simple physical models for some of the gas dynamics and *feedback processes.

Successful models must not only agree with the observed statistical properties of galaxies, such as their *luminosity function and *clustering as a function of galaxy morphology and colour, but must explain the existence of the main components of the *Milky Way Galaxy, namely the central bulge, a rotating, thin stellar disk, and the Galactic halo. Rotating disks form naturally if the proto-galactic gas clouds have some angular momentum, i.e. if they are slowly rotating, as might be expected from tidal forces due to the gravitational pull of other, nearby clouds. As the clouds collapse, angular momentum must be conserved and so the rotation speeds up. This rotation will cause a previously spherical cloud to bulge outward around its equator (just as the Earth's equator bulges slightly). Subsequent collapse will then proceed more quickly along the rotation axis of the cloud and a rotating, pancake-like object is expected to form.

In addition to physical processes occurring within a *dark matter halo, neighbour-ing halos will be gravitationally attracted and may merge together. In a rapidly changing *gravitational potential, stellar orbits will be mixed up via violent relaxation and the gas will be shock heated. It is believed that elliptical galaxies may result from mergers of two or more progenitors of comparable mass, resulting in a randomization of stellar orbits, whereas spiral galaxies form in isolated halos, and have only accreted small amounts of matter, if any. Such a model naturally explains the morphology–density relation (*see* GALAXY CLASSIFICATION).

galaxy groups

Galaxy groups are associations of *galaxies occupying similar volumes to *clusters of galaxies, but containing fewer members—typically less than 100. There is a continuum of galaxy associations, from pairs and groups with as few as three members, to richer groups, poor clusters, and rich clusters. Like clusters, richer groups also contain hot gas and may emit in *X-rays, but only the closest, richest groups have so far been catalogued in X-ray surveys. The majority of galaxies in the *Universe are found in groups; there are few isolated galaxies. Our own Galaxy is a member of the *Local Group.

Compact groups are distinguished by the distances between the member galaxies being comparable to their size (see for example, Fig. 1). They are thus a good place to study interactions between galaxies.

Fig. 1. The compact galaxy group number 66 in Paul Hickson's catalogue, as imaged by the Sloan Digital Sky Survey.

galaxy rotation curves

*Spiral galaxies are observed to rotate about an axis perpendicular to their disk. The rotation curves of spiral galaxies show how the rotation speed varies with distance from the centre of the galaxy, and is an important tool in estimating the distribution of mass in spiral galaxies. An example is shown in Fig. 1.

The rotation of galaxies was discovered in 1914 by Vesto Slipher, who observed the galaxies M31 and the Sombrero galaxy using a long-slit spectrograph. In a two-dimensional spectrogram plotting intensity as a function of wavelength and position along the spectrograph slit, Slipher noticed inclined absorption lines. He correctly interpreted these inclined lines as due to varying systematic velocities within the galaxy, as would be expected if a partially edge-on galaxy was rotating: those stars in that half of the disk rotating away from us would have a larger *Doppler shift than stars in the other half of the disk rotating toward us.

By assuming circular rotation about the centre of the galaxy and a spherically symmetric mass distribution, one can balance the centripetal acceleration by the gravitational force to show that

$$v^2 = \frac{GM(R)}{R}.$$

Here v is the circular velocity at radius R, G is Newton's gravitational constant, and $M(R)$ is the mass contained within radius R. While

only strictly true for a spherical mass distribution, it is still a good approximation for a flattened mass distribution, as we might expect within a disk galaxy.

Note that rotation can only be measured for galaxies that are partially edge-on to us, so that a Doppler shift can be detected. One needs to estimate the inclination of the galaxy in order to correct a measured line-of-sight velocity to a true orbital velocity.

It was noticed as early as 1939 by Horace Babcock and 1940 by Jan Oort that the rotation curves of galaxies indicated higher mass at large radii than would be expected from the distribution of starlight. Radio observations of the 21 cm line of neutral hydrogen in the 1950s were able to probe galaxy rotation curves to larger radii than optical spectra.

The typical spiral galaxy rotation curve shows a steep, roughly linear rise from the centre of the galaxy out to a couple of kiloparsecs or so (depending on the size of the galaxy), with a 'flat' (i.e. approximately constant) velocity of ~ 100 km/s to the maximum detectable distance. Note that these flat rotation curves show that galaxies do not rotate as rigid bodies, such as compact disks, as in this case we would have $v \propto R$. Instead, galaxies exhibit **differential rotation**, whereby stars further from the centre of the galaxy take longer to complete one orbit than close-in stars.

By the early 1970s, flat rotation curves were routinely detected. It was not until the late 1970s, however, that the community was convinced of the need for *dark matter halos around spiral galaxies. A minor industry grew up modelling the rotation curves of spiral galaxies. By assuming reasonable mass-to-light ratios for the various visible components of a galaxy (the bulge, thin disk, and thick disk) and measuring the amount of gas present in the gas disk, one could estimate the contribution to the rotation curve for each component. Even by assuming the largest reasonable mass-to-light ratios for the disks, known as the 'maximal disk' model, these models were unable to predict the flatness of the observed rotation curve beyond the stellar disk (see Fig. 1). The inescapable conclusion, assuming that Newton's law of *gravity holds on cosmological scales, was that the visible galaxy was embedded in a much larger dark matter halo, that contributes roughly 50–90% of the total mass of a galaxy.

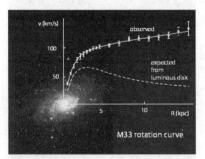

Fig. 1. Observed rotation curve of the galaxy M33 obtained from radio observations of neutral hydrogen superposed on a visible image of the galaxy to the same scale. If all of the mass in this galaxy was associated with the visible matter, the rotation curve would be expected to follow the dashed line.

Some have interpreted extended galaxy rotation curves as being explained by a modification to Newton's laws, known as *modified Newtonian dynamics (MOND). However, this is currently very much a minority viewpoint.

Sofue, Y. and Rubin, V. *Annual Review of Astronomy and Astrophysics, 39, 137*, 2001.

Faber, S. M. and Gallagher, J. S. *Annual Review of Astronomy and Astrophysics, 17, 135*, 1979.

galaxy surveys

A galaxy survey is a systematic record of *galaxies that have been found using at least reasonably well defined selection criteria.

Only a small number of galaxies can be seen with the naked eye, most notably the Large and Small Magellanic Clouds visible from the southern hemisphere and, rather less easily, the Andromeda Galaxy in the northern sky. With the invention of the telescope in 1608 by Hans Lipperhey, and its first astronomical use by Galileo Galilei in 1610, many other *nebulae (diffuse sources of light) could be seen. The first person to systematically catalogue nebulae was Charles Messier who published a catalogue of 110 nebulae in 1784. Forty of these nebulae are in fact galaxies, the remaining objects are mostly clusters of stars too faint to be individually resolved, within our own Galaxy. These nebulae still bear the designation 'M' today; for example the Andromeda Galaxy is also known as M31, being the 31st nebula in the *Messier catalogue.

More nebulae were catalogued in the 18th and 19th centuries, notably by William Herschel and his son John. By 1907, John Dreyer's *New General Catalogue and Index Catalogue* contained 13 000 nebulae and star clusters. At this time, only positions on the sky were catalogued; distances to the nebulae were unknown, and opinion was divided on whether the nebulae were small systems within our Galaxy or separate 'island Universes'.

The existence of distinct galactic systems was proved beyond doubt by Edwin *Hubble in 1923, when he measured a distance to M31 of at least 300 kpc using *Cepheid variable stars. A few years later, Hubble discovered a linear relation between the recession velocity of a galaxy, which is easily obtained from the *Doppler shift in its spectrum, and distance, thus allowing estimates of distance to large numbers of galaxies.

Since then, many catalogues of galaxies have been published, some containing complete *redshift information, others providing two-dimensional maps of the galaxy distribution. Here we list some of the more significant galaxy catalogues which detected galaxies from images of the sky.

Shapley–Ames Catalogue—The first reasonably well defined catalogue of galaxies was published by Harlow Shapley and Adelaide Ames in 1932, and revised by Allan Sandage and Gustav Tammann in 1981. It contains 1 246 galaxies over the whole sky to an approximate *magnitude limit $m \approx 13.2$, although the magnitude limit is now known to be rather fuzzy, with the catalogue being about 50% complete at $m \approx 12.7$.

Zwicky Catalogue—The catalogue published in six volumes by Fritz *Zwicky and collaborators (1961–8) contains 27 837 galaxies in the northern sky to a blue magnitude limit $m_B \approx 15.5$, thus going more than two magnitudes deeper than the Shapley–Ames Catalogue. Zwicky's catalogue formed the basis for the *CfA Redshift Survey.

*Lick Galaxy Catalogue—Published in 1967 by Charles Donald Shane and Carl Wirtanen, this catalogue does not contain parameters for individual galaxies, but contains counts of one million galaxies to magnitude $m \approx 18.9$ in 10 arcminute cells. This was by far the largest catalogue of galaxies made before the advent of machine-produced catalogues.

Uppsala Galaxy Catalogue—The Uppsala Galaxy Catalogue (UGC), published in 1973 by Peter Nilson, contains 12 939 galaxies brighter than $m = 14.5$ and/or larger than 1 arcminute in diameter on blue prints from the Palomar Observatory Sky Survey (northern hemisphere).

ESO Survey—This is essentially a southern hemisphere counterpart to the UGC, containing 18 422 galaxies, star clusters and nebulae with diameter 1 arcminute or larger on survey plates taken with the European Southern Observatory (ESO) 1 m Schmidt telescope at La Silla, Chile.

Reference Catalogue of Bright Galaxies—Gerard de Vaucouleurs and collaborators have published a series of three catalogues, starting in 1964, each superseding the last. The latest, abbreviated RC3, was published in 1991 and contains 23 022 galaxies. About half of these meet the joint conditions: diameter greater than 1 arcminute, B magnitude brighter than 15.5 and recession velocity below 15 000 km/s.

IRAS Point Source Catalogue—The *Infrared Astronomical Satellite (IRAS) surveyed the sky in four passbands at 12, 25, 60, and 100 microns in 1983. Several groups selected galaxy catalogues from the IRAS Point Source Catalogue based on flux ratios (galaxies are redder than stars in these passbands due to the presence of dust). The IRAS data have the advantage of being unaffected by Galactic obscuration, and thus enabled galaxies to be catalogued closer to the plane of the Galaxy (the *zone of avoidance) than in optical surveys.

*APM Galaxy Survey—The Automated Plate Measurement (APM) Galaxy Survey utilized plate scanning technology to digitize 269 photographic plates taken in the b_j passband with the United Kingdom Schmidt Telescope (UKST) at Siding Spring, Australia. Completed in 1990, two million galaxies were catalogued, enabling the first reliable measurement of the galaxy *correlation function on angular scales above 2 degrees. These results provided some of the first evidence that the then-popular *critical density cold *dark matter model was in error. The APM Galaxy Survey also provided the source catalogue for the *two degree field galaxy redshift survey.

The Edinburgh–Durham Southern Galaxy Catalogue, completed around the same time and finding consistent results, was based on the same plate material as the APM survey, but digitized with the COSMOS measuring machine in Edinburgh.

Two Micron All Sky Survey—The *Two Micron All Sky Survey (2MASS) provided the first complete sky coverage in the near-infrared part of the *electromagnetic spectrum around 2 microns in wavelength, using two dedicated 1.3 m telescopes, one in Arizona and one in Chile.

Millennium Galaxy Catalogue—The Millennium Galaxy Catalogue, completed in 2003, is a medium-deep survey covering 37.5 square degrees, taken with the Wide Field Camera on the Isaac Newton Telescope. Although much smaller in area than the Sloan Digital Sky Survey, it has a fainter isophotal detection limit of 26 mag arcsec^{-2}.

*Sloan Digital Sky Survey—The first (and to date the only) optical galaxy survey to use a dedicated telescope, the Sloan Digital Sky Survey (SDSS) is a combined imaging and redshift survey.

(See REDSHIFT SURVEYS for redshift surveys based on the imaging surveys listed here.)

Gamow, George (1904–68)

George Gamow (Fig. 1) played a key role in the early development of the hot *big bang theory. He was born in Odessa, Russia, in 1904, and received his PhD from the University of Leningrad where he had worked primarily on radioactivity, and in particular the alpha particle decay of atomic nuclei. After occupying research positions in Copenhagen and Cambridge he returned to the Soviet Union, but in 1933 he fled first to Europe and then to America, where he held the Chair in Physics at George Washington University from 1934 to 1956. He moved to the University of Boulder in Colorado in 1956, where he remained until his death in 1968 and where a building is named in his honour.

Gamow had scientific interests across a broad range of astronomy and even genetics, but in *cosmology his most famous work concerns the early stages of the big bang. He realized that the hot early stages of the expansion would permit the formation of light atomic nuclei, such as helium, offering an alternative to creation via nuclear fusion within stars. Gamow first published his ideas in 1946, but the theory was developed quantitatively in collaboration with Ralph Alpher. (Alpher, in collaboration with Robert Herman, would later publish the first prediction of the existence of the *cosmic microwave background, a prediction often erroneously attributed also to Gamow.) This resulted in a famous paper authored by Alpher, Bethe, and Gamow in 1948; Hans Bethe did not contribute to the work, but Gamow insisted on his inclusion to make the author list a pun on the first three Greek letters alpha, beta, gamma. While over subsequent years calculations of the origin of the light elements, dubbed cosmic *nucleosynthesis, have seen considerable refinement, the basic ideas of these early papers remain intact and in excellent agreement with observations.

Gamow is also well known as a writer of popular science, often featuring the character of Mr Tompkins used as a foil to enable complex scientific ideas to be explained in the

Fig. 1. George Gamow.

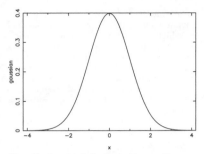

Fig. 1. The Gaussian distribution, in this case with zero mean and standard deviation σ equal to one. A general Gaussian has the same shape, with the x-axis labelled as $(x - x_0)/\sigma$ and the y-axis divided by σ.

simplest possible way, using maths only when absolutely essential. Such is their enduring appeal that some of his books remain in print today.

Gamow, G. *Mr Tompkins in Paperback*, Cambridge University Press, 1993.

Alpher, R. A. and Herman, R. *Genesis of the Big Bang*, Oxford University Press, 2001.

Gaussian

In the subject of statistics, where one is interested in the probability that a quantity takes some value or range of values, the most important function by far is the Gaussian distribution. This distribution is named after its discoverer Carl Friedrich Gauss, the most important mathematician of the 19th century (and arguably ever). The Gaussian distribution has a characteristic bell shape, seen in Fig. 1, characterized by a central value (here taken to be zero) and a width.

The mathematical form of the Gaussian is

$$p(x) = \frac{1}{\sqrt{2\pi}\,\sigma}\,\exp\left[-\frac{(x - x_0)^2}{2\sigma^2}\right].$$

Here x_0 is the mean value and σ, known as the standard deviation or dispersion, measures the width. The probability that x lies within

the range x to $x + dx$ is given by $p(x)dx$. The prefactor ensures that the integral $\int p(x)\,dx$ equals one, so that the total probability is one.

The importance of the Gaussian stems from a key theorem in statistics known as the **central limit theorem**, which states that if one adds independent random numbers from any type of source (subject to the technical condition that their distributions have finite variance, which is almost always true in practice), then the distribution of the sum tends to a Gaussian distribution. Many physical phenomena, such as instrumental measurement errors, tend to originate in the cumulation of small-scale randomness and hence follow the Gaussian distribution. Measurement errors are commonly quoted under the assumption that they are Gaussian, with *confidence limits quoted in terms of number of standard deviations.

For a Gaussian distribution, 68.3% of the probability lies with one standard deviation of the mean value, and 95.4% within two standard deviations. Rounded to 68% and 95% respectively, these are the most commonly used confidence intervals, and are often the ones quoted even if the distribution of a quantity is not Gaussian.

In addition to the uncertainty from measurement errors, some key features of the *Universe itself can only be described in probabilistic terms. The Universe possesses *density perturbations whose origin is believed to be random, and hence described by a probability distribution rather than precise values. Thus far all observations have been consistent with the initial perturbations (before their amplification via *gravitational instability) obeying Gaussian statistics. The simplest models of *cosmological inflation predict that the perturbations should be of this form. Searches for violations of Gaussianity are known as *non-Gaussianity tests, and are often applied to *cosmic microwave background anisotropy measurements in particular.

See also SIGNIFICANCE.

general relativity

General relativity is a theory of *gravity. Its basic premise is that gravity is caused by *curvature of *space-time, that curvature itself being induced by the presence of matter. The theory was devised by Albert *Einstein and published in 1915, its full title being the general theory of *relativity. It supersedes Isaac Newton's theory of gravity, though in most circumstances it very closely reproduces the predictions of Newton's theory. The most dramatic applications of general relativity are to *cosmology and to *black holes.

Motivation

Einstein's motivations for general relativity came from several sources. One was the shortcomings of his own theory of special relativity. Special relativity united space and time into a single concept, space-time, but space-time was still an arena in which events took place rather than an active participant. Moreover, special relativity selected out observers moving with constant relative velocity (called inertial observers) as special observers for whom the laws of physics took on a simple form, but why just them? Finally, special relativity did not allow the incorporation of gravity.

Einstein's great insight was that the last two are linked. By widening the requirement so that all observers, not just inertial ones, saw the same laws of physics, he was forced to include a description of gravity. There was no choice about it. Moreover, the gravitational force was in fact due to space-time being curved.

Einstein was very fond of thought experiments, where he would imagine various physical situations enacted, and of adopting fundamental principles that were to be encoded in his physical laws. In devising general relativity, one of the most important principles he

wished to respect is one known as the equivalence principle.

The equivalence principle states that the laws of physics experienced by an observer freely falling in a gravitational field are the same as those experienced by an inertial observer (as in special relativity) without a gravitational field. The thought experiment that goes with this is shown in Fig. 1. Consider one physicist stuck in a box somewhere in outer space, and another one in a box freely falling towards the Earth (ignoring effects of the atmosphere). The equivalence principle says that the two physicists can carry out any experiment they wish, with lasers, atoms, magnets or whatever, but they will not be able to tell from those experiments which of them is in which box. They will always get exactly the same answers.

A slight variation on this set-up is also helpful. Apply an equal acceleration to each box, so that the freely falling one is held stationary. This is most easily achieved by letting it rest on the ground; the ground provides a force, and hence an acceleration, that cancels out that coming from gravity, so that it stays where it is. We learn that the physics in a box resting on the Earth's surface is the same as in a box accelerating through empty space.

The equivalence principle is an assumption, but one that appears to be true of our *Universe. It has been tested to high precision without violations ever having been seen. It is also in accord with our everyday experience; we know for example that being accelerated upwards in a lift feels just as if the strength of gravity has been temporarily increased, and can imagine that the lift plunging down the shaft would leave us temporarily in free fall within it. We therefore need to formulate physical laws in a manner consistent with the equivalence principle, and in doing so we will have to incorporate gravity.

Even without progressing to a full theory incorporating it, the equivalence principle makes important predictions. If a light ray is sent upwards in a gravitational field, it will be *redshifted. To see this, compare a lab stationary on the Earth's surface with an equivalent accelerating lab (the lower figures of Fig. 1); as the light travels upwards the far end of the lab accelerates away and is moving faster by the time the light is received, meaning a greater gap between wave crests.

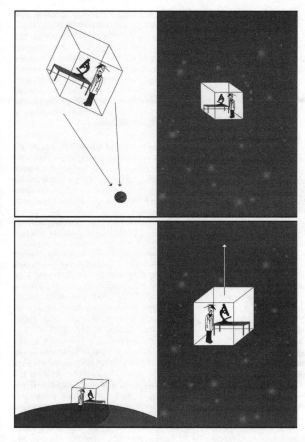

Fig. 1. Two schematic illustrations of the equivalence principle. In the top two diagrams, we have one laboratory free-falling under Earth's gravity and a second floating in free space. The equivalence principle states that the scientists in those labs cannot conduct any experiment (without looking out, of course) which would tell them which lab they were in. In the lower two diagrams an acceleration has been applied to each box; the left-hand one (not drawn to scale!) is now resting on the Earth's surface, while the right-hand one is undergoing a constant acceleration. Those two labs also experience the same laws of physics as each other (though not the same as the top pair).

It also predicts light will bend in a gravitational field (the same set-up, but with the light now fired horizontally—the acceleration makes the actually straight light ray appear parabolic to the observer). At the surface of the Earth this effect is too small to measure, but later we will see that the effect from the Sun can be measured. A related argument (interpreting the crests of the waves as the ticks of a clock), shows that clocks within a gravitational field run more slowly the deeper within the gravitational field they are. This is known as gravitational time dilation, and is particularly important near the *event horizon of a black hole.

Space-time curvature as gravity

That light bends under the influence of a gravitational field is clear evidence that space-time can be curved. This immediately presents considerable difficulties in visualization, four-dimensional space-time alone already being a challenge for most of us without worrying about it being curved. Accordingly, physicists like to think by analogy by reducing the number of dimensions from four to perhaps two, and perhaps additionally thinking only of curvature of space.

This leads to one of the simplest pictures of general relativity, known as the rubber-sheet analogy (see Fig. 2). Simply stretch a rubber

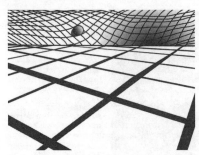

Fig. 2. A distorted rubber sheet gives a simple way of visualizing general relativity. Moving objects follow the geometry and appear to be drawn towards the massive objects which create the indentations.

sheet, and then place a massive object, say a bowling ball, at its centre. The bowling ball will distort the sheet in its vicinity, indenting it, and might for instance represent the Sun. A lighter ball, such as a golf ball, can represent a passing planet or comet. If rolled on a trajectory close to the bowling ball, its path will be curved around the point of closest approach. The slower the golf ball moves, or the closer it passes, the greater the deviation in its path. If sent on the right path, the golf ball can even, ignoring friction, be placed into a circular or elliptical orbit around the model Sun.

This analogy certainly has shortcomings, but still captures some of the essence of general relativity. Massive objects distort the geometry of space-time, and other objects move on paths influenced by this geometry.

The mathematics of general relativity, known as tensor calculus, is a complex subject, way beyond the scope of this book. Combined with the major conceptual leaps required by relativity, the theory became notorious for its impenetrability. In a famous story from 1919, British astrophysicist Arthur *Eddington was asked whether it was true that he was one of only three people in the world who understood general relativity; Eddington paused, and, on exhortation that modesty was unnecessary, explained 'I am trying to think who the third person is'. Fortunately the theory has become much more accessible over the years, though teaching of general relativity is normally reserved until close to the end of an undergraduate physics degree.

Ultimately, the relation between matter and space-time curvature is encoded in

*Einstein's equation. Einstein derived this equation under the considerations that it had to ensure conservation of energy and momentum, and had to give the same results as Newtonian gravity in situations where the gravitational force is weak. This final point is an example of the **correspondence principle**, which states that new physical laws should be good approximations to the old ones they supersede in situations where the original laws are well established.

Observational tests of general relativity

There are four classical tests of general relativity which have been carried out within the Solar System. The gravitational fields of the Earth and Sun are weak as far as relativistic effects are concerned, but precision measurements can be carried out which unveil subtle differences between Newton's gravitational theory and Einstein's. Those tests are

Gravitational redshift—The equivalence principle indicates that photons climbing out of a gravitational field should lose energy, and hence be redshifted to lower frequencies. This effect was measured in the Earth's gravitational field by Robert Pound and Glen Rebka in 1960, using delicate measurements of gamma rays fired up and down a 22-metre tower.

Perihelion shift of Mercury—In Newton's theory, the orbit of one body about another is given by a perfect ellipse, retracing the same path over and over again. Within the Solar System, this simple pattern is spoiled because of interactions with other planets, which cause the perihelion (the closest point of the orbit to the Sun) to drift. For Mercury, calculations around Einstein's time using Newtonian gravity had shown that this drift should be 5558 arcseconds per century, but measurements showed that the actual rate was 5601 arcseconds per century. The discrepancy of 43 arcseconds per century seems small, but was seen as a deep mystery. Einstein was able to show that general relativity predicted deviations from Newton's theory giving an additional perihelion shift of 43 arcseconds per century, in stunning agreement with observations.

Bending of light by the Sun—The equivalence principle already predicts that light paths should bend near massive objects. Unfortunately the effect is very small, and can be seen only for light rays passing very close to the surface of the Sun. However the Sun itself is blindingly bright. Fortunately, during Solar eclipses the Sun is shielded by the Moon; by happy coincidence the Moon's apparent size is almost exactly the same as that of the Sun, so it obscures the Sun itself while still allowing us to see rays passing close to the Sun's surface. General relativity predicts the bending of rays skimming the Sun's surface to be 1.75 arcseconds, while Newton's theory also predicts a deflection but of only half this magnitude. A famous trip led by Eddington in 1919 sought to confirm this, and in doing so launched Einstein's career as a global celebrity.

Radar echo delays—The final classic test was discovered much later, in 1964 by Irwin Shapiro, who calculated that a signal reflected from a planet or satellite will be delayed if the signal passes near a massive body. He was able to carry out his own experiment. This effect includes both the bending of the light rays and time dilation near a massive body. Nowadays this method has led to the most accurate confirmation of general relativity, using a signal reflected from the Cassini satellite when it passed behind the Sun in 2002 en route to an observing mission at Saturn.

In addition to these Solar System tests, which all apply in weak gravity regimes, general relativity is now tested in the strong gravity regime by observations of pulsars, which are rapidly rotating neutron stars. These tests are described under the following heading.

General relativistic phenomena

A range of astrophysical phenomena are distinctively relativistic, and hence require general relativity. Most are dealt with elsewhere in this book. An incomplete list is

- The global evolution of the Universe.
- *Cosmological perturbation theory (i.e. the evolution of irregularities in the Universe).
- Black holes.
- *Gravitational waves.

One specific topic worthy of mention is binary pulsars, which give the most powerful tests of general relativity in the strong gravity regime. A pulsar is a neutron star, an extremely dense relic left after a star explodes

in a *supernova, which emits detectable radio waves. In 1974 Russell Hulse and Joseph Taylor discovered a binary pulsar, meaning a system of two closely orbiting neutron stars where one was a pulsar. They were awarded the 1993 *Nobel Prize in physics for that discovery. This system has been observed for over 30 years now, and the precision observations of its properties give an excellent vindication of general relativity. In particular, the two neutron stars are seen to spiral together at a rate predicted from the expected rate of gravitational wave radiation from the system. This agreement is spectacular enough to remove any serious doubts about whether gravitational waves exist, although it doesn't amount to a direct detection.

In 2003, an even more exciting system was discovered where *both* neutron stars could be detected as pulsars, enabling the physical properties of the system, such as the neutron star masses, to be pinned down even more precisely. Although it is early days, this new system promises to raise the accuracy of tests of relativity in the strong-gravity limit.

Alternatives to general relativity

General relativity is a highly successful theory which has withstood all observational tests for almost a century. Nevertheless, like Newton's theory (which stood for over 200 years), it may yet prove to be of only limited validity and be itself superseded by a more fundamental theory.

Various examples have been considered over the years, mostly taking the basic structure of general relativity and adding modifications. For instance, the *Jordan–Brans–Dicke gravity theory allows the strength of gravity, normally given by Newton's constant G, to vary in time and space. Although no direct evidence has been found for any such theory, their study remains useful as a way of quantifying just how well general relativity does perform. It is also well worth remembering that the laws of gravity that we see in the Universe today may well not have applied during the earlier epochs that cosmology seeks to explore.

Ultimately, general relativity is predicted to break down at very small length scales of order the *Planck length (equivalently, extremely high energies), due to a deep-seated incompatibility between general relativity and quantum mechanics. It is thought that in that regime general relativity will be replaced by a theory of *quantum gravity, of which the best current candidates are *superstrings and *M-theory.

Einstein, A. *Relativity: The Special and the General Theory*, Crown Publications, 1995 (reprinted edition).

Ellis, G. F. R. and Williams, R. M. *Flat and Curved Space-times*, Oxford University Press, 2000 [Moderately technical].

Weinberg, S. *Gravitation and Cosmology*, John Wiley & Sons, 1972 [Technical].

geometry of the Universe

Far back in history, intellectual battle was waged between competing ideas as to the geometry of the Earth. Did it possess a flat geometry, extending to infinity in all directions? Or did it possess a spherical geometry, corresponding to the surface of a sphere?

While that argument was fought and won long ago, a remarkably similar battle has been fought in recent years concerning the geometry of the *Universe. The Universe as we perceive it has three dimensions of space (up–down, left–right, forwards–backwards), meaning that we require three numbers (coordinates) to uniquely specify a position. In addition it has one time dimension. Albert *Einstein's theory of *general relativity unites these into a single entity, four-dimensional *space-time. According to Einstein's theory, the force of *gravity is caused by space-time being curved, with that *curvature caused by whatever material bodies (e.g. stars) are within the space-time. While we find it easy to visualize a curved two-dimensional surface, such as the surface of the Earth, it is an extreme challenge even just to imagine four-dimensional space-time itself, far less it being curved; nevertheless it can readily be described using the mathematics of differential geometry introduced by Bernhard Riemann in the 1800s and exploited by Einstein.

The geometry of the Universe refers to the shape of the three spatial dimensions, usually referring to an idealized Universe where all material is spread evenly, thus obeying the perfect *cosmological principle where all points of the Universe are equivalent. However, it turns out that this requirement is *not* sufficient to uniquely determine the geometry, but leaves open three possibilities. Intriguingly, despite Einstein's

assurance that the presence of gravity curves four-dimensional space-time, one of these possibilities is that the three-dimensional space alone is flat.

The three possible geometries are flat, spherical, and hyperbolic. A flat geometry is one where the usual laws of geometry apply, for instance that the internal angles of triangles add up to 180 degrees, and the circumferences of circles are 2π times their radii. The geometry can be traced by light rays, which always follow a straight line (defined as the shortest path between two points); in a flat geometry two parallel light rays will maintain a constant separation.

A spherical geometry for the Universe is the three-dimensional analogue of the two-dimensional surface of the Earth. Here parallel light rays converge together, exactly as ships heading northwards from the equator will find their paths converging to the North Pole. And like the finite surface of the Earth, such a Universe would possess a finite volume, and a traveller setting out in a straight line would, given enough time, travel around the Universe to return to their starting point.

The final possible geometry is the hyperbolic one, by far the hardest of the three to picture, in which parallel light rays diverge from one another. Its two-dimensional analogue is sometimes illustrated as a saddle-shaped surface. Such a Universe would, most likely, extend forever (*see* TOPOLOGY OF THE UNIVERSE).

The curvature of space-time, and hence in particular the curvature of space, is determined by the material content of the Universe. A determination of the amount of matter in the Universe would lead to a prediction for its geometry. If the total amount of material (including any contribution from a *cosmological constant) is precisely the *critical density, the geometry will be flat. Any more than that, and the extra gravity will curve the geometry to the spherical geometry. Any less, and the geometry will be hyperbolic.

As recently as the mid 1990s, there was considerable dispute amongst cosmologists as to the favoured geometry, with almost everyone either supporting a flat geometry or a hyperbolic one. Eventually, however, it became possible to determine the geometry quite directly by studying the *cosmic microwave background (CMB) radiation left over from the hot *big bang. This radiation has been travelling uninterrupted since the Universe was very young, and hence comes from a great distance making it ideal for testing the geometry. By studying the characteristic size of patterns in the microwave background—the *CMB anisotropies—cosmologists were able to deduce the geometry. Early experiments provided hints but the definitive measurements were made by the *Boomerang experiment, reported in 1999 and 2000. These demonstrated conclusively that while a flat Universe gave a good description of the observed patterns, a significantly hyperbolic Universe could not.

It is impossible to ever be sure that the Universe has precisely the flat geometry, as there would always remain the possibility of a curvature too small to be detected. However, from the advances of the last few years we know that curvature is either absent or plays only a minor role in the dynamics of the Universe. This lends support to the idea that the Universe underwent an *inflationary expansion during its early stages, as such models predict that the Universe was expanded to such a large size that any residual curvature should be indistinguishably small. By ensuring this, inflation is said to solve the *flatness problem.

Geometry alone is not a complete specification of the properties of space. One needs also to consider the topology of the Universe.

globular clusters

Globular clusters are compact, gravitationally bound systems of between roughly 10 thousand and 1 million stars. See Fig. 1 for a nearby example. They are spherical in shape, ranging from of order 10 to 100 parsecs in diameter.

The earliest-known globular clusters belong to our *Milky Way Galaxy. By measuring distances to the known globulars, and by assuming that they are distributed isotropically about the centre of the Galaxy, in 1917 Harlow Shapley was able to estimate the distance of the Solar System from the Galactic centre. He substantially overestimated the distance due to an error in his distance calibration, but he was able to show that we do not lie close to the Galactic centre, as had previously been thought.

About 150 Milky Way globular clusters are now known. Our larger neighbour, Andromeda, may contain around 500. Some giant ellipticals have around 10 000 globular clusters, found at distances up to 100 kpc

Fig. 1. The globular cluster NGC288.

from the galaxy centre. Studying the motions of these globular clusters allows one to estimate the total mass of their host galaxies.

Globular clusters are formed of the oldest known stars: their spectra show almost none of the heavy elements formed through stellar *nucleosynthesis in younger stars such as the Sun. Age estimates of globular clusters, obtained by comparing the observed colours and luminosities of their stars with stellar evolution models, put nearly all of them to be in the range 12–16 billion years. The largest source of uncertainty in determining the age is that in distances to the clusters. The most recent estimates of the oldest clusters give ages 13.5 ± 2 Gyr. They are thus useful **chronometers**: allowing one to set lower limits on the age of galaxies, and hence, the Universe.

An independent estimate of the *age of the Universe, 13.7 ± 0.2 Gyr from the *WMAP satellite shows that globular clusters must have formed very early in the Universe's history.

gluons

Gluons are a type of elementary particle. They are responsible for interactions between *quarks, in particular playing the role of preventing *protons and *neutrons breaking up into their constituent three quarks. The interaction caused by gluons is usually called the strong nuclear force or the strong interaction. Unlike the electromagnetic or gravitational interaction, its strength does not reduce with distance, and this property prevents isolated quarks from existing. (For more about gluons, *see* STANDARD MODEL OF PARTICLE PHYSICS.)

grand unification

During the 1970s, particle physicists had established the *standard model of particle physics, which provided a description of the three non-gravitational *fundamental forces of Nature, namely electromagnetism, and the weak and strong nuclear forces. However this model simply stuck together separate theories describing each of these interactions. The next step was to try to properly unify the theory of interactions into one coherent structure. This goal became known as grand unification, and the corresponding theories grand unified theories or GUTs.

Motivation

The main idea for grand unification came from the standard model, which lays out a classification of particles and describes the unification of two of the fundamental forces, electromagnetism and the weak interaction, into a single force known as the electroweak interaction. According to the standard model, at energies above the mass-energy of the W and Z bosons (about 100 GeV in *particle physics units), electromagnetism and the weak interaction are indistinguishable facets of the single electro-weak force. However such high energies are rarely seen on Earth, being just within reach of the largest particle accelerators, and due to the action of the *Higgs boson, the two forces become distinct at low energies.

Might the same also be true of the strong interaction? Further clues come from the change in interaction strength with particle energy (technically known as 'running of the coupling constants'), which indicates a convergence of the interaction strengths at a very high energy, around 10^{16} GeV, not much short of the *Planck scale. (Incidentally, this convergence appears to work precisely only if *supersymmetry is taken into account, and it is seen as indirect evidence in support of supersymmetry.) The interpretation, then, is that at energies this large and above, the strong interaction merges with the electroweak interaction to behave as a single unified force. At around that energy, a *phase transition similar to the *electro-weak phase transition, and featuring its own high-energy Higgs bosons, would take place giving the strong force its separate identity.

A second motivation for grand unification is simplification. The current standard model features a large number, around 20, adjustable parameters, meaning fundamental constants which are not predicted by the structure of the theory but instead must be measured. Particle physicists find it unnatural and unaesthetic that their theory has so much freedom, as it corresponds to a lack of predictability. A hope for grand unification is that these 20 parameters might be predictable from a smaller number of fundamental parameters associated with grand unification.

Types of grand unification

Unfortunately there is no unique theory of grand unification. The guiding principle of such theories is to choose a model—known as **symmetry group**—which is large enough to encompass the three forces. The simplest such theory carries the technical name $SU(5)$ [this is the name of the type of symmetry around which the theory is based, the symmetry being large enough to contain the whole standard model symmetry $SU(3) \times SU(2) \times U(1)$], but unfortunately it is ruled out by observations (see below). There are an infinite number of more complicated models that can be constructed. The idea of grand unification offers no route out of this impasse, and progress will require guidance from a yet more fundamental theory.

Predictions of grand unification

Although grand unification takes place at an energy scale utterly out of reach of any conceivable particle accelerator, it nevertheless makes testable predictions which are particularly important for cosmology.

A first clear prediction is that *protons are not absolutely stable particles, and can decay, for example into a positron plus photons. The standard model, by contrast, predicts protons are absolutely stable due to conservation of *baryon number. Obviously any short lifetime would be rather disastrous for us (and in blatant contradiction to the fact that the present Universe has survived full of protons for over 10 billion years), but typically the prediction is for very long lifetimes indeed, of order 10^{30} years or more. Perhaps surprisingly, this is within the realm of experimental test; obviously you don't watch one proton for 10^{30} years to see if it decays, but if you watch 10^{30} protons (that's about a tonne's worth) for a year then by chance one should decay if the lifetime is right. Modern experiments are in fact much more sensitive even than that, and proton decay has never been seen, the Japanese Super-Kamiokande experiment now reporting a lower limit of 10^{35} years. If the actual value is close to the lower limit, the chance of one single proton in your entire body decaying during your lifetime is about 1 in 10 000. The limits on the lifetime are strong enough that the simplest GUT, the $SU(5)$ model which predicts the shortest lifetime, has now been ruled out.

The second clear prediction is that the grand unification phase transition will produce objects known as *topological defects, and in particular a type known as *magnetic monopoles are inevitable. These are potentially rather troublesome, and some care is required to make grand unification consistent with current cosmological models.

Cosmologists working on early Universe physics often take the grand unification scale as a guideline for when interesting physical processes might take place. This energy was achieved only around 10^{-34} seconds after the *big bang! For instance, the grand unification energy is often taken as an estimate of the energy scale of *cosmological inflation. It turns out that inflation taking place at around that energy scale is liable to produce *cosmic microwave background anisotropies at around the observed level.

Beyond grand unification

Grand unification still ignores the gravitational force. More ambitious theories such as *superstring theory and *M-theory seek to include it also, and in doing so posit even larger symmetry groups. The precise relation of string theory to grand unification is rather unclear. It may be that string theory motivates a particular theory of grand unification, which then at low energy leads to the standard model as described above. However it may also be that string theory bypasses grand unification altogether, with its *compactification leading directly to the standard model.

gravitational field
See GRAVITATIONAL POTENTIAL.

gravitational instability
Gravitational instability is the mechanism by which structures, such as *galaxies, formed in the *Universe from an initial state that was almost *homogeneous. Regions which are overdense exert extra gravitational attraction on their surroundings, drawing material in and enhancing the initial unevenness. Accordingly, while a perfectly homogeneous Universe will always remain so, any perturbations introduced lead to an instability whereby they are exaggerated by gravitational attraction. Ultimately, the growth of perturbations leads to *structure formation and to *galaxy clustering. The process is sketched in Fig. 1.

The simplest type of *density perturbation is an *adiabatic perturbation, whereby each type of material in the Universe contributes to a total perturbation in the density. In such cases gravitational instability is able to act immediately. Alternatively, in the case of *isocurvature perturbations, the total density is initially unperturbed (with the perturbations of different particle types cancelling each other out), and only after the perturbations begin to convert to adiabatic ones can

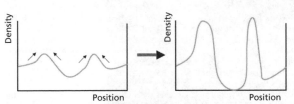

Fig. 1. A sketch of gravitational instability in action. In the left panel there are small irregularities in the density. The overdense regions provide greater gravitational attraction on neighbouring particles, pulling them towards the high-density regions. Some time later, we see that the initial irregularities are enhanced, as shown in the right panel. If this plot were on the scale of individual galaxies, we can consider it as showing how the material to form each galaxy is assembled. If it is on a much larger scale, we can consider it as showing why individual galaxies tend to be clustered close to other galaxies.

gravitational instability induce structure formation.

As far as the *dark matter is concerned, *gravity might well be the only relevant force, and the detailed process of gravitational instability can be followed for instance using *N-body simulations. However a complete analysis of structure formation, including *galaxy formation, needs to consider other interactions as well as gravity, for instance radiation pressure and energy injection from supernovae. Nevertheless, gravity can be considered the principal reason why structures formed in our Universe. Gravitational instability can however only act provided there are indeed initial perturbations in the Universe, which might for instance be created during a period of *cosmological inflation.

gravitational lensing

According to *Einstein's *general relativity, a large mass distorts the local *geometry of *space-time causing nearby light rays to be deflected around the mass. This prediction was observationally confirmed by Arthur *Eddington in 1919 when he was able to measure the predicted deviation in the apparent positions of stars close to the eclipsed Sun. The deflection of light rays by mass is known as gravitational lensing.

Suppose that a large mass, such as a *cluster of galaxies, the **lens**, lies very close to the line of sight to a distant object, the **source**. We will then detect light from the source that has been bent around the gravitational lens. This has two observational consequences. First, the source will be amplified in brightness, since some of its light that would otherwise have passed either side of us will be received by us. Second, the shape of the source may be distorted by the lensing process, just as viewing the world through a wine glass distorts the shape of everyday objects.

Gravitational lensing may be divided into three regimes, depending on the strength of the effect and the mass of the lens.

Strong lensing

Strong lensing describes the situation when a single object is discernibly brightened or distorted by a gravitational lens. Strong lensing requires exquisite alignment of observer, lens, and source and is thus a relatively rare phenomenon.

If the observer, lens, and source are perfectly aligned, and if the lens mass distribution is spherically symmetric, then the source would appear as a ring of light, known as an Einstein ring. In practice, the alignment is not perfect, and so rather than seeing a continuous ring, one may observe arcs of light, as in Fig. 1, or multiple images of the same source, as in Fig. 2. When multiple images are seen, one should bear in mind that the light has travelled to us along different paths (from the same source) for each of the images. Some paths may be longer than others, and so a time-delay may be detected between the images if the source varies in brightness (a common occurrence with *quasars).

Fig. 1. A spectacular Hubble Space Telescope image of the Abell cluster 2218. Most of the galaxies in this image are members of the relatively nearby cluster, but also visible, as faint elongated arcs, are the distorted images of background galaxies that have been gravitationally lensed by the foreground cluster.

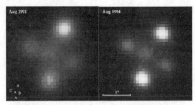

Fig. 2. Quadruple image of the quasar Q2237+0305, known as the Einstein Cross, taken on two different dates. The central object is the lensing galaxy at a redshift 0.04. The four outer objects are lensed images of the same redshift 1.7 quasar. The change in relative brightness of the images is due to microlensing.

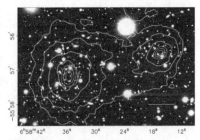

Fig. 3. The reconstructed mass distribution of the merging cluster 1E0567–558 from a weak lensing analysis is shown as contour lines superposed on an optical image. The white bar near the lower right indicates a scale of 200 kpc at the distance of this cluster.

The length of the time-delay scales with the *Hubble constant H_0, and observations of multiply-lensed quasars have been used to place constraints on H_0 with the result $H_0 = (67 \pm 13)$ km/s/Mpc.

Weak lensing

Weak lensing is a much more frequent occurrence, and describes a situation where the observational effects of lensing are too weak to detect for a single source, but where the effects are statistically discernible for a large number of sources.

Weak lensing was first applied to lensing of background *galaxies by clusters of galaxies. Few clusters exhibit such spectacular lensing as Abell 2218 in Fig. 1, but by analysing the shapes of many hundreds of background galaxies, the mass distribution within the cluster may be determined (see Fig. 3). If one can measure the shape and orientation of the images with sufficient accuracy, then one finds that they are elongated and aligned tangentially around the lens. Since galaxies are not necessarily circular, this effect of weak lensing is undetectable for a single galaxy of unknown intrinsic shape, but can be detected given a sufficient number of galaxies with a range of intrinsic shapes and orientations.

Since the amplitude of the lensing effect depends on the source–lens and lens–observer distances, then if one knows (or can estimate) the *redshifts of the lensed sources, a three-dimensional map of the lensing mass may be constructed. This technique allows one to estimate the fluctuations in the three-dimensional distribution of matter in the Universe.

In fact, any line of sight we choose will be affected by weak lensing to some extent,

since each *photon's path is affected by matter inhomogeneities along or near its path. It is just a matter of how accurately we can measure shapes and fluxes of background sources. In particular, lensing of source galaxies by galaxy lenses (known as galaxy–galaxy lensing) has now been statistically detected, providing further constraints on the mass distribution in *galactic halos.

Microlensing

Microlensing is the name given to gravitational lensing by dark objects of less than or of order of the mass of the Sun. Although these are many orders of magnitude less massive than entire galaxies or clusters of galaxies, their compact size means that they may pass exquisitely close to the line of sight to a background source, in this case usually a star. On these small distance scales, the microlenses have a significant proper motion, and so microlensing is a transient phenomenon, whereby a source may increase in brightness as the lens approaches the line of sight, only to fade again as the lens moves on. For close enough alignments, the image of the source may be brightened by more than a factor of ten for a duration of several days, and so the term microlensing refers to the mass of the lens, not to the amplitude of the magnification. Microlensing has been successfully used to detect the presence of *MACHOs in the halo of our Galaxy.

It may also occur simultaneously with strong lensing to give rise to changes in the relative brightness of lensed images (as seen in Fig. 2).

Cosmological applications

The great power of gravitational lensing is that it is due to the total mass of the lens, and so probes dark as well as luminous matter. The physics (Einsteinian gravity) is also extremely well understood, and so observations may be interpreted in a straightforward manner.

Applications of lensing relevant to cosmology have included:

- Mapping the mass distribution within clusters of galaxies.
- Use as a 'cosmic telescope' to detect faint objects at high redshift that would be undetectable if they were not magnified by lensing.
- Constraining the MACHO population within the Galactic halo using microlensing experiments.
- Constraining the mass distribution in other galactic halos using galaxy–galaxy lensing.
- Measurement of the time-delay in strong lens systems to constrain the Hubble constant.
- A direct measurement of the clustering of mass in the Universe.

Now that telescopes and instruments are becoming more and more sensitive, to some extent *all* astronomical observations are affected by weak lensing, and the effects of lensing need to be taken into account when counting the frequency of very faint, high-redshift sources.

http://relativity.livingreviews.org/Articles/
lrr-1998-12/

gravitational potential

Rather than describing the gravitational force exerted by a body, given for instance by Newton's law of *gravity, it is often convenient to use a quantity known as the gravitational potential, which measures the amount of potential energy that objects have through their presence in a gravitational field. This gravitational potential energy is, for instance, available to be converted into kinetic energy as particles move in response to the gravitational field.

The gravitational potential is usually denoted by the symbol Φ (a Greek capital phi), and is a function of position and time. Forces always act so as to reduce the potential, and so the force is in the direction where the potential reduces quickest. Mathematically this can be written

$\vec{a} = -\nabla\Phi$ where $\vec{a}$ is the vector acceleration, equal to the force per unit mass, and ∇ is the gradient operator taking the derivative in each direction. Note that this does not guarantee that the object will move in that direction; this law gives the acceleration, not the velocity. For instance, a planet moving in a circular orbit does so under a force always directed to the centre of the orbit, at right angles to the direction of motion.

The acceleration does not depend on the mass of the body within the gravitational field (the gravitational force is proportional to the body's mass, but the acceleration that a force induces is inversely proportional to the mass). Accordingly, the way in which the object moves does not depend on its mass. At the Earth's surface, we experience this in the law that, neglecting air resistance, all objects fall at the same rate. However, the mass will decide how much the body itself contributes to the gravitational potential, and thus its effect on other objects that might be present.

In many applications, Newton's theory of gravity is perfectly usable even if technically it is only an approximation to *Einstein's *general relativity. However to describe large-scale *cosmological perturbations, it is necessary to use general relativity, which is rather more complex. Even the simplest and most important type of perturbations, responsible for the formation of *cosmic structure, need to be described by two separate potentials. Under some circumstances (the technical jargon is 'in the absence of anisotropic stresses'), particularly in the very young *Universe, these two potentials become equal to each other, and generalize the single gravitational potential of Newton's theory. For this reason, they are sometimes known as the **Newtonian potential** even though they correspond to Einstein's theory rather than Newton's.

The gravitational potential is widely used in astrophysics and *cosmology to describe gravitational interactions. For instance, it is used to describe the gravitational field of the *Milky Way Galaxy or, more generally, of other *galaxies and galaxy *clusters, to calculate the gravitational forces in an *N-body simulation, to define the *density perturbations that give rise to structure in the Universe, and to compute the origin of large-scale *cosmic microwave background anisotropies via the *Sachs–Wolfe effect.

gravitational waves

Gravitational waves are ripples in *space-time, which propagate as waves at the *speed of light. They are predicted by Albert *Einstein's theory of *general relativity. They have never been directly observed, but their existence has been inferred indirectly through studies of binary pulsar systems. In the context of *cosmology, they are sometimes referred to as *tensor perturbations.

Properties of gravitational waves

Gravitational waves are described in Einstein's theory as wave-like configurations in the *curvature of space-time. Unlike conventional space-time curvature around massive objects, they can exist even if the space-time region contains no matter. Something has to be responsible for causing these waves in the first place, but once formed they can freely travel through space, typically expanding and losing amplitude just like a water wave on a pond after a rock has been thrown in. Any object, such as the Earth, which happens to be in a region where a gravitational wave is passing will be briefly influenced by the wave, before it continues on its way leaving no trace.

A passing gravitational wave induces quite a complex motion. Suppose a wave travels vertically upwards through the floor. All the motion will be generated in the horizontal plane transverse to the direction the wave is travelling. First you will be stretched in two opposing directions, say left–right, and simultaneously squashed in the forward–backward direction. As the wave continues, you will briefly be restored to normality, and then instead be squashed in the left–right direction and stretched in the forward–backward direction before returning once again to your starting shape. This cycle will repeat as long as the wave is present.

Gravitational waves interact only extremely weakly with matter, as a consequence of the weakness of the gravitational force. [If gravity seems strong to you, bear in mind that it takes the entire mass of the Earth to provide the force that holds you to its surface, and your leg muscles are easily strong enough to overcome that force when you jump in the air.] This weakness operates in two directions; it is very hard to generate significant gravitational waves by moving objects, and it is very hard to detect the effects of any gravitational waves that have been generated. This weakness is compounded by the fact that gravitational radiation has a quadrupole nature, rather than the dipole nature of electromagnetic radiation. This means that generation of gravitational waves can only take place if there are departures from spherical symmetry, and unfortunately most astrophysical objects tend to be near spherical.

In principle any motion complicated enough to have a varying quadrupole moment generates gravitational waves, for instance you waving your arms above your head. But in practice such gravitational waves are far too weak to ever be detected. In order to get significant amounts, astrophysically large masses must be in rapid motion. The classic example is a close binary system containing compact objects such as neutron stars or *black holes. The waves generated by such a system are shown schematically in Fig. 1, spreading out from the source in a spiral pattern.

Indirect detection

The gravitational waves in a binary system are carrying energy away from the system, and so even if the waves themselves cannot be detected, it may be possible to observe the leakage of energy from the system, which will slowly alter the orbital parameters such as orbital period and separation. Indeed, this measurement has been achieved in the famous binary pulsar PSR B1913+16, discovered by Russell Hulse and Joe Taylor in the 1970s, and for which they won the *Nobel Prize for Physics in 1993. This system, in which a pulsar orbits a second neutron star, has been monitored for decades now, and the rate of energy loss is perfectly consistent with being due to gravitational wave emission. Accordingly, there is no serious doubt that gravitational waves exist, even though at time of writing they have not been directly detected.

(Most or even all neutron stars are thought to emit strongly-directional beams of radio waves from their magnetic poles, and those whose radio beams are visible from the Earth are known as pulsars. As the pulsar spins, the radio beams repeatedly intersect the Earth giving an extremely regular sequence of pulses. Other neutron stars are probably no different from pulsars—it just happens that their radio emission is not pointed towards the Earth.)

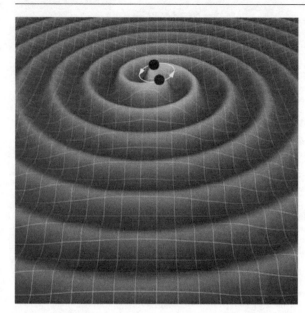

Fig. 1. An impression of the gravitational wave pattern generated by two compact objects (for example neutron stars) in close orbit.

Direct detection

Direct detection of gravitational waves is one of the key goals of *particle astrophysics. According to general relativity, the Earth is continually bathed in low-amplitude gravitational waves, generated by close-orbiting compact binaries, by mergers of neutron stars and/or black holes, and perhaps in the early stages of the *big bang. However the physical effect of these waves is extraordinarily small; a passing wave will alter the distance between two objects by perhaps only one part in 10^{21}. Put another way, if two objects were placed 1000 kilometres apart, the distance between them would vary by about 10^{-15} metres, roughly the width of a single *proton.

Measuring such a small variation is, of course, extraordinarily challenging, but is just entering technical feasibility using laser interferometry. A laser beam is split and the two halves are sent at right angles to one another through a hard vacuum, reflected back, and combined to form an interference pattern. Such a pattern is very sensitive to changes in the lengths of the 'arms' of the interferometer. A passing gravitational wave may, for instance, slightly shorten one arm and lengthen the other as the laser beam is crossing, causing a shift in the interference pattern.

The above experiment is greatly complicated by the need to distinguish real gravitational waves from other sources of length variation, including thermal fluctuations of the mirror surfaces and seismic noise. To reach the required sensitivity, the experiments need to be very large, of order kilometres in size, with the sensitivity enhanced by bouncing the laser beams up and down the arms many times (requiring very high efficiency mirrors as well as creating challenging vacuum system requirements) before recombining them. In the 1990s, in recognition that the technology had become mature enough, funding was secured to build five large detectors. The largest project is LIGO in the US, comprising two separate 4km detectors in Washington State and Louisiana. A French–Italian collaboration has built a 3km detector called VIRGO, a British–German collaboration a 600m detector called GEO600, and a Japanese collaboration a 300m detector called TAMA300. At time of writing all the construction is complete and they have begun operations, but will need time to push

Fig. 2. One of the two Laser Interferometer Gravitational-wave Observatory (LIGO) detectors, this one near Livingston, Louisiana. The two interferometer arms, one of which is fully visible here, are four kilometres long. Laser beams are reflected up and down the enclosed vacuum tubes many times before being brought together to form a sensitive interference pattern.

their sensitivity towards that needed to detect gravitational waves. Fig. 2 shows one of the two LIGO detectors.

The presence of multiple detectors plays an important part in the strategy for distinguishing gravitational waves from experimental noise. Genuine gravitational waves, emitted for instance in the final merger of two orbiting black holes, can be detected simultaneously at several different detectors, leading to strongly correlated signals, while the various noise sources act independently at each detector. By comparing the data streams from coincident runs of the different detectors, one can seek much weaker signals than would otherwise be possible.

The next generation of gravitational wave detectors, beyond those already operating, is likely to be space-based. This is in order to increase the effective arm length, while also doing away with the need to have actual arms since the laser beams can simply be sent through the vacuum of space. The Laser Interferometer Space Antenna (LISA), a joint NASA/ESA mission, is planned for launch around 2015, and will consist of three separate space-craft exchanging laser signals from positions 5 million kilometres apart.

As well as using the detected waves to learn more about general relativity, the ultimate goal of such studies is to create a new scientific field of gravitational-wave astronomy, whereby gravitational waves are used to learn about astrophysical processes in the same way we already use light.

Cosmological implications

All the above experiments, including LISA, are principally targeting specific astrophysical events such as black hole and neutron star mergers, and hence are not directed specifically towards cosmological questions. Nevertheless, those results may have direct impact on cosmology, as distant sources may act as *standard candles, and the mergers of *supermassive black holes may shed light on the *galaxy evolution process.

Cosmological gravitational waves, known as a stochastic gravitational wave background or as primordial gravitational waves, may also exist, and indeed the theory of *cosmological inflation predicts that they do. Gravitational waves are necessarily created during inflation by the same quantum process believed responsible for generating *density perturbations. Different models of inflation predict different amplitudes for the gravitational waves, and hence they provide one method for distinguishing between different models of inflation.

Inflationary gravitational waves are not well suited to detection by interferometers as described above, because on the length scales where those operate they are expected to be totally swamped by the cumulative signal from merging black holes. However on much longer length scales they are in principle detectable via their effect on the *cosmic microwave background (CMB); if gravitational waves are already present at the time the CMB forms, the distortion to the space-time gets imprinted on the pattern of *CMB anisotropies. This signal is in addition to the one expected from primordial density perturbations, with observations showing us the total effect of the two contributions. One can attempt to untangle them as the two sources have a different dependence on angular scale, with the gravitational wave contribution becoming negligible on scales much smaller than one degree.

There is currently no positive indication that gravitational waves exist, but at the same time present limits are quite weak; it is still possible that up to one-third of the large-angle anisotropies seen by the *WMAP satellite might be due to gravitational waves, rather than density perturbations as is the usual assumption. Ultimately, the strongest probe of primordial gravitational waves will be via *polarization of the CMB rather than the temperature anisotropies. In particular, gravitational waves generate a type of polarization known as B-mode polarization, which is not generated by density perturbations except on small angular scales via *gravitational lensing. Observation of large-angle B-mode polarization would strongly indicate the presence of primordial gravitational waves.

Gravitons

In quantum theory, waves have the interpretation of corresponding to collections of individual particles. Just as an electromagnetic wave is, at a microscopic level, made up of *photons, a gravitational wave is made up of large numbers of particles known as *gravitons.

LIGO Home Page:
    http://www.ligo.caltech.edu/
LISA Home Page:
    http://lisa.jpl.nasa.gov/

gravitons

In quantum theory, waves have the interpretation of corresponding to collections of individual particles. Just as an electromagnetic wave is, at a microscopic level, made up of *photons, a gravitational wave is made up of particles known as gravitons. These particles are *bosons and have spin-2, corresponding to the quadrupolar nature of gravitational radiation, and come in two polarization states. Since the gravitational force is a long-range force, they must have zero mass. [A subtle argument, known as

the van Dam–Veltman–Zakharov discontinuity, indicates that the mass must be precisely zero, not just extremely small. In a theory of *gravity with massive gravitons, the predictions of general relativity are not reproduced even if one takes the limit of vanishingly small mass.]

In principle gravitons should be the building blocks of a theory of *quantum gravity, acting as mediators of the gravitational force just as photons do for the electromagnetic force. Standard quantization of the graviton however leads to irretrievably infinite results, the jargon being that the theory is nonrenormalizable. This may however be overcome in *superstring and *M-theory. In superstring theory, the graviton corresponds to a particular type of oscillation of a closed loop of string. In a *braneworld scenario, most particles are described by strands of string whose ends are confined to the *brane which corresponds to our visible Universe. By contrast, the closed loop representing a graviton can move freely away from the brane, indicating that gravitational forces can penetrate the *extra dimensions (known as the bulk).

Individual gravitons have never been detected, and indeed some physicists have argued that such detection is impossible, due to the uncertainty principle and the extraordinary weakness of the gravitational force. Certainly no presently conceivable experiment has any chance of doing so, and so they are set to remain as hypothetical particles for the foreseeable future.

gravity

Of the four *fundamental forces of Nature, gravity is the one that has the most visible influence on our everyday existence. In *cosmology too, gravity is the dominant force in most circumstances, being responsible for the large-scale evolution of the *Universe as well as shaping the distribution of *galaxies.

You might be surprised to learn that gravity is, however, by far the *weakest* of the four fundamental forces (the others being electromagnetism, and the weak and strong nuclear forces). Indeed, it takes the entire mass of the Earth, a mere 6×10^{24} kilogrammes, to provide the gravitational attraction that prevents you drifting off into space. Yet your own muscles have little difficulty in overcoming this force to lift objects, and even a tiny magnet can do likewise. In fact, the electromagnetic force from the *proton on the *electron in a hydrogen atom is nearly 10^{40} times stronger than the gravitational force.

For large objects, however, gravity wins, the reason being that it is always attractive. The electromagnetic force, by contrast, can be either attractive or repulsive, depending whether the charges are of the same or opposite sign. Opposite charges attract, and having done so tend to cancel each other out so that there is no net long range force.

The first theory of gravitation was due to Isaac Newton, whose great insight was that the force that held objects to the surface of the Earth might be the same force responsible for the orbit of the Moon. He is said to have been inspired around 1664/5 by an apple falling from a tree near his birthplace at Woolsthorpe Manor in Lincolnshire (that it actually hit his head is almost certainly apocryphal). His law was an inverse square law, meaning that the strength of gravity reduced in proportion to the square of the distance. The force between two objects of masses M and m is given by

$$F = \frac{GMm}{r^2} ,$$

where G is a fundamental constant of Nature, known as Newton's constant, which measures the strength of gravitational attraction. This theory predicts that planets move on elliptical orbits, as had previously been determined, but not explained, by Johannes Kepler.

[The identity of the original apple tree has been the subject of some investigation and controversy. The standard opinion is that it died around 1820 following a storm, but there are claims that it survives to this day. Several trees are said to be propagated from the original, including one planted at the Isaac Newton Institute for Mathematical Sciences in Cambridge. A small log said to be from the original was presented to the Royal Astronomical Society in 1912 and is displayed in its library.]

The acceleration a on the object with mass m generated by this force is given by $F = ma$ (force equals mass times acceleration), and is *independent* of the object's mass m which cancels out between the two equations. Therefore objects of different masses experience the same acceleration under gravity; in the absence of air resistance a feather and a cannonball would fall at the same rate. Likewise, the orbits of planets around the Sun do not depend on the planets' masses (in the limit where their mass is much less than that of the Sun).

Newton's theory of gravity stood for over 200 years, but was eventually superseded by Albert *Einstein's theory of *general relativity, published in 1915. Einstein radically rewrote the interpretation of gravity. According to his theory, massive objects cause *space-time itself to become curved, and what we perceive as the force of gravity is actually due to the effect of this curvature on the motion of objects. If gravity were the only force acting, objects such as planets simply follow straight lines (known as **geodesics**) in the curved space-time.

Despite the radical change in viewpoint that general relativity brings, in situations where the gravitational field is weak, such as that of the Earth or Sun, general relativity and Newton's theory give extremely similar answers, though detailed measurements are able to tell the difference in favour of general relativity. The main differences between the theories come in the strong gravity regime, as in the vicinity of *black holes. The hot *big bang cosmology (*see* 'Overview') is based upon general relativity, and uses its equations both to follow the average (homogeneous) evolution of the Universe, and also to follow the growth of *density perturbations.

General relativity is the standard theory of gravity used in modern physics, and detailed calculations usually assume its validity. However other theories of gravity have been proposed; one example is *modified Newtonian dynamics, and another the *Jordan–Brans–Dicke gravitational theory in which Newton's constant G is allowed to vary with location and time. In situations where gravity is well tested, such as in the Solar System, such theories have to make predictions extremely similar to those of general relativity, otherwise they would be in violation with observations. However one should bear in mind that many cosmological applications, particularly in the very young Universe, are to situations where general relativity has not been tested, and so its validity remains an open question. Indeed, cosmological observations are the only way of testing whether general relativity might not have applied during the Universe's earliest epochs.

Will, C. *Theory and Experiment in Gravitational Physics*, Cambridge University Press, 1994 [Technical].

Some theories about Newton's apple tree can be found in an article by Keesing, R. G., *Contemporary Physics, Vol. 39*, p. 377, 1998.

Gunn–Peterson effect

The Gunn–Peterson effect (also sometimes called the 'Gunn–Peterson trough' or 'Gunn–Peterson test') is a probe of the *ionization history of the *Universe. It was proposed and applied in 1965 by American astronomers James Gunn and Bruce Peterson. For several decades after the test became known, the key observational result was that the effect they proposed was *not* seen in observations. This implied that the Universe was in a state of high ionization, to a distance as far as the furthest known *quasars. Eventually, in 2001, the effect was discovered in extremely distant quasars detected by the *Sloan Digital Sky Survey.

The effect concerns interactions of relatively high-energy radiation as it traverses the *intergalactic medium (IGM) towards us. It is normally applied to radiation from distant quasars, as they provide bright point-like sources which emit strongly at high energies. Typically light can propagate through the IGM over vast distances with little chance of an interaction, as it is such a low-density environment. However, *photons which happen to have an energy corresponding to *electron transitions between atomic energy levels have a considerably enhanced interaction rate. Of particular importance is the *Lyman alpha transition of hydrogen, which transfers an electron between the ground state and the first excited state. Photons with this characteristic energy (in *particle physics units 10.2 electron-volts) strongly interact with atomic hydrogen clouds, giving rise to what is usually called Lyman-alpha absorption. [More properly, the interaction should be called resonant Lyman-alpha scattering, as the absorbed photon is swiftly re-emitted but in a different direction from its original path. The word 'resonant' indicates the matching of

the photon energy to the electron energy level difference. To us however it appears as if it were absorbed.] Study of quasar spectra to examine the interactions in the IGM is known as the study of *quasar absorption systems.

What Gunn and Peterson realized is that light coming towards us from distant quasars redshifts to longer wavelengths as it travels, and as it does so its wavelength may pass through the wavelength corresponding to the Lyman alpha transition. At that point, if there is any neutral (i.e. unionized) hydrogen gas in its vicinity it will be scattered, preventing us from seeing the light. Such scattering is so efficient that we should see no light at all that has passed through this key wavelength during its travels towards us. However, although quasars do show absorption features (in particular the *Lyman alpha forest), some such light is detected from distant quasars. This means that there must be essentially no neutral gas in the IGM—the gas must instead be in a highly ionized state (the interactions between Lyman-alpha photons and free electrons being around a million times less efficient than those with atoms). This observation indicated the need for *reionization of the Universe.

For almost 40 years, all searches for the Gunn–Peterson effect gave a null result, covering quasars out to *redshifts around 5. This was therefore a lower limit on the epoch of reionization. In 2001, the Sloan Digital Sky Survey discovered new quasars at redshifts approaching 7, and for the first time saw evidence of the Gunn–Peterson effect. The implication is that the reionization process was completing around those epochs.

Now that the effect has been detected, astronomers seek to measure separate effects due to neutral hydrogen and neutral helium, which theory predicts to have reionized at slightly different epochs as different energies are required to liberate electrons from each.

hadrons

The word hadron refers to any particle which is made from *quarks. Examples include *protons and *neutrons.

halo

See DARK MATTER HALO.

halo occupation distribution

The halo occupation distribution (HOD) model is a simple way of assigning *galaxies to *dark matter halos. The HOD is a function that describes how many galaxies are expected within a dark matter halo of given mass. It predicts that the form of the galaxy *correlation function $\xi(r)$ (or the projected correlation function $w_p(r_p)$) is not a pure power-law (that is a straight line in a plot of $\log \xi$ against $\log r$) but should show a 'shoulder' at scales $r \sim 10$ Mpc as the clustering signal transitions from being dominated by galaxy pairs in the same halo (one-halo contribution) to galaxy pairs in two different

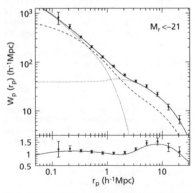

Fig. 1. Upper panel: projected galaxy correlation function $w_p(r_p)$ measured from the Sloan Digital Sky Survey, along with the best fit from the halo occupation distribution model. The two dotted lines show the one- and two-halo contributions to $w_p(r_p)$ which sum to give the continuous line. The dashed line shows a prediction for the matter correlation function. Lower panel: data and model divided by the best-fit power law.

halos (two-halo contribution). This feature is indeed seen in galaxy clustering in the *Sloan Digital Sky Survey: see Fig. 1.

Note that the HOD assigns galaxies to halos in a purely statistical way. In contrast, *semi-analytic models incorporate real, if simplified, physics in assigning galaxies to halos.

Hammer–Aitoff projection

The Hammer–Aitoff projection is a way of representing an all-sky map in two dimensions, for example on a page or computer screen. It was invented by Ernst Hammer based on an earlier idea by David Aitoff, and is commonly referred to simply as the Hammer projection by geographers. Confusingly, astronomers commonly call this projection simply the 'Aitoff projection', while cartographers use the term 'Aitoff projection' to refer to something slightly different (and which actually was invented by Aitoff).

The information we receive from the sky can be thought of as originating from the surface of a sphere, sometimes called the celestial sphere. However for display in books and papers it is necessary to translate the information on that sphere into a two-dimensional plane, a process known as projection. It is inevitable that some compromises in representation, particularly shape and area, must be made when making a projection.

The problem is exactly the same as the one of representing a map of the complete surface of the Earth, and so has a long history. The most famous projection, formerly popular in atlases, is the Mercator projection invented by Gerardus Mercator in 1569. This was designed with sea navigation in mind, such that lines of constant compass bearing correspond to straight lines in the projection. However it has the famous drawback of seriously distorting the areas of land masses far from the equator, for instance making Greenland look much larger than it is, and generally biasing the appearance of the world in favour of the northern hemisphere countries as they are further from the equator. In recent years it

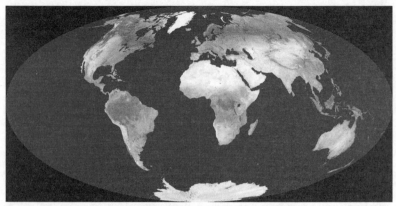

Fig. 1. A map of the surface of the Earth using the Hammer–Aitoff projection, centred on the Greenwich meridian. Lines of constant latitude and longitude are shown. *Compare* MOLLWEIDE PROJECTION.

has fallen from favour, with atlases switching to more sophisticated projections such as the Winkel Tripel projection, which compromise between distortion of shape and area.

There are well over thirty projections in common use (see the weblink below), making different compromises according to need. For astronomy maps, the most important criterion is usually that the projection should preserve areas, so that two regions of equal area on the sky appear also with equal area in the map. The Mercator projection does not have this property, but the Hammer–Aitoff projection does (at a cost of a distortion in shape). It is therefore said to be an equal-area projection.

The Hammer–Aitoff projection projects the sphere into an ellipse, with the equator lying horizontally and the poles in the centre at the top and bottom, as seen in Fig. 1 showing a Hammer–Aitoff projection of the Earth's surface. While the map preserves areas throughout, the shapes of landmasses near the poles can be highly distorted.

The orientation of an astrophysical Hammer–Aitoff projection can be freely chosen, but most commonly it is shown in Galactic coordinates, which are defined so that the Galactic plane (seen in the sky as the *Milky Way) lies along the equator with the centre of the Galaxy at the zero of longitude. In such maps, features due to our Galaxy appear as a central horizontal band,

for instance in maps of *cosmic microwave background anisotropy.

http://www.geometrie.tuwien.ac.at/karto/

Harrison–Zel'dovich spectrum

The Harrison–Zel'dovich spectrum is a particular type of *density perturbation in the *Universe, in which the perturbations are of *adiabatic form, and where the typical size of perturbations on all length scales is the same. This corresponds to a *spectral index of one. It was first considered by Edward Harrison in 1970, and by Yakov *Zel'dovich in 1972. It is sometimes also referred to as a *scale-invariant spectrum, though this latter terminology is also used with more general meaning.

In fact one needs to be a little more precise as to just what quantity is independent of scale; it is the perturbations to the *metric describing the spatial geometry that are constant. Yet more precisely, the *power spectrum of spatial *curvature perturbations on a uniform-density slicing of *space-time is a constant independent of scale. The density perturbation itself can be defined in a variety of different ways, and may appear to have scale-dependence in some of those.

A Universe with Harrison–Zel'dovich perturbations has no sense of scale; if you were given a map of the perturbations you would not be able to tell whether the map was one megaparsec across, or the size of the whole *observable Universe. Or, for that matter, a zillion times larger than the observable Universe.

The term Harrison–Zel'dovich spectrum always refers to the initial perturbations. Evolution, including *gravitational instability, *pressure forces, etc., modifies the initial form, breaking the scale-invariance. In order to judge whether the initial spectrum was of Harrison–Zel'dovich form given the presently observed perturbations, one has to untangle those evolutionary effects.

The Harrison–Zel'dovich spectrum is a very popular choice amongst cosmologists, because it gives a good fit to observational data, and because the simplest models of the *inflationary cosmology predict that the perturbations are close to the Harrison–Zel'dovich form.

hierarchical structure formation

There have been two competing pictures of how the *cosmic structure in the *Universe formed. In the first picture, known as *top-down structure formation or the 'monolithic collapse' model, large things such as *clusters and *superclusters of galaxies form first, and then subsequently fragment to yield individual *galaxies. Such a mode of structure formation is expected in a Universe dominated by hot *dark matter: the near-relativistic velocities of the hot dark matter particles prevent small structures from forming early on. The second, and now widely accepted, picture of structure formation is the hierarchical, or 'bottom-up' scenario. Here, small things, such as *globular clusters and *dwarf galaxies form first, which later merge to form giant galaxies which in turn conglomerate to form clusters and superclusters of galaxies. The hierarchical model of structure formation arises naturally in a Universe dominated by *cold dark matter.

The growth of structure in cold dark matter Universes may be followed by running *numerical simulations. These simulations model galaxies as halos of dark matter particles and allow one to trace the growth of *dark matter halos with time. One finds (e.g. Fig. 1) that the present-day halo of a typical galaxy is formed from the past merger of less massive halos. This merging is an ongoing process: it is perhaps inevitable in the future that the *Milky Way will eventually merge with other members of the *Local Group.

The theory of hierarchical structure formation makes several observational predictions.

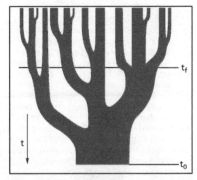

Fig. 1. Schematic representation of a 'merger tree' depicting the growth of a massive halo as the result of a sequence of mergers. Time increases from top to bottom; the widths of the branches of the tree represent the masses of the individual halos. Time t_0 denotes the present; it is possible that further sub-halos may merge with the parent halo in future. Time t_f denotes the formation time, the time at which a parent halo containing at least half of the mass in the halo today was first created.

- Small objects should be older than large objects. This is observationally confirmed to some extent, for instance galaxies tend to be dynamically 'relaxed' (that is they have reached a state of equilibrium), whereas groups and clusters of galaxies are less obviously relaxed. For example, many clusters are elongated and may show more than one central concentration of galaxies: the *Coma cluster is a prominent example of this. This is even more true for superclusters, which tend to be flattened or even filamentary structures.

  However, when one studies different types of galaxies, the observational picture is not quite so consistent with hierarchical structure formation. Dwarf galaxies, which one would expect to be amongst the oldest galactic systems, actually appear to be relatively young, as evidenced by their high star formation rates. Conversely, massive elliptical galaxies, which might be thought to be the result of a long history of mergers, show only old stellar populations. This discrepancy is likely to arise due to the distinction between the ages of the stars which comprise a galaxy, and the age at which they join a galactic system. It also seems likely that star formation is an inefficient process in relatively isolated systems such as dwarf galaxies, and so is still ongoing today, whereas a rapid burst of star formation may occur in merging systems, leaving an evolved stellar population relatively soon after the merger event.

- The cold dark matter model makes a prediction for the shape of the galaxy *correlation function which is well matched (assuming the effects of a *cosmological constant are included) with observations. A top-down formation model predicts relatively more clustering on large scales than is seen.

- Amongst the first things to form in a hierarchical picture are gas clouds. These have been detected along the lines of sight to *quasars via the *Lyman alpha forest. One would not expect to see this forest of hydrogen clouds in a top-down model.

Higgs boson

The Higgs boson is a fundamental particle responsible for the breaking of symmetries in the *Universe. It is named after Scottish physicist Peter Higgs who proposed the mechanism of symmetry breaking, the Higgs mechanism, in the 1960s, drawing on ideas from the study of *phase transitions in laboratory materials.

The original Higgs boson is the one in the *standard model of particle physics, described further under that topic. Its role is to give a separate identity to the electromagnetic and weak nuclear forces, which otherwise would behave as a single electro-weak force. In doing so, the Higgs boson is responsible for giving a mass to fundamental particles.

The term 'Higgs boson' is also used to describe the particles breaking other symmetries; in a *grand unification theory there would be grand-unified Higgs particles responsible for differentiating the strong and electro-weak forces. *Supersymmetry breaking may also have its Higgs particles. It was originally hoped that the grand-unified Higgs particles might be good candidates to have caused *cosmological inflation, but that hope has never been properly realized.

homogeneity

The Universe is homogeneous if, at a given time, all points within it are equivalent. This requirement is usually coupled with *isotropy, which demands that the *Universe looks the same in all directions. If both conditions are met, the Universe is described by the *Robertson–Walker metric.

Homogeneity does not imply isotropy. An example of a scenario which is homogeneous but not isotropic is a Universe pervaded by a uniform electric or magnetic field. Each point is the same as any other, but the field lines will select out a special direction. Likewise, isotropy does not imply homogeneity; any spherically symmetric distribution is isotropic as seen from its centre. Only if there is isotropy around *every* point is homogeneity implied.

Homogeneity is not respected exactly by our Universe (locations on the surface of the Earth are not equivalent to those within the Sun, for instance), but is a good approximation on sufficiently large scales. Accordingly, a usual starting point to describe the Universe is to assume a homogeneous distribution giving the Robertson–Walker metric, and then to add deviations from that, which are known as **inhomogeneities**. The evolution of such *density perturbations is one of the key observable properties of the Universe.

Such inhomogeneities are usually assumed to obey what is known as **statistical homogeneity**, meaning that differences between locations arose by chance, and that any statistical measure of the inhomogeneities such as the *power spectrum is independent of position. As an illustration, consider two people rolling dice a large number of times. Although their individual sequences will differ, they will still share statistical properties such as the mean value thrown and its standard deviation.

The essence of homogeneity is that there is nothing special about our location in the Universe. This assumption is known as the *cosmological principle.

horizon distance

The horizon distance is the maximum distance from which we can receive a signal from a distant object or phenomenon in the *Universe. Because the Universe has a finite *age, and because the *speed of light is a universal speed limit, the horizon distance is finite. The region enclosed within the horizon distance is known as the *observable Universe.

It is impossible to know what the Universe might be like beyond the horizon distance. In particular, the Universe might well be infinite in extent, but if so we will never know it. Some theories of the early Universe predict that the Universe might look very different at distances much larger than the horizon distance, but again it is impossible for us to verify that prediction.

A simple estimate of the horizon distance is given by multiplying the age of the Universe by the speed of light. Even in the simplest cosmological models this is an underestimate, because the light began its travels when the Universe was much smaller and so could make progress more easily (an equivalent viewpoint is that the region the light traversed when the Universe was young has subsequently increased in size due to the expansion, so when we measure it today we get a bigger answer). An accurate calculation needs to compute the distance light can travel taking into account the expansion of the Universe, and for the *standard cosmological model it is a few times bigger than one would get by multiplying the age of the Universe by the speed of light.

The horizon distance can be spectacularly modified if the Universe experiences a period of very rapid expansion in its early stages, known as *cosmological inflation. This rapid expansion can help resolve some otherwise puzzling features of the Universe, one being the *horizon problem. Because the properties of the Universe at such early times are uncertain, often the horizon distance is computed without including the contribution from early times.

horizon entry and exit

At any given moment in its evolution, the *Universe is characterized by one important length scale, the *Hubble length. This is the distance that light is able to travel in a *Hubble time, where the Hubble time is characteristic of the expansion rate of the Universe. When considering *perturbations in the Universe, it is very useful to known whether their length scale is smaller or greater than the Hubble length. For some reason, in this context the Hubble length is often referred to as the 'horizon', even though the *particle horizon is a rather different quantity.

Let's consider a perturbation on a particular length scale; this might for instance be the length scale relevant for *galaxy formation, or

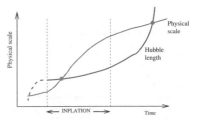

Fig. 1. The evolution of a physical scale in comparison with the Hubble length in an inflationary cosmology. Whether there exists an epoch before inflation is uncertain, but at the early stages of inflation the physical scale of interest is well within the horizon. The one shown exits the horizon at the first dot, and re-enters much later at the second dot long after inflation has ended. Other physical scales could be shown as lines shifted upwards or downwards of the one plotted. The Hubble length evolution stays the same. For instance, shifting the line downwards to consider a smaller scale gives a later epoch of horizon exit and an earlier one of horizon re-entry.

a length scale on which we observe *cosmic microwave background anisotropies. During the early stages of its evolution the perturbation has a small amplitude and evolves according to *cosmological perturbation theory. At this time the dominant effect is stretching of the perturbation due to the expansion of the Universe, which increases its length in proportion to the *scale factor of the Universe. An especially interesting epoch is the one where it is equal to the Hubble length, an epoch known as **horizon crossing**. Depending on whether the perturbation is becoming bigger or smaller than the Hubble length, this phenomenon is known as horizon exit and horizon entry (or re-entry) respectively. Different length scales undergo horizon crossing at different times.

According to the *inflationary cosmology, all the scales on which we observe *cosmic structure have crossed the horizon twice, as shown in Fig. 1. During inflation, physical scales are increasing much more rapidly than the Hubble length, and so scales are constantly exiting the horizon. The length scales on which we observe structures begin well inside the Hubble length early on, with inflation causing them to rapidly expand and cross the horizon. It is at this point that *inflationary perturbations are imprinted on the Universe, giving the large-scale irregularities that ultimately seed structure formation.

After inflation ends, physical length scales are growing more slowly than the Hubble length and scales begin to re-enter the horizon. This happens in a last-out, first-in fashion; those scales which exited the horizon towards the end of inflation re-enter long before those which exited earlier on. Those scales which are only re-entering the horizon around the present epoch, and which are observed via large-angle cosmic microwave background anisotropies, would have exited the horizon around 50 *e-foldings before inflation ended.

horizon problem

The horizon problem is one of the classic problems of the hot *big bang cosmology (*see* 'Overview') which motivated the development of the *inflationary cosmology. The hot big bang model (without inflation) predicts that our present *observable Universe would have been made up of a large number of separate regions which would be unaware of each other's existence, as light would not have had time to travel between them. However, observations indicate that these regions must have been almost identical in their properties. How can that have arisen if the regions cannot have communicated?

The problem

The simplest manifestation of the horizon problem is in the *cosmic microwave background (CMB). The temperature of the microwave background is almost perfectly uniform across the sky, and a natural explanation of this would be to say that the Universe is in a state of near *thermal equilibrium, where the different regions have exchanged energy to reach a balance. Exactly that argument explains, for instance, why the opposite walls of the room you are sitting in are at almost the same temperature.

Unfortunately, this is not possible for the Universe. Consider two *photons of light coming to us from opposite directions, as shown in Fig. 1. They have travelled uninterrupted, at the fastest speed possible, since the epoch of *decoupling, and hence have travelled most of the way across the present observable Universe. Since the photon coming from one direction has only been able to reach us, it is clear that no photon, and hence nothing of any sort, could possibly have gone from the point of origin of one photon to the point of origin of the other. Those two regions

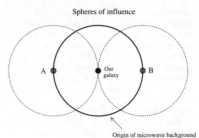

Spheres of influence

Origin of microwave background

Fig. 1. An illustration of the horizon problem. We receive microwave radiation from points A and B on the opposite sides of the sky. These points are well separated and would not have been able to interact at all since the big bang—the dotted lines indicate the extent of regions able to influence points A and B by the present—far less manage to interact by the time the microwave radiation was released.

have had no way of exchanging information, and are said to be **causally disconnected**.

A common objection to this argument is that at times closer and closer to the big bang, the different points in the Universe become closer and closer, and indeed would have been coincident at the instant of the big bang. Surely they could interact then?! It turns out not. Although the points are closer the nearer to the big bang we are, the amount of time available to send a signal is also less. By taking the limit carefully, one can show that the distance a signal can be sent is always finite and small, however close to the instant of the big bang it set out.

In fact, the situation is much worse, because the above argument says that those two regions could not even have communicated by today. However, if interactions are to explain why they are at almost identical temperatures, they would have to have already taken place by the time of *decoupling, only 400 000 years or so after the big bang. Even regions which appear quite close together to us now would not be able to interact by that time; in fact, any regions on the *last-scattering surface which are more than about one degree apart were causally disconnected at the time of decoupling.

While the CMB is the most direct version of the horizon problem, it also arises in studies of cosmic *nucleosynthesis, which took place when the Universe was about one second old. The abundances of elements formed at this stage are very sensitive to the precise density of material present. Any significant variations, and element production will take place differently in each region, and when all the regions in our observable Universe are added together the net result would be different from the successful predictions of nucleosynthesis in a *homogeneous Universe. The number of separate causally disconnected regions is even larger at nucleosynthesis than at decoupling, so in some ways this version of the problem is even more severe.

The horizon problem is also sometimes referred to as the homogeneity problem: why is our Universe so close to homogeneity when no physical mechanism within the big bang cosmology exists to create it? This highlights yet another aspect of the horizon problem, which is how to account for the fact that the Universe is not perfectly homogeneous, but rather includes *density perturbations which give rise to *structure formation. The same reasoning that says there is no possibility of smoothing the Universe before decoupling also implies that there is no way of creating density perturbations, which also requires the ability to move matter/energy around. However, *CMB anisotropies are observed on scales much larger than the one degree scale of causal disconnection.

Within the standard hot big bang model, therefore, it is impossible to produce a physical theory capable of explaining why the Universe is, to a first approximation, homogeneous on very large scales. It is then also impossible to produce a physical theory explaining the origin of the large-scale density perturbations which are superimposed upon this homogeneity. Instead, one is forced to assume that the Universe was born with those properties already in place at the instant of the initial big bang.

The earliest known mention of the horizon problem was by Ralph Alpher, James Follin, and Robert Herman in a 1953 paper during their studies of cosmic nucleosynthesis, where they noted that the correct element abundances were only achieved if the many causally separated regions at that time shared the same properties. The problem was known to exist, though not much discussed, throughout the following decades, but no ideas were available to address it until 1981, when Alan Guth developed the inflationary cosmology.

Solution by inflation
Guth found a solution to the horizon problem with a radical postulate—that the hot big bang era did not extend back all the way to the initial *singularity. Instead, he suggested that during its earliest stages the Universe underwent a burst of very rapid, indeed accelerated, expansion which he named the inflationary epoch. In doing so, he was motivated by the horizon problem along with two other problems known as the *flatness problem and *monopole problem. Inflation was able to simultaneously resolve these problems.

Inflation solves the horizon problem by taking small regions which are already in causal contact, and expanding their size so dramatically that our entire observable Universe lies comfortably within one such region. This requires that during the inflationary era, the Universe undergoes expansion by a factor of at least 10^{27}! In fact, many models of inflation predict even more expansion than that.

Guth, A. H. *The Inflationary Universe*, Jonathan Cape, 1997.

hot big bang cosmology
See 'Overview: the hot big bang cosmology'.

hot dark matter
See DARK MATTER.

Hoyle, Fred (1915–2001)
The British astronomer Sir Fred Hoyle (Fig. 1) is probably best remembered today for being a leading proponent of the *steady-state theory of *cosmology, which invoked continuous creation of matter in order to avoid the existence of the *big bang, a term Hoyle himself coined in mockery. Unfortunately, the steady-state theory is almost certainly wrong, yet

Fig. 1. Sir Fred Hoyle.

Hoyle still made extremely important contributions to modern cosmology, most notably his work on *nucleosynthesis.

Born in Yorkshire in 1915, Hoyle showed innate mathematical ability from a very young age: he was able to write out the twelve-times table by the age of four. He read mathematics at Cambridge, graduating in 1936, whereupon he studied for a Masters degree in Physics. After obtaining his Physics MA and a fellowship at St John's College in 1939, inspired by Raymond Lyttleton, Hoyle focused his attention on astrophysics.

During the war, Hoyle worked alongside Hermann Bondi and Thomas Gold in the development of radar. At the same time, Hoyle, Bondi, and Gold developed the steady-state theory of cosmology. This theory was published in 1948, and two years later popularized in Hoyle's book *Nature of the Universe*. Although the rival big bang hypothesis was essentially confirmed in the 1960s with the discovery of the *cosmic microwave background radiation, Hoyle continued to push his steady-state theory.

After the war ended in 1945, Hoyle became a lecturer in mathematics at Cambridge, and in the following year submitted his first research papers on the synthesis of heavier elements from hydrogen and on the origin of *cosmic rays. Hoyle developed his theory of nucleosynthesis in collaboration with William Fowler and Geoffrey and Margaret Burbidge, resulting in 1957 in the famous Burbidge, Burbidge, Fowler, and Hoyle paper *Synthesis of the Elements in Stars*. This seminal paper provided the first comprehensive account of how the elements are produced in the interiors of stars, and is probably Hoyle's greatest contribution to astrophysics. Prior to this paper, it had been widely assumed that all of the elements were produced in the hot primordial universe. It is generally recognized that Fowler and Hoyle did most of the core work on this paper, with the Burbidges providing the observations. In 1983, Fowler was awarded the *Nobel Prize in Physics for this work. Controversially, and possibly due to his other unpopular theories on the origin of life (see below), Hoyle was not included in the Prize. Fowler himself acknowledged in his autobiography that 'the grand concept of nucleosynthesis in stars was first definitely established by Hoyle in 1946'.

Hoyle joined the staff of the Mount Wilson and Palomar Observatories in 1956. In 1958 Hoyle returned to Cambridge to become Plumian Professor of Astronomy and Experimental Philosophy. Hoyle was the founding director of the Institute of Theoretical Astronomy (now known simply as the Institute of Astronomy, or IoA) at Cambridge in 1967. He became vice-president, then president, of the Royal Astronomical Society and was active in the Science Research Council in promoting the case for a 4-metre telescope in the Southern Hemisphere. He subsequently presided over the inauguration of the Anglo-Australian Telescope in 1974.

He was knighted in 1972, but around this time became disillusioned with the Cambridge system and resigned from his formal UK appointments and chose to live in the Lake District. He took up visiting appointments at the California Institute of Technology and Cornell University in the US, as well as at the Universities of Cardiff and Manchester in the UK, and began his interest in the origin of life. In books co-written with Chandra Wickramasinghe, Hoyle argued that evolution of life on Earth was continually being affected by the arrival on the Earth's surface of organic molecules from space. These ideas were dismissed by almost all biologists and physicists at the time, and are still largely rejected today.

In his later years, Hoyle concentrated on writing popular works of fiction, while continuing to promote his controversial and unpopular theories on steady-state cosmology and extra-terrestrial origins of life. He died at his home in Bournemouth on 20 August 2001.

While controversial as a scientist, Hoyle was widely respected for the conviction of his opinions and for his work in popularizing science, particularly astronomy, through his books, articles, radio and television appearances, and always-packed public lectures. He was awarded six prizes or medals, including the Crafoord Prize from the Royal Swedish Academy of Sciences in 1997 which was perhaps in partial recompense for his missing out on a share of the 1983 Nobel Prize. He authored more than 500 scientific papers, half a dozen non-fiction books, and 15 works of fiction. Hoyle's greatest scientific legacy, however, will almost certainly be his pioneering early work on nucleosynthesis of the elements inside stars.

Hoyle, F. (autobiography) *Home Is Where the Wind Blows: Chapters from a Cosmologist's Life*, Oxford University Press, 1994.

Mitton, S. *Conflict in the Cosmos: Fred Hoyle's Life in Science*, Joseph Henry Press, 2005.

HST
See HUBBLE SPACE TELESCOPE.

Hubble, Edwin (1889–1953)
Hubble's greatest contribution to *cosmology was his discovery in 1929 that the *Universe is expanding. It is no exaggeration to say that this observation completely revolutionized 20th century cosmology and provided the first major piece of evidence in support of the *big bang.

Edwin Powell Hubble (Fig. 1) was born in 1889 in Missouri. He moved to Illinois in 1898 where he became a renowned athlete before obtaining a degree in mathematics and astronomy from the University of Chicago. He then spent three years in Oxford, obtaining a Masters degree in law. On returning to the US, he spent several years as a high school teacher in Indiana and served in the First World War.

He returned to astronomy after the war, studying at Yerkes Observatory of the University of Chicago, obtaining a PhD in 1917, with a dissertation entitled 'Photographic Investigations of Faint Nebulae'. In 1919 he moved to Mount Wilson Observatory in California, where he spent the rest of his life until his death in 1953. It was at Mount Wilson Observatory that Hubble made his most important discoveries. These include:

Fig. 1. Edwin Hubble.

- Measurements of *distances to *galaxies using *Cepheid variable stars. This discovery, announced in 1924, proved that galaxies were not part of the *Milky Way Galaxy, but separate systems in themselves.
- With Milton Humason, discovery in 1929 of the *redshift–distance relation for galaxies, now known as the *Hubble law. The present rate of expansion is known as the *Hubble constant.
- A morphological classification scheme for galaxies, still in use today. (*See* GALAXY CLASSIFICATION.)

The *Hubble Space Telescope, launched in 1990, was named in his honour, and accurately determining the Hubble constant was one of its key objectives.

Christianson, G. E. *Edwin Hubble: Mariner of the Nebulae*, University of Chicago Press, 1995.

Hubble constant
The Hubble constant H_0 measures the present expansion rate of the Universe. It is named after Edwin *Hubble, who discovered the *Hubble law for the expansion of the Universe in 1929. In general the expansion rate depends on the age of the Universe and is given by the *Hubble parameter $H(t)$; Hubble's constant can be considered as the present value of the Hubble parameter $H_0 = H(t_0)$, and is thus only constant in space, not in time.

The Hubble constant relates the *redshift of a distant object to its distance. Since distances to the vast majority of extra-Galactic objects are inferred from their redshifts, the value of the Hubble constant sets the distance scale for the Universe, and hence has a direct bearing on the inferred luminosities, masses, and sizes of distant objects. The Hubble constant also sets the *age of the Universe, $t_0 \approx H_0^{-1}$, and an estimate for the size of the *observable Universe, $R_0 \approx ct_0$, where c is the speed of light. (The actual value is closer to $R_0 \approx 3ct_0$, see *horizon distance.) The *critical density ρ_c of the Universe is also dependent on H_0, $\rho_c = 3H_0^2/8\pi G$.

One of the longest-standing challenges in astronomy has been to accurately measure the Hubble constant. Hubble himself grossly overestimated it, ending up with a value around 500 km/s/Mpc, due to incorrect assumptions about the luminosities of *Cepheid variable stars yielding underestimated distances. Since the age of the Universe is inversely related to the value of the Hubble constant, this implied that the Universe was very young, incompatible with the Earth's geological record.

Subsequent estimates of H_0 revised its value downwards by a factor of 5 or 10. One group, led by Allan Sandage of the Carnegie Institutions, derived a value for H_0 around 50 km/s/Mpc. The other group, associated with Gerard de Vaucouleurs of the University of Texas, obtained values that indicated H_0 to be around 100 km/s/Mpc. The uncertainties estimated by each group precluded the other group's result, an indication that both groups were underestimating systematic uncertainties in their analyses, and an illustration of the difficulty of measuring distances on cosmological scales.

A long-term key programme for the *Hubble Space Telescope (HST) was to refine the value of the Hubble constant. Between 1991 and 2000, the HST Key Project team and others determined distances to 25 *galaxies using Cepheid variables in order to calibrate five secondary distance indicators (*see* DISTANCE LADDER): the *Tully-Fisher relation, the fundamental plane of elliptical galaxies, Type Ia *supernovae, Type II supernovae, and *surface brightness fluctuations. Once calibrated with Cepheid distances, each method can be applied to more distant galaxies and, by taking into account a model for the *peculiar velocity field, used to obtain an estimate of the Hubble constant. The reason Cepheid distances cannot be used directly to estimate the Hubble constant, is that peculiar velocities are relatively large compared to the Hubble expansion at the small maximum distances (~ 25 Mpc) to which Cepheids can be resolved, even using HST. The inferred value of H_0 would thus be extremely sensitive to the assumed velocity field. By pushing out to greater distances (60–400 Mpc) with the secondary distance indicators listed above, one's estimate of H_0 is much less sensitive to the assumed velocity field.

By combining the H_0 estimates and their associated uncertainties from the five secondary distance indicators, the HST Key Project team arrived in 2001 at the culmination of a series of 30 papers at an overall estimate for the Hubble constant of $H_0 = 72 \pm 8$ km/s/Mpc.

More recently, in 2006, the *WMAP satellite team independently arrived at an estimate

for the Hubble constant of $H_0 = 73 \pm 3$ km/s/Mpc, in excellent agreement with the HST key project. The agreement is particularly striking given that the WMAP analysis used completely different observations and model assumptions to the HST key project.

Given the until-recently large uncertainties in the value of the Hubble constant, it has become customary to express any results that depend on the value of the Hubble constant in terms of the parameter h, where $H_0 = 100h$ km/s/Mpc. So, for example, if a distance is quoted as $200h^{-1}$ Mpc, and you believe that $H_0 = 72$ km/s/Mpc, then the actual distance would be $200/0.72 \approx 278$ Mpc.

Freedman, W. et al., *Astrophysical Journal*, 553, 47, 2001.

Hubble Deep Field (HDF)

The Hubble Deep Field refers to a pair of very deep images taken with the second Wide Field Planetary Camera (WFPC2) on the *Hubble Space Telescope (HST). The original HDF was observed in the northern hemisphere, and is now sometimes referred to as HDF-N to distinguish it from a subsequent deep field observation made in the south: HDF-S.

The HDF concept was the brainchild of the then-director of the Space Telescope Science Institute (STScI), Robert Williams, in 1995. He wanted to put a significant allocation of Director's Discretionary Time into a project that would have maximum scientific impact and legacy value.

The location of HDF-N on the sky (approximately $12^h 36^m$, $+62° 12'$ in J2000 equatorial coordinates) was chosen to be optimally placed for observing by HST and to avoid regions affected by bright stars, large nearby *galaxies, and Galactic *dust. Images were taken for 10 consecutive days (approximately 150 orbits of HST) through four passbands in December 1995. Roughly 35 hours were spent observing through each of three optical filters with central wavelengths of 450, 606 and 814 nm, and 50 hours was spent with the less sensitive *ultra-violet filter around 300 nm. Shorter 1-orbit images through each filter were also obtained of the fields immediately adjacent to the primary field in order to facilitate spectroscopic follow-up by ground-based telescopes, which typically have a much wider field of view than HST's WFPC2.

Fig. 1. The Hubble Deep Field (North) image. Almost all objects visible in this image are galaxies.

Images of comparable depth (to *magnitude $V \approx 30$) have been made from the ground. What makes the HDF so successful is the combination of depth, superior image quality, and multi-passband nature of the data.

The most significant results from the HDF relate to the evolution of galaxies. In particular, the steep number–magnitude counts found in shallower ground-based surveys, indicative of galaxies being brighter in the past, appears to flatten off around $B \approx 25$. It is also striking how much blank sky the HDF-N image (Fig. 1) contains, in spite of its great depth. This means that galaxies were physically small (≈ 3 kpc) in the past, in agreement with *hierarchical structure formation. One also sees a larger fraction of irregular and multi-component galaxy images in HDF compared with shallower surveys, suggesting that star formation has been rapidly declining since *redshift $z \sim 1$.

Due to the great scientific success of the original HDF, a second deep field (HDF-S) was observed in the southern sky in 1998. Rather than randomly locating HDF-S, it was deliberately centred on a redshift $z \approx 2.24$ *quasar so that correlations between *quasar absorption system redshifts and the redshifts of galaxies in the field could be studied. Simultaneous, parallel observations were made with the three HST instruments WFPC2, the Space Telescope Imaging Spectrograph (STIS), and the Near Infrared Camera and Multi-Object Spectrometer (NICMOS), the latter two of which were not available when HDF-N was observed.

When the Advanced Camera for Surveys (ACS) instrument was installed on HST during the 2002 servicing mission, with its factor of 10 improvement in efficiency compared with WFPC2, it was decided to make an even deeper image of the Universe, known as the Hubble Ultra Deep Field (UDF, see Fig. 2). This field was also observed with NICMOS. The total exposure time was 1 million seconds (11.3 days), broken down into 800 exposures taken during the course of 400 orbits between September 2003 and January 2004. Nearly 10 000 galaxies are visible in this image, which covers an area of sky just one-tenth of the diameter of the full Moon. About 100

Fig. 2. The Hubble Ultra Deep Field image.

of the smallest, reddest galaxies are probably among the most distant known, existing when the universe was just 800 million years old, corresponding to a redshift $z \sim 7$. Most of the fainter galaxies show a disturbed, irregular appearance, compared with the spiral and elliptical galaxies we see today, indicating that they are still in the process of formation.

http://hubblesite.org/

Hubble diagram

In 1931, Edwin *Hubble and Milton Humason plotted the distances to several nearby *galaxies, determined by observations of *Cepheid variables within the galaxies, against their recession velocity, determined by the *redshift of their spectra (Fig. 1). They noticed that the points on this plot, now known as the Hubble diagram, lay roughly along a straight line, telling Hubble and Humason that distance is linearly proportional to recession speed (a galaxy at twice the distance will also have twice the recession speed). Since it is much easier to measure the redshift of a galaxy than its distance directly, the Hubble diagram provides an easy way to estimate distances to galaxies.

Although the first plot of recession velocity against distance was published in 1931, the first reference to a relation between these two quantities appears in a leaflet authored by Hubble, published by the Astronomical Society of the Pacific in July 1929.

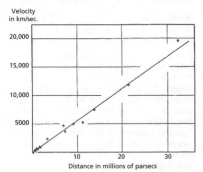

Fig. 1. A plot published by Hubble and Humason in 1931 of the mean recession velocity for galaxies in clusters or groups against the estimated distance to each cluster or group.

The slope of the best-fit line on the Hubble diagram is known as the *Hubble constant, H_0, and recession velocity v and distance d are mathematically related by the equation $v = H_0 d$. Due to a calibration error in Hubble's Cepheid distance measurements, he initially thought that H_0 had a value of roughly 500 km/s/Mpc. Today, the favoured value, as determined by the *WMAP satellite team, is $H_0 = 73 \pm 3$ km/s/Mpc.

Hubble expansion

By obtaining the spectra of a number of nearby *galaxies, Edwin *Hubble was able to measure their *redshifts and hence recession velocities. He found that nearly all galaxies showed a positive recession velocity, that is they are moving away from us, and thus inferred that the *Universe is expanding. This expansion is named after its discoverer, and is referred to as Hubble expansion or Hubble flow.

By also measuring distances to the same galaxies, by observing *Cepheid variable stars within them, Hubble was able to demonstrate that recession velocity is proportional to distance, a relation known as the *Hubble law.

It is important to note, however, that each galaxy has a *peculiar velocity relative to the Hubble flow, and so when one uses redshift as an estimate of distance, this estimate will be biased by the line-of-sight component of the peculiar velocity. Also, systems bound by *gravity, such as stars and individual galaxies, do not grow with the Hubble expansion, but remain of fixed size in *proper coordinates.

Hubble law

The linear relation between recession velocity v and distance d of a remote object,

$$v = H_0 d,$$

where H_0 is known as the *Hubble constant. This law is the unique form such that the same law is seen from *every* point within the *Universe—all observers see that *galaxies recede radially with velocity proportional to their distance. The Hubble law therefore respects the *cosmological principle, which states that there are no special observers in the Universe.

(For further information on the discovery of this law, *see* HUBBLE DIAGRAM and HUBBLE EXPANSION.)

Hubble length

The Hubble length (also known as the Hubble distance) is the characteristic length scale of the *Universe, and is obtained by multiplying the *Hubble time by the *speed of light. It therefore estimates the distance that light could have travelled since the beginning of the Universe, known as the *horizon distance.

Perhaps surprisingly, the Hubble length typically underestimates the distance that light can have travelled. The reason is that the Universe has been expanding, and so was smaller during the early stages of the light's travels, whereas the Hubble length measures the total distance in today's units. The present Hubble distance is in fact only about one-third of the horizon distance. Its value, from multiplying the Hubble time by the speed of light, is

$$cH_0^{-1} = 3000h^{-1}\mathrm{Mpc},$$

where $H_0 = 100\,h\,\mathrm{km/s/\,Mpc}$.

In considering distances much less than the Hubble distance, the cosmological expansion can be ignored. For instance, if we receive light from the Andromeda Galaxy around one megaparsec away, this is a very small fraction of the Hubble length and so the Universe will hardly have expanded at all while that light travelled to us.

As with the Hubble time, the Hubble length can be a useful concept during earlier stages of the Universe's evolution, when it will be much smaller than today. Ordinarily it gives a good indication of the distance over which causal interactions could have taken place by a given time, though this correspondence is broken during epochs where the Universe undergoes accelerated expansion, such as *cosmological inflation. Indeed, this breakdown is crucial for the inflationary solution to the *horizon problem.

Hubble parameter

The Hubble parameter measures the expansion rate of the *Universe as a function of time. Its present value is known as the *Hubble constant, which is one of the key *cosmological parameters that cosmological observations seek to determine.

Edwin *Hubble's original discovery of the expansion of the Universe in 1929 concerned the present expansion rate of the Universe. *Hubble's law relates the velocity v with which a *galaxy is moving away from us to its distance d, the law being a direct proportionality

$$v = H_0 d,$$

where H_0 is Hubble's constant. So, an object twice as far from us is receding with twice the velocity. This law corresponds to a uniform expansion of the Universe; if one observer sees all galaxies moving radially away according to the Hubble law, then *every* observer will see the same law, regardless of which galaxy they reside in. If galaxies are distributed evenly throughout the Universe, then they remain so.

The expansion rate has been changing as the Universe expands, and the above equation can be generalized to apply at any epoch (in doing so, the objects considered have to be close enough that the expansion rate does not change significantly as the light propagates from one to another), writing

$$v = H(t)d.$$

The quantity H is now known as the Hubble parameter and is a function of time, its present value being the Hubble constant. This definition can still be used even before galaxies form in the Universe, by applying it to individual particles.

The Hubble parameter can be related to the *scale factor of the Universe $a(t)$, which measures the size of the Universe as a function of time. In the language of calculus, the relation is

$$H \equiv \frac{da/dt}{a} = \frac{d\ln a}{dt}.$$

In a *cosmological model, the Hubble parameter is determined from the *Friedmann equation. This shows that the expansion rate is related to the *density of material in the Universe and to the spatial *curvature. Solving the Friedmann equation under various physical circumstances, in order to determine the Hubble parameter's evolution, is one of the most common tasks faced by cosmologists in aiming to test their models.

Hubble Space Telescope (HST)

The Hubble Space Telescope (Fig. 1) has been the most successful, as well as the best-known, optical telescope to operate from space. HST is a cooperative programme of the European Space Agency (ESA) and the National Aeronautics and Space Administration (NASA), with the Space Telescope Science Institute (STScI) in Baltimore, Maryland, responsible for conducting and coordinating science operations. The 2.4-metre reflecting telescope was deployed by the crew of the space shuttle *Discovery* in low-Earth orbit at an altitude of 600 kilometres in April 1990.

HST's current complement of science instruments includes three cameras, two spectrographs, and fine guidance sensors (primarily used for accurate positional observations). Because of HST's location above the Earth's atmosphere, these science instruments can produce high-resolution images of astronomical objects. The majority of ground-based telescopes can seldom provide resolution much better than 1 arcsecond, except momentarily under the very best observing conditions. HST's resolution is about 10 times better, at around 0.1 arcseconds.

It was originally conceived that HST would be brought back to Earth every five years for refurbishment. However, concerns over contamination and structural stresses led NASA to decide on a three year cycle of on-orbit servicing. Soon after launch, it was realized that HST suffered from spherical aberration due to a design error (the primary mirror was 2 microns too flat at its edge), severely compromising imaging quality. Fortunately, the refurbishing mission of December 1993 (Fig. 2) successfully cured the effects of spherical aberration and restored the full functionality of HST.

The following instruments are currently available on HST:

Wide Field Planetary Camera 2—The original Wide Field/Planetary Camera (WF/PC1) was replaced by WFPC2 during the 1993 refurbishing mission. Mirrors within WFPC2 are designed to compensate for the spherical aberration of the telescope. WFPC2 is actually four cameras: an L-shaped trio of wide-field sensors and a smaller, high resolution ('planetary') camera tucked in the square's remaining corner. For a famous image taken with this camera, *see* HUBBLE DEEP FIELD.

Space Telescope Imaging Spectrograph—A spectrograph spreads out the light gathered by a telescope so that its intensity can be

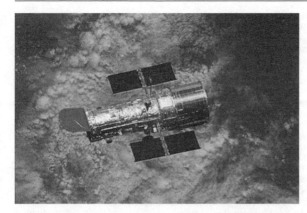

Fig. 1. The Hubble Space Telescope as seen against the Earth after a week of repair and upgrade by Space Shuttle *Columbia* astronauts in 2002.

analysed as a function of wavelength. The Space Telescope Imaging Spectrograph (STIS) covers a spectral range from the *ultra-violet (UV; 115 nanometres) through the visible, red and the near-*infrared (IR; 1000 nanometres). The field of view is 25×25 arcsec in the UV and 50×50 arcsec in the optical–near-IR. Because of HST's excellent spatial resolution, STIS can record the spectrum at many locations within a single object. This has been particularly useful in studying *supermassive black holes at the centre of galaxies.

Fig. 2. A photograph of Hubble's first servicing mission in 1993. The telescope is docked with Space Shuttle *Endeavor*.

Near Infrared Camera and Multi-Object Spectrometer—The Near Infrared Camera and Multi-Object Spectrometer (NICMOS) provides the capability for infrared imaging and spectroscopic observations of astronomical targets. NICMOS detects light with wavelengths between 0.8 and 2.5 microns—beyond the limit of perception of the human eye. The sensitive mercury-cadmium-telluride (HgCdTe) arrays which comprise the infrared detectors in NICMOS must operate at very cold temperatures. NICMOS keeps its detectors cold inside a cryogenic dewar (a thermally insulated container much like a thermos bottle) containing frozen nitrogen ice. The dewar cools the detectors for years, much longer than any previous space experiment. NICMOS is HST's first cryogenic instrument.

Advanced Camera for Surveys—The ACS comprises three instruments in one:

- a Wide Field Channel (WFC), with a field of view of 202×202 arcsec covering the optical–near-IR wavelength range 370–1100 nm.
- a High Resolution Channel (HRC), with a field of view of 26×29 arcsec covering the near-UV–near-IR wavelength range 200–1100 nm.
- a Solar Blind Channel (SBC), with a field of view of 31×35 arcsec, covering the UV wavelength range 115–170 nm.

The primary design goal of the ACS Wide Field Channel is to achieve a factor of 10 improvement in discovery efficiency, compared to WFPC2, where discovery efficiency is defined as the product of imaging area and instrument throughput. The other two channels provide new ultra-violet imaging capability.

Fine Guidance Sensors—The Fine Guidance Sensors (FGS) in addition to being an integral part of the HST Pointing Control System (PCS), provide HST observers with the capability of precision astrometry and milliarcsecond resolution over a wide range of magnitudes ($3 < V < 16.8$).

Whenever possible, two scientific instruments are used simultaneously to observe adjacent regions of the sky. For example, while a spectrograph is focused on a chosen galaxy or *quasar, the WFPC can image a region of sky offset slightly from the main

target. During observations the Fine Guidance Sensors track their respective guide stars to keep the telescope pointed steadily at the right target.

The HST is operated as an international research centre; any astronomer worldwide may apply for observing time. Competition is fierce due to high demand, and only one of every ten proposals is successful.

The observatory has had a profound impact on the field of cosmology due to its uniquely high-resolution imaging and spectroscopy. Two of the most important results from HST are the **Hubble Key Project** to determine the *Hubble constant and the discovery of supermassive black holes at the centres of most nearby galaxies.

http://www.stsci.edu/hst

Hubble time

The Hubble time is the characteristic timescale of expansion of the *Universe. It is related to the *Hubble parameter. The present value of the Hubble time gives an estimate of the *age of the Universe. It is somewhat uncertain as the value of the *Hubble constant is not yet precisely known.

The Hubble constant has units of inverse time, though this is somewhat concealed if, as is common, it is written as a velocity divided by a distance. One often sees it expressed as

$$H_0 = 100h \, \text{km/s/Mpc},$$

where h is a constant whose value is about 0.7; this form is convenient for reading off that an object which is, for instance, one megaparsec away will (ignoring *peculiar velocities) be receding from us at about 70 kilometres per second. However, one can convert the kilometres to megaparsecs, or vice versa, to cancel them off and write

$$\frac{1}{H_0} = 9.77h^{-1} \times 10^9 \, \text{years}.$$

The left-hand side of this expression has the dimensions of time, and is known as the Hubble time. Substituting in the observed value $h \approx 0.7$ gives the Hubble time as about 14 billion years. This is in excellent agreement with more precise age determinations giving the age as 13.7 billion years, though in fact the close correspondence is somewhat of a coincidence.

The Hubble time indicates the timescale on which it is necessary to take the expansion of the Universe into account. If the elapsed time is much less than it, the expansion can be ignored; for instance you don't worry about the Universe expanding during a trip to the chemist. Roughly speaking, if you were able to wait around for a Hubble time, you can expect the Universe to have approximately doubled in size by the end of it, though this type of approximation works less well during epochs where the Universe is undergoing *acceleration, as it is believed to be at the present.

The Hubble time is also a useful concept at earlier stages of the Universe's evolution, once again giving a characteristic timescale on which the expansion of the Universe cannot be ignored. During the early stages the Universe expanded much more rapidly, meaning that the Hubble parameter was much larger and hence the Hubble time shorter. During either *radiation-dominated or *matter-dominated eras, the Hubble time is always a good estimate of the age of the Universe at that time, though this is not necessarily true during a period of *cosmological inflation.

hydrodynamical simulations
See SMOOTHED PARTICLE HYDRODYNAMICS.

ICM
See INTRA-CLUSTER MEDIUM.

IGM
See INTERGALACTIC MEDIUM.

IMF
See INITIAL MASS FUNCTION.

inertial frame
An inertial frame is the reference frame of an observer who is not subject to any external forces. Such an observer moves with a constant velocity. Inertial frames have particular significance in the theory of special relativity, which states that all inertial observers experience the same laws of physics. (For more details, *see* RELATIVITY.)

*General relativity goes beyond the concept of inertial observers, who no longer have special status once *gravity is brought into play. Nevertheless, in general relativity there is still a concept of a local inertial frame, with the laws of special relativity applying at least within a small *space-time region around any observer.

inflation
See COSMOLOGICAL INFLATION.

inflationary cosmology
See COSMOLOGICAL INFLATION.

inflationary perturbations
The inflationary cosmology was introduced by Alan Guth in order to explain a number of otherwise puzzling features about the large-scale *Universe—the *flatness problem, *horizon problem and *monopole problem. However it was subsequently realized that inflation also provided a mechanism capable of creating primordial *density perturbations in the Universe, which could be responsible for *structure formation. This idea has strong support from observations, and is the most powerful evidence for the inflationary hypothesis.

Observed perturbations
To a first approximation, the Universe is usually described as smooth, with a uniform *density of material within it. Such an idealized Universe is *homogeneous and *isotropic and can be described by the *Robertson–Walker metric. However the real Universe has variations in density from point to point, which in the present Universe are quite pronounced, for instance as seen in the *clustering of galaxies. Only by averaging over very large regions can the present Universe still be approximated as homogeneous.

According to the theory of *gravitational instability, the initial variations in the density were rather small, and later grew under their gravitational attraction to cause structure formation. The small initial variations are known as *perturbations. They can be seen quite directly in the form of *cosmic microwave background (CMB) anisotropies, the small point-to-point variations in the temperature of the cosmic microwave background.

In a cosmological model without inflation, no physical mechanism is capable of generating large-scale perturbations. The argument is essentially exactly that of the horizon problem; because the *speed of light is finite it is not possible to move material over large distances and hence create inhomogeneities. This limitation is known as *causality. In the CMB, we see clear evidence of anisotropies on large angular scales (a few degrees or larger) which could not have been created by the time the CMB was released at *decoupling. In the absence of inflation, the initial perturbations would have to already exist as part of the initial conditions of the Universe, and no physical theory predicting them could be derived. (A possible escape route from this argument is if the speed of light is not a universal constant, but instead might have changed as the Universe expanded and have been much larger earlier on. *See also* VARYING FUNDAMENTAL CONSTANTS.)

Perturbation generation
Two key ingredients make it possible for inflation to generate density perturbations in the Universe. One is the extremely rapid

expansion, characterized by the *scale factor growing more rapidly than the *Hubble length. The second is that the Universe is *quantum mechanical, with everything in it subject to quantum processes governed by Werner Heisenberg's famous Uncertainty Principle.

The rapid expansion is essential for circumventing the causality argument above. The scale factor determines how physical scales are stretched as the Universe expands, while the Hubble length governs the scale of causal interactions. If perturbations are generated on small scales, much less than the Hubble length and hence consistent with causality, the inflationary expansion can stretch them to much larger scales, comparable to or even much greater than our present *observable Universe.

However if our Universe obeyed only the laws of classical (i.e. non-quantum) physics, even this would not be enough, because classical physics would say that if the Universe started out homogeneous, then it would have to stay so since each location would evolve in the same way according to those laws. Fortunately, in the quantum world such irregularities can be generated spontaneously thanks to the chance element of quantum processes. These are known as **quantum fluctuations**, and all materials are subjected to them.

Density/curvature perturbations

The most important type of perturbations are density perturbations, as these are responsible for structure formation. In the inflationary cosmology, these are generated from quantum fluctuations in the material driving inflation, normally called the *inflaton. The quantum fluctuations get caught up in the inflationary expansion and stretched to very large scales, where ultimately they become 'frozen in' as density perturbations.

Perturbations are quantified by two properties, their length scale (distance from peak to peak, for example) and their amplitude (the difference between the density in the peaks and troughs). Perturbations on the largest length scales are generated first, as they are the ones which receive the greatest inflationary expansion after their creation.

In the simplest inflation models with only one scalar field, the perturbations are of a type known as *adiabatic perturbations, corresponding to genuine point-to-point variations in the density. Observations favour this type of perturbation. Such perturbations are also known as *curvature perturbations, since the variations in density induce variations in the geometric *curvature of the Universe as according to *Einstein's equation. Indeed, it is typically preferable to describe these perturbations geometrically using the curvature, rather than by using the density.

Inflation predicts that there will be density perturbations across a very wide range of length scales, ranging from those much bigger than the observable Universe, generated in the early stages of inflation, to those much smaller than the observable Universe generated near the end. However observations of structure formation are sensitive only to those perturbations on scales ranging from typical galaxy separations (a megaparsec or so) up to the size of the observable Universe. Those perturbations were generated some time before inflation ended, when the Universe was roughly 10^{-20} of the size it would reach by the end of inflation.

The amplitude of perturbations on a given scale depends on the inflation model under consideration, i.e. on the *potential that governs it. The amplitude is quite well measured by CMB anisotropy experiments, and can typically be matched in any inflationary model by adjusting the energy scale (i.e. the epoch) at which it takes place. More interesting is how the amplitude of perturbations varies with length scale, which is quantified by the *spectral index of the density perturbations, usually indicated by the symbol n. Inflation models tend to make specific predictions for n, which can be tested against observations, and indeed many inflation models have been ruled out as viable models because they predict a spectral index at odds with observation.

Although the perturbations start out as quantum fluctuations, by the end of inflation their quantum nature is no longer important and they can be thought of as perturbations obeying classical physics laws. The process of conversion from quantum to classical is sometimes called **decoherence**. It has a laboratory analogue in the creation of particles in strong electric fields, and mathematically is also very similar to the process of Hawking evaporation of particles from *black holes.

Tensor perturbations

*Tensor perturbations, also known as primordial *gravitational waves, are a second kind

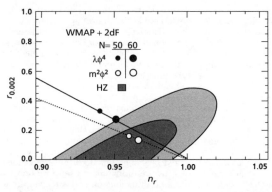

Fig. 1. The constraints on the inflationary parameters n and r from three years of observations by the WMAP satellite. The dark shaded region shows the 68% confidence limit, and the light shaded region 95%.

of perturbation automatically generated by inflation. Rather than corresponding to quantum fluctuations in the material driving inflation, they instead are generated by quantum fluctuations in the *metric of space-time itself, which is also not immune to the uncertainty principle.

Tensor perturbations do not lead to structure formation, but do generate CMB anisotropies. Observations, such as those of the *WMAP satellite, have seen no evidence of the existence of tensor perturbations, but the upper limits set so far are quite weak and permit as much as one third of the large-angle anisotropies to be due to tensors. Future experiments including the *Planck satellite will either discover the existence of tensors, or significantly tighten the observational bounds.

Tensor perturbations are inevitably produced by inflation, but there is no definite prediction of how significant they will be, which varies from model to model. In general they are suppressed with respect to the density perturbations, and in some models their amplitude is too small ever to be observed, while other models predict a level easily accessible by upcoming observations. Evidence for tensor perturbations would be extremely strong support for the inflationary scenario.

Tensor perturbations are usually quantified by the tensor-to-scalar ratio r (*scalar perturbations being yet another name for density/curvature perturbations). Inflation models usually make a definite prediction for this quantity along with the spectral index n. Testing inflation models therefore amounts

to predicting the values of n and r from the model, and then assessing which values of these parameters are allowed. Fig. 1 shows the current constraints on n and r from the WMAP satellite.

If tensor perturbations are detected, there is a further powerful test that can be attempted, known as the **consistency equation**. The simplest models of inflation predict that the density perturbations and the tensor perturbations are related to one another, as they are both generated by the same inflationary potential energy. The consistency equation applies independently of the actual form of the potential. Verifying it however requires rather accurate measurement of the tensor perturbations, determining not just their amplitude but also how that amplitude varies with length scale (known as the tensor spectral index). This will be difficult to achieve even if they are detected.

Isocurvature perturbations

The simplest inflation models, based on a single scalar field, predict adiabatic density perturbations, tensor perturbations, and nothing else. However, more complicated models of inflation which may feature more than one scalar field can give rise to additional perturbations known as *isocurvature perturbations. This is because there will be independent perturbations in each scalar field, so that in addition to perturbations in the total density there may be perturbations in the relative densities of the scalar fields which leave the total density unperturbed.

How such perturbations affect observations depends on how the density in the scalar fields is converted into the materials present in the later Universe, e.g. one field might decay into *dark matter and the other into *baryons, *photons and *neutrinos. There is presently no observational evidence that isocurvature perturbations exist in our Universe, and observational bounds are becoming quite strong.

A particular variant on isocurvature perturbations is the *curvaton scenario, in which the perturbations in the inflaton field itself are negligible and all the observed structures are postulated to have arisen instead from perturbations in a second scalar field. This model relies on a conversion of isocurvature perturbations generated during inflation into adiabatic perturbations sometime afterwards.

inflaton

The word 'inflaton' is used as a shorthand for the material responsible for causing *cosmological inflation. Usually inflation is thought to be caused by a *scalar field, and then the word may be used either for the field itself, or for the fundamental particles that the field is describing. The origin of the word comes from particle physicists' fondness for naming particles to end in -on, examples being *electron, *neutron, and *proton.

Presently the identity of the inflaton is unknown, though ultimately it may be associated with a particular field predicted by some fundamental theory such as *M-theory.

infrared (IR)

The term infrared refers to radiation in that part of the *electromagnetic spectrum which has wavelength in the approximate range 1 mm to 1 micron. It is familiar to us on Earth as the heat felt from the Sun or from a toaster. The radiation from evolved stars peaks in the near-infrared (around 2 microns). At longer wavelengths, infrared light is a valuable tracer of *dust that has been heated by starlight.

InfraRed Astronomical Satellite (IRAS)

The InfraRed Astronomical Satellite was a joint project of the US, the UK, and the Netherlands. Launched in January 1983, and with a mission lifetime of 11 months, the IRAS mission performed an unbiased, sensitive, nearly all-sky survey in four mid–far *infrared passbands centred on 12, 25, 60, and 100 μm (see Fig. 1).

IRAS increased the number of then-catalogued astronomical sources by about 70%, detecting about 350 000 infrared sources. By surveying in the infrared part of the *electromagnetic spectrum, IRAS was able to observe galaxies behind the plane of the *Milky Way, and thus provide the first nearly whole-sky *galaxy survey. An infrared-selected sample of galaxies also provides a different perspective: galaxies that are bright in the mid-infrared tend to be young, dusty, and star-forming compared with those selected in the optical. The *selection function of IRAS galaxies peaks at relatively low *redshift, as the survey is not very deep, but includes a number of infrared-luminous galaxies to quite high redshift.

A *redshift survey based on the IRAS Point Source Catalogue, the IRAS PSCz Redshift Survey, mapped out the three-dimensional distribution of 15 000 galaxies over 83% of the sky. Some important results from this redshift

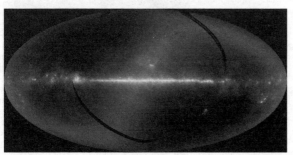

Fig. 1. Almost the entire sky as seen by the IRAS satellite, in a Hammer-Aitoff projection in Galactic coordinates. The horizontal band across the centre of the image is the Galactic plane. The diffuse S-shaped feature is emission from dust in the zodiacal plane of the Solar System. Black stripes are regions of the sky that were not observed before the satellite ran out of liquid helium coolant.

survey include measurements of the *cosmic dipole and the *power spectrum of galaxies.

http://irsa.ipac.caltech.edu/IRASdocs/iras.html

inhomogeneities

Material in the Universe is, on the largest scales, distributed smoothly, corresponding to *homogeneity and *isotropy. However there are deviations from this perfect smoothness, and they are known as inhomo-geneities or *density perturbations. Such inhomogeneities are typically enhanced by *gravitational instability, whereby overdense regions exert extra gravitational force on their surroundings and draw more material in, leading to *structure formation.

initial mass function (IMF)

The initial mass function describes the distri-bution of the masses of stars when they first start fusing hydrogen to helium inside their cores. An assumed IMF is a required input into models of *galaxy formation and evolu-tion, such as *semi-analytic galaxy formation models.

Observationally, it is very hard to deter-mine. All one can do is to measure the present-day stellar *luminosity function and, using models of how we think stars evolve, work backwards to estimate how many stars were made at each mass.

The simplest IMF is known as the Salpeter IMF, after its proposer, Edwin Salpeter, a sim-ple power-law of the form:

$$\xi(M) = \xi_0 M^{-2.35}.$$

Glenn Miller and John Scalo proposed alter-native forms for the IMF, known as the Scalo and Miller–Scalo IMFs, in which the power-

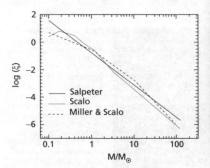

Fig. 1. Three commonly used initial mass functions.

law index varies according to the mass range being considered (see Fig. 1).

initial singularity
See BIG BANG.

integrated Sachs–Wolfe (ISW) effect
See SACHS–WOLFE EFFECT.

intergalactic medium (IGM)

Except for those *galaxies within a com-pact *galaxy group or near the core of a *cluster of galaxies, a typical galaxy is 10–100 times smaller than the distance to its near-est neighbour. The vast majority of space is thus empty of stars and galaxies, and indeed the average *density of the *Universe is less than one atom per cubic metre. However, intergalactic space is not entirely empty: a rarefied gas with a filamentary structure is thought to connect the high-density regions of the Universe where galaxies lie. This gas, which mostly comprises ionized hydrogen, has a density around 10-100 times the average density of the Universe and is known as the intergalactic medium (IGM). See Fig. 1 for the distribution of hydrogen gas in a simu-lation. Despite the extremely low density of the IGM, the huge volume of 'empty' space means that it weighs more than the com-bined mass of all of the known stars and galaxies.

As we attempt to look further and fur-ther out into space, we are looking through more and more of this gas. Fortunately for

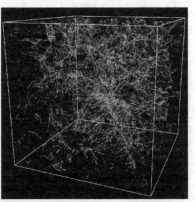

Fig. 1. The density of hydrogen gas in a simulated region of the Universe $25h^{-1}$ Mpc on a side at redshift $z = 4$. Note the filamentary distribution of the gas.

observational *cosmology, the IGM is highly ionized; that is the hydrogen atom is split into its constituent *proton and *electron. In this ionized state, the IGM absorbs very little light, allowing us to see objects at great distances.

The IGM is not uniformly ionized, however. Some regions, often known as 'clouds', are not so highly ionized. This variation in ionization is due to ionizing *ultra-violet radiation from both *quasars and giant bursts of star formation in the young Universe. The neutral hydrogen within the clouds absorbs light from background quasars at wavelengths that depend on the *redshift of the cloud, leading to a characteristic absorption-line spectrum. If the light passes through several clouds at different redshifts, one sees a 'forest' of absorption lines at different wavelengths known as the *Lyman alpha forest. Such clouds are known as *quasar absorption systems.

By observing in the ultra-violet part of the *electromagnetic spectrum, one can also detect helium in the IGM. (Possessing two electrons, it takes more energy to completely ionize a helium atom.) On the whole, helium is detected at the same locations as neutral hydrogen. However, there exist some more highly ionized parts of the IGM where only helium is detected. The distribution of these partially ionized regions of the IGM match theoretical models of *structure formation in the young universe.

Miralda-Escudé, J. *Science*, *293*, *5532*, p. 1055, 2001.

http://fuse.pha.jhu.edu/wpb/sci_h2347.html

intra-cluster medium (ICM)

The intra-cluster medium refers to the general environment within a *cluster of galaxies, usually particularly referring to the hot gas which has failed to collapse to form stars, but which is held in place by the gravitational attraction of the cluster environment.

This gravitational attraction (known as the cluster *gravitational potential well) is primarily due to the presence of *dark matter in the cluster, which however cannot be observed directly. The gas, by contrast, can be observed as it gains so much energy falling into the cluster that it is heated to many millions of degrees, leading to emission in the *X-ray part of the *electromagnetic spectrum which is detectable by space observatories such as *XMM-Newton and *Chandra. Observations can measure both the temperature of the gas (from the spectrum of the emission) and the *density (from the total emitted luminosity given the spectrum). The gas offers a simple way of estimating the total mass of the galaxy cluster, including the dark matter; by assuming it is in a static configuration one can balance the gravitational attraction against the gas pressure inferred from measuring its temperature. Such mass estimates play an important role in using galaxy clusters to probe *cosmological models.

One particular ICM property of interest is the *cluster baryon fraction, which measures the fraction of the total cluster mass due to *baryons, i.e. stars and gas. As galaxy clusters are so large, this provides a good estimate of the Universal balance between baryonic and dark matter.

As observations have improved, it has become necessary to model the ICM evolution much more carefully. Modern treatments will include the effect of cooling of the gas as it radiates its X-rays, and heating of the gas due to *supernovae and *active galactic nuclei (effects collectively known as *feedback). *Cosmic rays may also play a significant role in heating the ICM. *Numerical simulations of cluster formation and evolution attempt to model these processes as accurately as possible.

Another important property of the ICM is its *metallicity (where in astronomer-speak 'metal' means any element heavier than helium). Such metals are believed to be formed only within stars, and released via stellar winds and supernovae. However, the gas in the ICM is observed to be enriched with metals (both because of characteristic emission lines due to metals in the X-ray spectra, and from absorption features seen in the spectra of quasars lying behind clusters), with an overall metallicity of about one-third that of the Sun. The exact mechanism of this enrichment is unclear, but it does indicate that material from within galaxies can be recycled into the ICM with quite a high efficiency.

ionization history

The ionization history of the *Universe measures the extent to which the atoms in the Universe are ionized, as a function of epoch. The lowest energy state of atoms is always a neutral state, where they contain the same number of *electrons as *protons to give electrical neutrality. Ionization refers to the possibility that atoms may lose one or more of their electrons when in high-energy environments, for instance bathed in hot *black-body radiation. A state of complete ionization is often called a **plasma**.

The early Universe is an environment of high enough energy to ensure full ionization. The cooling of the Universe first allows atoms to form at the epoch of *recombination at an age of about one hundred thousand years, shortly before the release of the *cosmic microwave background (CMB) radiation. During this epoch, the Universe makes a rapid transition from the fully ionized state to one of near neutrality, though a small amount of *residual ionization is left over due to the final electrons being unable to find their partner nuclei. This near-neutral state persists to the much more recent past, when the process of *reionization is induced by the first structures to form. This returns the Universe to a highly ionized state. In the *intergalactic medium this situation remains right up to the present, though within *galaxies material has cooled sufficiently to restore much of it to the neutral atomic state.

The state of ionization of a cloud of gas determines how readily radiation can travel through it. Low-energy *photons such as those of the CMB are unable to interact at all with atoms, as they lack the energy to excite the electrons between energy states. They therefore pass unaffected through regions containing neutral gas. However after reionization takes place, they can interact with the population of free electrons thus created, meaning that some will scatter. This effect is quantified by the reionization *optical depth, which is measured to show that approximately 10% of CMB photons scatter before reaching us.

For higher-energy photons the situation may be reversed; if their energies happen to correspond to particular transitions between energy levels the photons can interact much more strongly with neutral atoms than with free electrons. An example is *Lyman-alpha photons, corresponding to electron transitions between the ground state and the first excited state. Such transitions are seen in *quasar absorption systems, where the features seen in *quasar spectra are sensitive to the amount of neutral gas present but are not measurably affected by the presence of ionized gas. The overall state of ionization of

the Universe can be probed out to the most distant known quasars through a technique called the *Gunn–Peterson effect.

While simple treatments may just follow the overall ionization state, more sophisticated ones will allow for the possibility of different types of atom being in different ionization states, at least separately considering hydrogen and helium-4 which are the most abundant elements in the Universe. Having two electrons, helium has two separate ionization states corresponding to the loss of one or both electrons. The energy required to establish either of those states is higher than that needed to ionize hydrogen, so during some epochs of the Universe one may find that hydrogen is fully ionized while helium-4 is either in its neutral state or only singly ionized. Very detailed calculations of *CMB anisotropies need to follow each species separately through the recombination process.

The ionization history determines a related quantity known as the *visibility function of the CMB, which indicates the epoch at which a given photon, seen by us at the present, last scattered on its trip towards us.

IRAS
See INFRARED ASTRONOMICAL SATELLITE.

irregular galaxies
See GALAXY CLASSIFICATION.

isocurvature perturbations
Isocurvature perturbations are a particular type of *perturbation to the *density of material in the *Universe. In an isocurvature perturbation, the spatial variations in the density of two types of material (e.g. *cold dark matter and radiation in the form of *photons) are initially equal and opposite magnitude, so that the total density remains constant even though each species separately has spatial density variations. There are several different types of isocurvature perturbation, as the balance can be between any two types of material.

The name arises because the spatial *curvature is initially unperturbed ('iso' means 'same', so different regions of the *Universe have the same spatial curvature). There is however still curvature of the full four-dimensional *space-time. In older cosmological literature a concept of **isothermal perturbations** was sometimes discussed, meaning perturbations where the temperature stayed constant, but such perturbations can always be re-expressed as isocurvature perturbations and the phrase is seldom seen nowadays.

Because initially the total density is constant, there is no gravitational force. However the different rates of *redshifting of the materials will change this, partially converting the isocurvature perturbations to *adiabatic perturbations which will then grow under *gravitational instability. However *structure formation with isocurvature perturbations is much less efficient than with adiabatic ones.

There is presently no observational evidence in favour of isocurvature perturbations, adiabatic being the preferred paradigm. Nevertheless, it remains possible to have a mixture of both types provided the isocurvature ones are subdominant, and some types of *inflation models predict this.

See also DENSITY PERTURBATIONS.

isotropy

The term isotropy means that the *Universe looks the same is all directions. The requirement is usually coupled with *homogeneity, which requires that all points in the Universe are equivalent to each other. If both conditions are met, the Universe is described by the *Robertson–Walker metric.

Homogeneity does not imply isotropy. An example of a scenario which is homogeneous but not isotropic is a Universe pervaded by a uniform electric or magnetic field. Each point is the same as any other, but the field lines will select out a special direction. Likewise, isotropy does not imply homogeneity; any spherically symmetric distribution is isotropic as seen from its centre. Only if there is isotropy around *every* point is homogeneity implied.

Our Universe is not precisely isotropic, but rather shows deviations from it, for instance there being *galaxies in some directions and not in others. Nevertheless isotropy is quite closely respected on large scales; any two regions of the sky of say one square degree contain a similar number of galaxies of a given brightness.

Deviations from isotropy are known as *anisotropies, one of the most important types being the variations in the intensity of the *cosmic microwave background radiation from different directions.

James Webb Space Telescope (JWST)

The James Webb Space Telescope, named after a former NASA Administrator, and previously known as the Next Generation Space Telescope, is a large, *infrared-optimized space telescope, scheduled for launch in 2013. It is seen as a natural successor to the *Hubble Space Telescope. JWST will have a large mirror, 6.5 metres in diameter, and a sunshield the size of a tennis court (see Fig. 1). Neither the mirror nor sunshield will fit into the rocket fully open, so both will fold up and open only after reaching orbit, about 1.5 million km from the Earth.

JWST will carry four main instruments:

Mid-InfraRed Instrument (MIRI)—An imager/spectrograph covering the wavelength range 5 to 27 microns.

Near-InfraRed Camera (NIRCam)—An imager combining a large field of view with high angular resolution, covering a wavelength range 0.6 to 5 microns.

Near-InfraRed Spectrograph (NIRSpec) —Capable of observing spectra over a wavelength range of 1 to 5 microns of more than 100 objects simultaneously in a 9-square-arcminute field of view.

Fine Guidance Sensor (FGS)—Incorporates a wide-field camera that provides narrowband imagery with a filter that can be tuned over a wavelength range of 1.6 to 4.9 microns (with a gap between 2.6 and 3.1 microns).

The JWST science goals are divided into four themes, two of which are directly relevant to *cosmology. The key objective of **The End of the Dark Ages: First Light and Reionization** is to identify the first luminous sources to form and to determine the *ionization history of the young Universe. The key objective of **The Assembly of Galaxies** is to determine how *galaxies, and the *dark matter, gas, stars, metals, morphological structures, and *active galactic nuclei within them, evolved from the epoch of *reionization to the present day. The other two themes are **The Birth of Stars and Protoplanetary Systems** and **Planetary Systems and the Origins of Life.**

http://www.jwst.nasa.gov/

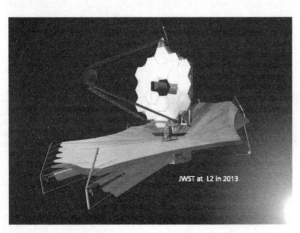

Fig. 1. Artist's impression of the James Webb Space Telescope in orbit. The segmented primary mirror points to the left. Beneath this is the large sunshield which protects the sensitive instruments from direct sunlight.

Jordan–Brans–Dicke gravity theory

The vast majority of cosmological research assumes that Albert *Einstein's theory of *general relativity is a valid description of the gravitational force. Certainly that theory has proven extremely successful in every environment it has been tested in. Nevertheless, some of the physical regimes of *cosmology are more extreme then those in which relativity has been tested, and there remains the possibility that Einstein's theory of *gravity is modified in those circumstances.

The best known, and simplest, modification to Einstein's theory was described by Carl Brans and Robert Dicke in the 1960s, and was long known as the Brans–Dicke theory. It was subsequently discovered that the idea had already been discussed by Pascual Jordan in 1955, and the theory is now commonly known as the Jordan–Brans–Dicke (JBD) theory. The idea behind the theory is that the strength of gravity, indicated by Newton's constant G, becomes a variable, and its value can be influenced by the distribution of matter just in the way the *curvature of space can. That is to say, the strength of the gravitational force can vary both with position and with time. The theory introduces a new dimensionless parameter, called ω (Greek omega), which determines the extent to which the strength of gravity can vary, and in the limit where this parameter becomes very large Einstein's theory is recovered.

JBD theory has implications both in cosmology and for the gravity around massive objects such as our Sun. The Sun distorts the strength of gravity, which becomes slightly weaker close to the Sun. Various experiments have been carried out in order to test whether there is indeed such a variation, without any detection being made. The strongest current limit comes from an experiment where radio waves were bounced off a satellite called Cassini when it passed behind the Sun (as seen from Earth) en route to carry out observations of the planet Saturn. Those observations indicated that the value of ω must be at least 40 000, implying that if JBD theory is valid it must be a version whose predictions are extremely close to those of Einstein's theory. [One might reasonably argue whether 40 000 really is close to infinity, as required to recover general relativity, since the difference between those numbers is still infinity. However what really matters is not the value of ω, but rather the value of $1/\omega$, and $1/40\,000$ is indeed rather close to zero.]

In a cosmological context, JBD theory opens up the possibility that the strength of gravity varies with time, the typical effect being that gravity was weaker in the past. Limits on this variation can be obtained from the way this modifies the *gravitational instability process which leads to *structure formation, but such limits are presently much weaker than the Cassini limit above. This implies that the predictions of JBD theory are essentially identical to those of general relativity as far as cosmology is concerned.

More complex gravity theories are also possible, often called **extended gravity theories**. For example, if the Brans–Dicke parameter ω is itself allowed to vary with position we have what is called a scalar–tensor theory. One possibility that such theories allow is that ω was much smaller in the past while still satisfying the present Cassini bounds, so that nonstandard gravity effects could have been much more significant in the early Universe. There is however no indication from observations that such modifications to gravity did indeed take place.

Will, C. *Theory and Experiment in Gravitational Physics*, Cambridge University Press, 1993.

JWST

See James Webb Space Telescope.

K correction

The K correction is a correction that is applied so that the absolute magnitudes of objects at different *redshifts are measured in the same restframe passband. The term 'K correction', and its definition, was introduced in a monumental 74-page paper published in 1956 by Milton Humason, Nicholas Mayall, and Allan Sandage of Mount Wilson Observatory.

K corrections come about because the wavelength of light emitted by an object at redshift z will have increased by a factor $(1 + z)$ by the time the light reaches the observer. This means that light detected at optical wavelengths may have been *emitted* in the *ultra-violet (UV) part of the *electromagnetic spectrum. Most *galaxies emit less light in the UV than the optical (see Fig. 1), and so the inferred luminosity will be underestimated, unless one applies a K correction. This correction depends on the object's *spectral energy distribution (SED) s, its redshift z, and the observing passband b. The correction is applied in the sense $M_{corr} = M_{obs} - k(s, b, z)$. For an early-type Sa *spiral galaxy observed in blue light at moderate red-shift, $z \lesssim 0.5$, $k(z) \approx 3z$. As a rule of thumb, moderate-redshift K corrections are larger for blue than red passbands, and for 'early-type' (e.g. elliptical) spectra than late-type (e.g. Sc) spectra. When observing in the *infrared, beyond the peak in a galaxy's SED, the K correction can be of the opposite sign, thus correcting a magnitude that would otherwise be too *bright*. The is also true of an optical-band K correction for an extremely blue object, such as the one in the top-left panel of Fig. 1.

Kaluza–Klein theories

See EXTRA DIMENSIONS.

Keck Observatory

The Keck Observatory, named after the W. M. Keck Foundation which provided grants totalling more than 140 million US dollars, comprises a pair of 10-metre telescopes on the summit of Mauna Kea, in Hawaii (see Fig. 1, over page). Each of the identical telescopes, the largest individual telescopes in the world, is over eight storeys tall and weighs about 300 tonnes. The primary mirrors are each composed of 36 hexagonal segments which can be individually angled to a precision of 4 billionths of a metre (4 nm) to compensate for deformations caused by the pull of *gravity as the telescope tracks objects across the sky. Telescope operators control the telescopes at the summit while observers acquire data remotely at the observatory headquarters in Waimea, Hawaii. The first telescope was commissioned in 1993, with the second following three years later.

A suite of optical and *infrared instruments are available, including four optical spectrographs (which are also capable of taking direct images); adaptive optics systems, using both natural and laser guide stars, which enable much of the image degradation due to the Earth's atmosphere to be removed; near-infrared cameras and spectrograph; and a mid-infrared interferometer. The last of these combines light from the two Keck telescopes, providing the effective

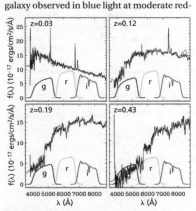

Fig. 1. The spectra of four galaxies at various redshifts as labelled, along with the Sloan Digital Sky Survey g, r and i filter response functions.

Fig. 1. The twin Keck telescopes on the summit of Mauna Kea on the Big Island of Hawaii.

resolution of a telescope 85 metres in diameter. This enables Keck to measure the diameters of stars and to detect extra-solar planets.

The combination of a large light collecting area with high resolution imaging enables the Keck Observatory to observe the most distant *galaxies and *quasars in great detail.

The DEEP (Deep Extragalactic Evolutionary Probe) project used the Keck DEIMOS spectrograph to study the detailed kinematics of galaxies at redshift $z \approx 1$, providing powerful constraints on the formation of galaxies.

http://www.keckobservatory.org

large-scale structure
See COSMIC STRUCTURE.

last-scattering surface
The last-scattering surface is the location from which the *cosmic microwave background radiation originates. It delineates the edge of the *observable Universe.

During the early hot stages of the *Universe's evolution, the Universe was in an ionized state where *electrons were not attached to individual nuclei. During this epoch (*see also* COSMIC MICROWAVE BACKGROUND) any radiation present would interact strongly with the free electrons, bouncing around in a random walk. The Universe was therefore highly opaque. At the time of *recombination, the energy had dropped sufficiently that electrons began to join with nuclei to form atoms; since low-energy *photons cannot interact with atoms the Universe then became transparent.

For any individual photon of light, there would be a final interaction with an electron, after which the photon travelled freely, perhaps eventually to be detected by us. (An exception is that some photons may have scattered again in the nearby Universe after *reionization of the Universe had been induced by the early stages of structure formation.) This epoch is known as *decoupling. As it happened when the Universe was around one-thousandth of its present size, the photons have been travelling uninterrupted for most of the lifetime of the Universe. Those we receive presently on Earth therefore originated on the surface of an extremely large sphere centred at our location. There is nothing special about that particular sphere; every location in the Universe has its own last-scattering surface located around it.

If recombination of the electrons were instantaneous, the last-scattering surface would be very sharp, the sphere being of essentially zero thickness. This is not a bad approximation, but in practice the finite duration of recombination means that the last-scattering surface has a finite width, with some photons undergoing their last scattering sooner than others. This finite width is responsible for suppressing *cosmic microwave background anisotropies on very small angular scales, and is related to *Silk damping.

leptons
Lepton is a generic name for one of the types of fundamental particle. It refers to particles which interact via the electromagnetic and weak nuclear forces, but not the strong nuclear force. The most familar example is the *electron. The leptons occur in pairs, the electron's partner being the electron *neutrino, and there are three such pairs, sometimes called generations. The second and third pairs are essentially replicas of the first pair but with a higher mass.

According to the *standard model of particle physics, the fundamental constituents of Nature are leptons and *quarks, with further particles mediating the interactions between them.

Lick galaxy catalogue
The Lick galaxy catalogue, also known as the Shane and Wirtanen catalogue, is remarkable for containing roughly one million *galaxies compiled entirely by eye, without the aid of computer digitization. The catalogue is named after Lick Observatory in California, whose 20-inch Carnegie astrograph (a telescope designed for photographing wide areas of sky) was used to obtain the survey plates.

The catalogue includes 1 246 photographic plates covering the sky above declination $\delta = -20$ degrees, each plate covering 6 by 6 degrees of sky, with the plate centres separated by 5 degrees. This overlap between neighbouring plates allows the sensitivity of each plate to be approximately calibrated. Galaxy counts were made by mounting each plate in a frame with north–south motion. A microscope could move in the east–west direction, with scales marked

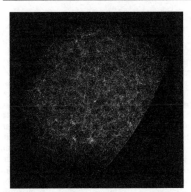

Fig. 1. A map of galaxy counts in the Lick catalogue centred on the north Galactic pole. The grey-scale represents the numbers of galaxies in 10 × 10 arcminute regions.

at one arcminute increments. Cross-hairs within the microscope marked out a square 10 arcminutes on a side.

In a truly heroic effort, Charles Donald Shane and Carl Wirtanen used this set-up to count the number of galaxies in each of the 36 × 36 10 arcminute cells on each of the 1 246 plates. (Nicholas Mayall counted galaxies on 5 plates early on in this programme.) Shane and Wirtanen estimated that the magnitude limit of their counts was approximately 18.4, while recognizing that some *low surface-brightness galaxies brighter than this would have been missed and that some plates would go fainter than others. Several corrections were made to the galaxy counts: an off-axis correction to account for vignetting and poorer image quality away from the centre of each plate, an east-west gradient in galaxy density on each plate of obscure origin, allowance for atmospheric extinction and exposure time, a factor for the type of photographic emulsion used, and a factor to calibrate the three observers to a consistent count rate. The overlap regions between neighbouring plates proved vital in arriving at these correction factors.

Contour maps of the galaxy density and tables of the numbers of galaxies in one square degree cells (made by simply summing the counts in 6 × 6 arrays of 10 arcminute cells) were published in a series of papers between 1954 and 1967. The catalogue was used by Shane and Wirtanen to determine the surface density of galaxies in galaxy *clusters and to investigate interstellar *extinction in the Galaxy. They arrived at a mean interstellar extinction towards the Galactic poles of 0.46 magnitudes, substantially larger than the value of 0.25 magnitudes assumed by Edwin *Hubble.

In 1977, Michael Seldner and colleagues at Princeton University re-reduced the original counts of Shane and Wirtanen, deriving new correction factors, and published the Lick galaxy counts at the full resolution of 10 arcminutes in a famous poster entitled 'One million galaxies' (Fig. 1). Ed Groth and Jim Peebles, also at Princeton University, measured the two- and three-point angular *correlation functions for the Lick galaxy catalogue. They found that the two-point function was well-represented by a power-law (a straight line in a log-log plot) on small scales, out to about 2.5 degrees, dropping sharply on larger angular scales. This sharp cut-off in large-scale power was cited as evidence in favour of the *critical density *cold dark matter model, in which the mass of the *Universe is dominated by massive, non-*baryonic particles.

This remained the definitive measurement of the galaxy angular correlation function until the *APM galaxy survey in 1990. With the aid of computer digitization and improved calibration data, it became apparent that part of the real large-scale *clustering signal had been removed by Seldner et al. in applying their correction factors to the Lick catalogue, and that the power-law in the correlation function at the depth of the Lick survey extends to nearly 10 degrees.

light cone

A light cone is a concept from *relativity theory, both special and *general relativity. It uses the fact that nothing can travel faster than the *speed of light to divide the *spacetime relevant to particular observers, such as ourselves, into separate sections, depending on whether signals can be sent or received by the observer. It illustrates the property of *causality.

Light cones are most easily illustrated in a space-time diagram. Such a diagram illustrates the properties of space-time, usually simplified by ignoring one or even two of the three space dimensions. Time is shown flowing up the page and distance across (and perhaps into) the page. The axis scales are

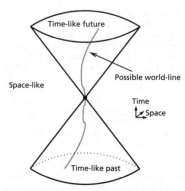

Fig. 1. A light cone shown in a space-time diagram, where one of the three spatial dimensions has been omitted. The observer sits at the apex of the two cones, which represent light rays travelling radially inwards and outwards of that point. The cones divide the space-time into three regions, and the observer is restricted to motion within the past and future light cones.

usually chosen so that a beam of light travels at an angle of 45 degrees, see Fig. 1. The space-time diagram is designed to show *events*; points in the diagram represent things happening both at specific places and specific times.

Consider our present location, both in space and in time, and imagine that we send out a beam of light in all directions. In a space-time diagram with one dimension suppressed, the rays of light trace out a cone, the light cone, with our location at the apex. Equally, suppose that we receive light which at some time in the past was sent towards us from all directions; this traces a mirror-image cone in the past, the two cones meeting at our space-time location. Note that these light cones are specific to us; someone else at a different location will have their own light cone. Indeed, as time passes our own light cone will evolve, the future light cone always shrinking and the past light cone always expanding.

These two cones partition space-time into three regions. Inside the forward cone is called **time-like future**. It consists of all regions of the space-time to which we are able to send a signal. If our signal travels at the speed of light it will follow the cone itself; if it is slower it will lie within the cone. The backward cone is **time-like past**, and consists of all the regions of space-time that could have sent

a signal to be received by us at the present time.

The remaining region, outside those two cones, is the **space-like** region. We are unable to communicate with those regions of space-time, and they with us; the regions are said to be **causally disconnected** from us. In thinking about this, be sure to remember that we are talking about events in space-time, not just space. For example, if you want to send a message to the star Alpha Centauri (four light years away) that gets there next Tuesday, you just cannot do it. You cannot even in principle do anything that will affect it that soon. That's not to say you can never communicate with Alpha Centauri, but if you do, you have to wait at least four years into the future, by which time it may be at the same location in space, but it is at a different one in space-time.

We can also describe our motion relative to the light cone. Since we ourselves can never exceed the speed of light, whatever we do we are always compelled to stay within our future light cone as we move into the future. Likewise, we must have reached our present point from some location in our past light cone. The path that we trace through space-time, known as our **world-line**, must always be at an angle steeper than 45 degrees when shown in a space-time diagram.

An important property of the standard *big bang cosmology is that sufficiently separated locations are causally disconnected; because the *age of the Universe is finite, the past light cone of an observer encloses a finite volume. This means we can see only a finite part of the Universe, known as the *observable Universe. This causal disconnection leads to a puzzle known as the *horizon problem.

likelihood

The likelihood, a shortened form of the phrase **likelihood function**, is the probability of a given dataset within a model. The model will typically contain several parameters which can be varied to give the best fit to the data, and the likelihood then is a function of those parameters. High values of the likelihood, indicating that the observed data were quite probable if that model is the true one, are interpreted as indicating that the data favour those parameter values. The likelihood is a key component of the *Bayesian inference framework, its role being to update the prior knowledge of parameter values (before the data was obtained) to that after the data is taken into account.

The method of **maximum likelihood** is one of *parameter estimation, which determines parameters by finding the values which give the maximum likelihood. In this method, the best estimate of a given parameter, in light of a dataset, is the one which maximizes the likelihood. In addition to knowing the maximum-likelihood value, one usually also wishes to know what range of parameter values, near the maximum-likelihood value, give fits to the data that are comparably good.

A drawback of the maximum-likelihood method is that it does not ask whether the fit to the data is good in an absolute sense, only which parameter values of the model give the best fit. Even if the model is a terrible fit to the data for all possible parameter values, there will still be a maximum-likelihood version. Overall goodness-of-fit tests have to be carried out separately.

The likelihood function is a property of a given dataset. When cosmologists want to include new datasets in their analyses, the first thing they need to know is the likelihood function for that data. They then make theoretical predictions from their chosen models for particular parameter values, and use this likelihood function to determine the corresponding likelihood of these parameters. This is then repeated for different parameter values.

In a cosmological context, maximum likelihood analyses are usually carried out using Markov Chain *Monte Carlo methods. These use random numbers to explore the shape of the likelihood function in the vicinity of its maximum, in order to set *confidence limits on parameters.

The likelihood function depends on all the parameters varied in fitting to observational data. If we are interested in a particular parameter, we can average over the effect of other parameters. This process is known as marginalization, and results in the marginalized likelihood function for one (or more) parameters. This process however throws away information on correlations between parameters.

The average likelihood over all parameters is known as the Bayesian evidence, and is the key statistic for carrying out *model selection analyses.

linear density perturbations
See DENSITY PERTURBATIONS.

linear perturbation theory
See COSMOLOGICAL PERTURBATION THEORY.

Local Group
The Local Group is the name given to the *galaxy group of which our own *Milky Way Galaxy is a prominent member. In addition to the Milky Way, the Local Group contains four objects from the *Messier catalogue: the Andromeda Galaxy M31 and its *satellites M32 and M110, as well as the Triangulum Galaxy M33. The Local Group also contains at least 30 less-luminous galaxies within a region of around 3 Mpc in diameter. Several nearby *dwarf galaxies have been discovered, which are satellites of the Milky Way, including the Large and Small Magellanic clouds, which are visible to the unaided eye in the Southern hemisphere. Much fainter dwarf satellites include those in Canis Major, Ursa Minor, Draco, Carina, Sextans, Sculptor, Fornax, and Leo.

http://www.seds.org/messier/more/local.html

LOFAR
LOFAR stands for LOw Frequency ARray for radio astronomy. It is an ambitious project to build an interferometric array of radio telescopes distributed across north-western Europe, with a total effective collecting area of up to 1 square kilometre. The processing of the data is done by a supercomputer, situated at the University of Groningen.

LOFAR started as a new and innovative effort to force a breakthrough in sensitivity for astronomical observations at radiofrequencies below 250 MHz. The individual detectors are simple dipole antennae, sensitive to radio signals from all directions. These are combined into a **phased array** using the aperture synthesis technique developed in the 1950s. The LOFAR design has concentrated on the use of large numbers of relatively cheap antennae, with the mapping performed using aperture synthesis software. The electronic signals from the LOFAR antennae are digitized, transported to a central digital processor, and combined in software in order to map the sky. The cost is dominated by the cost of electronics and will follow Moore's law, becoming cheaper with time and allowing increasingly large telescopes to be built. So LOFAR is an information technology telescope. The antennae are simple but there are a lot of them—25 000 in the full LOFAR design. To make radio pictures of the sky with adequate sharpness, these antennae are to be arranged in clusters that are spread out over an area of ultimately 350 km in diameter. (In phase 1, which is currently funded, 15 000 antennae and maximum baselines of 100 km will be built.) Data transport requirements are in the range of many Tera-bits/sec and the processing power needed is tens of Tera-FLOPS: more than 10^{13} FLoating-point Operations Per Second. (For comparison, at the time of writing, typical desktop computers are capable of a few Giga-FLOPS. The IBM Blue Gene/L computer has reached 280 TFLOPS.)

The mission of LOFAR is to survey the *Universe at radio frequencies from 10 to 240 MHz with greater resolution and greater sensitivity than previous surveys. LOFAR will be the most sensitive radio observatory until the next generation large array radio telescope, the *Square Kilometre Array (SKA), comes on line around 2020.

The sensitivities and spatial resolutions attainable with LOFAR will make possible several fundamental new studies of the Universe. In the very distant Universe, over the *redshift range $7 < z < 10$, LOFAR will search for the signature produced by the *reionization of neutral hydrogen. This crucial phase change is predicted to occur at the epoch of the formation of the first stars and *galaxies, marking the end of the so-called 'dark ages'. The redshift at which reionization is believed to occur will shift the 1 420 MHz line of neutral hydrogen into the LOFAR observing window. In the distant 'formative' Universe ($1.5 < z < 7$), LOFAR will detect the most distant massive galaxies and will study the processes by which the earliest structures in the Universe (galaxies, *clusters, and *active galactic nuclei) form, and probe the intergalactic gas.

LOFAR was first proposed in 1997 and a feasibility study carried out in 1999. In 2003, LOFAR's Initial Test Station became operational. This consists of 60 inverse V-shaped dipoles; each dipole is connected to a low-noise amplifier, which provides enough amplification of the incoming signals to transport them over a 110 m long coaxial cable to the receiver unit. In 2005, an IBM Blue Gene/L supercomputer was installed at the University of Groningen for LOFAR's data processing. In 2006 the first LOFAR station (Core Station 1, aka. CS1) was commissioned. A total of 96 dual-dipole antennae (the equivalent of a full LOFAR station) are grouped in 4 clusters, the central cluster with 48 dipoles and the other 3 clusters with 16 dipoles each. The clusters are distributed over an area of ~ 500 m in diameter.

http://www.lofar.org/

look-back time

Light travels at a finite speed, of roughly 300 000 km/s. The Moon has an average distance from us of about 385 000 km, and thus the light reflected from the Moon's surface has taken just over 1 second to reach us from there. We thus see the Moon as it was just over 1 second ago; this is the look-back time for the Moon. In general, look-back time is the *distance to an object, divided by the *speed of light, except that for distant objects we need to allow for the *expansion of the Universe.

Of course, we don't expect features on the Moon's surface to change on a timescale of order 1 second (although this time-delay was noticeable when communicating with Apollo astronauts by radio, whose signals also travel at the speed of light). For most other astronomical objects, however, look-back time is significant. For the Sun, it is about 8.3 minutes, and we say that the Sun is at a distance of 8.3 light-minutes. The next nearest star, Proxima Centauri, being about 268 000 times as far away as the Sun, has a look-back time of more than 4 years.

Once we look beyond the Galaxy, look-back times are in excess of tens of thousands of years and start to approach astronomical timescales. Particularly when looking at distant *galaxies and *quasars, it is important to bear in mind that we are seeing them when they were only a small fraction of their present age. Significant look-back times are what enable us to study *galaxy evolution: more distant galaxies will, on average, be at an earlier stage of their evolution than nearby galaxies.

In order to calculate look-back time from *redshift, we need to know the expansion history of the *Universe, and so it depends on the *cosmological parameters. In a Universe with non-zero *cosmological constant, look-back time cannot be written as a simple function of redshift without using calculus, but in the special case of an *Einstein–de Sitter Universe, with zero *curvature and cosmological

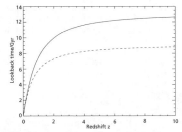

Fig. 1. Look-back time as a function of redshift assuming a concordance cosmology (continuous line; $\Omega_\Lambda = 0.7$, $\Omega_m = 0.3$) and an Einstein–de Sitter Universe (dashed line; $\Omega_\Lambda = 0.0$, $\Omega_m = 1.0$), both assuming a Hubble constant of 72 km/s/Mpc.

constant, look-back time may be written as

$$t = \frac{2}{3H_0}\left[1 - \frac{1}{(1+z)^{3/2}}\right],$$

where H_0 is the *Hubble constant and z is redshift. For this cosmology, the *age of the Universe may be obtained by letting redshift z tend towards infinity, yielding an age $t_0 = 2/(3H_0)$.

In Fig. 1, we plot look-back time as a function of redshift both for an Einstein–de Sitter Universe, and for the *standard cosmological model, both assuming a Hubble constant of 72 km/s/Mpc. Note that the age of the Universe in an Einstein–de Sitter Universe with $H_0 = 72$ km/s/Mpc is inconsistent with the ages of the oldest *globular clusters.

Light from the highest-known redshift object at the time of writing, a quasar with a redshift $z = 6.4$, was emitted when the Universe was only 800 million years old, just 6% of its current age.

low surface brightness (LSB) galaxies
Low surface brightness galaxies may be defined as those *galaxies whose *surface brightness is one magnitude or more fainter than the night-sky brightness ($\mu_B \approx 22.5$ mag arcsec^{-2} for a moonless sky). This property naturally makes them extremely difficult to detect and measure, and they are thus very under-represented in most *galaxy surveys. It seems likely that perhaps 90% of all galaxies are LSBs, only a tiny fraction of which have been catalogued.

There is a strong correlation between surface brightness and luminosity, and so most LSBs are also *dwarf galaxies. Most of their

*baryonic matter is in the form of neutral hydrogen gas, which typically forms a giant halo occupying a much larger volume than the visible stars. An even larger proportion of their mass, more than 95%, is in the form of non-baryonic *dark matter.

Compared with normal, high-surface brightness galaxies, LSBs are mainly isolated field galaxies, found in regions devoid of other galaxies. In their past, they had fewer tidal interactions or mergers with other galaxies, which could have triggered enhanced star formation. This may explain their small and dim stellar content.

luminosity density
The luminosity density gives the average total radiation emitted by all luminous sources per unit volume. Assuming that all of this radiation is accounted for by *galaxies, then the luminosity density j is equivalent to the product of luminosity L times the galaxy *luminosity function $\phi(L)$, integrated over all luminosities:

$$j = \int_0^\infty L\,\phi(L)\,dL.$$

If one has a reliable estimate of the mean mass-to-light ratio, then a measurement of the luminosity density immediately provides an estimate of the mean mass density of the Universe.

The luminosity density in near-*ultra-violet and blue optical passbands is dominated by recently formed stars, and so provides a good indication of the overall star-formation rate. Thus by measuring the near-UV luminosity density as a function of *redshift, one can constrain the *star-formation history of the *Universe.

In the local Universe, at redshift $z = 0.1$, the luminosity density in optical passbands is of order 10^8 solar luminosities per cubic megaparsec. In other words, a cubic region of space 100 parsecs on a side contains, on average, a luminosity equivalent to 100 million Suns.

luminosity distance
The luminosity distance is a way of expressing the *distance to a cosmological object. It is *not* the true distance, but rather the distance that the object appears to have, based on its luminosity, if one were ignorant of the cosmological effects of *redshift due to *expansion and of *spatial curvature.

The effect of redshift is always to reduce the luminosity, thus making the object appear to be further away than it really is and hence increasing the luminosity distance. Spatial curvature may either increase or decrease the luminosity distance. If the spatial *geometry is spherical then the intensity of light falls off more slowly than the usual inverse square law would imply; this increases the luminosity, and hences reduces the luminosity distance. Hyperbolic geometry has the opposite effect.

The power of the luminosity distance as a cosmological probe is due to it being a direct observable, at least provided the actual luminosity of the object being studied is known. Unfortunately that is rarely true in practice, but provided a class of objects can be considered a *standard candle (i.e. all members in the class have the same absolute luminosity or known variations from it) then the luminosity distance can still be used with the (possibly unknown) actual luminosity providing an overall multiplier. Combined with measures of the redshift of the objects, the luminosity distance can then be mapped out as a function of redshift, and compared with the predictions of cosmological models.

Closely related to the luminosity distance is the *angular-diameter distance, which is the distance objects appear to have based on their size. The two relations are connected by the **reciprocity relation**, that their ratio is $D_L/D_A = (1 + z)^2$ where z is the redshift. This relation is independent of spatial curvature, and is a simple consequence of the conservation of *photon number.

The most important current application of the luminosity distance is to type Ia *supernovae. The relation between luminosity and redshift of these supernovae is one of the most direct and strongest pieces of evidence for the existence of *dark energy in the *Universe.

luminosity function

The luminosity function, frequently abbreviated to LF and denoted by the Greek symbol ϕ (phi), describes the number density of a particular type of object (e.g. *galaxy or *quasar) as a function of intrinsic luminosity. It is an important cosmological tool for a number of reasons:

- It provides a way of testing theories of *structure formation and evolution, which, if successful, should correctly predict the observed luminosity function.

- It may be integrated to provide an estimate of the *luminosity density of the *Universe, thus providing constraints on the *star formation history.
- It is needed in order to calculate the *correlation function for a flux-limited sample of sources, via the *selection function.

In order to estimate a luminosity function, one needs a well-defined sample of objects with known fluxes and *distance estimates. In most cases, distance is estimated by measuring *redshifts. Knowing the flux and distance to each object, an intrinsic luminosity or absolute magnitude may be calculated for each, taking into account the *K correction.

When estimating the luminosity function from a sample of sources selected by observed flux (or equivalently, apparent magnitude), one needs to allow for the fact that luminous sources can be seen to a greater distance than dim ones. The simplest way to estimate the luminosity function is then to weight each object in the sample by the reciprocal of the volume defined by the maximum distance to which it could have been detected, this is called the '$1/V_{max}$' estimator. In this way, one compensates for the smaller fraction of faint objects that are included in the sample.

Such estimators are, however, susceptible to biases due to *cosmic structure within the sample. For instance, if a rich *cluster of galaxies happens to lie at low redshift within the survey volume, the relative density of faint galaxies will be over-estimated. Such biases may be overcome by surveying sufficiently large areas of sky or by employing more refined luminosity function estimators which are unaffected by density fluctuations within the sample.

The galaxy luminosity function estimated from the *Sloan Digital Sky Survey Data Release 5 (DR5) is shown in Fig. 1. The symbols show the number of galaxies per cubic megaparsec (vertical axis) in bins of absolute magnitude (horizontal axis). In most cases, the estimated uncertainty in density is smaller than the symbol size. The galaxy luminosity function is commonly fitted with a function proposed by Paul Schechter in 1976:

$$\phi(L)dL = \phi^* \left(\frac{L}{L^*} \right)^{\alpha} \exp \left(-\frac{L}{L^*} \right) d\left(\frac{L}{L^*} \right).$$

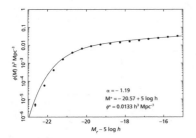

Fig. 1. The galaxy luminosity function measured from the Sloan Digital Sky Survey Data Release 5 (DR5). The symbols show estimates of the number density of galaxies in bins of r-band absolute magnitude, with luminosity decreasing from left to right. The line shows the best-fit Schechter function with parameters as indicated.

In this equation, L is luminosity and L^* is a characteristic luminosity at which the form of the Schechter function $\phi(L)$ changes from a power-law behaviour at fainter luminosities to an exponential decline at higher luminosities. The index α gives the slope of the faint-end power-law (a straight-line when $\log \phi$ is plotted as a function of magnitude) and ϕ^* sets the overall normalization (i.e. the amplitude) of the Schechter function. While the Schechter function, inspired by *Press–Schechter theory, provides a reasonable fit to most galaxy luminosity function estimates, it is now becoming clear, with larger and larger samples, that the Schechter function does not provide a perfect fit. In the example shown, a Schechter function is unable to simultaneously fit both the bright and faint ends of the luminosity function.

Estimation of the luminosity function for a cluster of galaxies is simplified by the fact that the galaxies within a cluster are at approximately the same distance from us, and therefore one needs only an estimate of the distance to the cluster as a whole, and not the distances to the individual galaxies. One must be careful in this case, however, in separating foreground and background galaxies from true cluster members.

Lyman alpha

Lyman alpha is the name given to the transition of an *electron from the first to the second energy level (or vice versa) of a hydrogen atom. According to *quantum mechanics, the electrons surrounding atomic nuclei may be in one of a number of discrete energy states. An electron may be promoted to a higher energy state when the atom absorbs a *photon whose energy is exactly equal to the difference in energy between the before and after states. Conversely, an electron may spontaneously drop from a higher to a lower energy state, accompanied by the emission of a photon carrying energy equal to the difference between the two states. Atoms can thus both absorb and emit radiation with a series of discrete energies.

The most abundant element in the *Universe is hydrogen. At relatively low temperatures, the single electron of the hydrogen atom is in its **ground** (lowest-energy) state. The series of transitions from this ground state to the first excited state, the second excited state, and so on, is known as the Lyman series after its discovery in 1906 by Theodore Lyman. The transitions within the series are labelled by the letters of the Greek alphabet, the first transition being referred to as Lyman alpha, often abbreviated to Lyα. This energy change corresponds to a photon of wavelength 1 216 Angstroms (one Angstrom being 10^{-10} metres), in the *ultra-violet part of the *electromagnetic spectrum.

In high-*redshift *quasars, Lyα is seen in emission as a strong, broad peak in the quasar's spectrum. This feature is due to hydrogen gas being heated by the infall of material onto the quasar's central *black hole. Thermal energy promotes electrons from the ground to the first excited state which is then re-emitted as a Lyα photon when the electrons return to their original state. The observed wavelength of the Lyα line is determined by the redshift of the quasar itself: $\lambda_{obs} = (1 + z)\lambda_{em}$.

As well as being useful in determining the redshifts of quasars, Lyα emission may be used to search for high-redshift ($z \approx 6 - 7$) galaxies. It is seen in absorption due to neutral hydrogen clouds along the line of sight to distant quasars: the Lyman alpha forest.

Lyman alpha forest

The Lyman alpha forest (Fig. 1 over page) is a series of sharp dips in the *spectrum of a distant *quasar due to Lyman alpha absorption of discrete quanta of radiation by clouds of hydrogen gas along the line of sight to the quasar.

In Fig. 1, the observed wavelengths have been divided by $(1 + z)$, where z is the

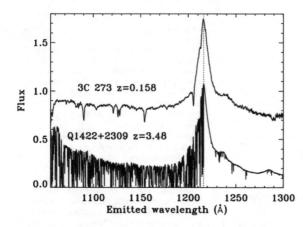

Fig. 1. The spectra of two quasars at low (z = 0.158) and high (z = 3.48) redshift. Observed features have been converted to the rest frame of each quasar, that is observed wavelengths are divided by $(1 + z)$, so that common features line up. The large peak in both spectra around wavelength 1 215Å is due to Lyman alpha emission from the quasars themselves. The series of dips at shorter wavelengths (the Lyman alpha forest) in the high redshift quasar are due to absorption by neutral hydrogen clouds at various redshifts along the line of sight to the quasar.

*redshift of the background quasar, so that the quasar restframe wavelength λ_{em} is plotted along the x-axes. In the case of the lower spectrum of the high redshift quasar, a 'forest' of absorption features is seen at wavelengths blueward of the Lyα emission feature. This absorption is due to clouds of neutral hydrogen gas between us and the quasar. It occurs at a whole range of different observed wavelengths because the clouds themselves have different redshifts, up to that of the background quasar.

There are a small number of clouds of hydrogen between us and the low-redshift quasar (it is much closer to us), and so only a few Lyα forest lines are seen. The broad absorption features seen in this spectrum are due to our own *Galaxy. It was only with the launch of the *Hubble Space Telescope that Lyα lines could be accurately measured in the ultra-violet part of the spectrum, and hence at low redshift.

The cosmological significance of the Lyman alpha forest is that it enables us to map out the distribution of small-scale density fluctuations in the Universe out to high redshift. Many more absorbing clouds are seen at high redshift than in the local Universe. Presumably most of the high-redshift clouds have either collapsed to form galaxies or otherwise have dispersed into the *intergalactic medium.

One may calculate the *power spectrum of the Lyα lines in order to determine

the *clustering of proto-galactic systems. The power spectrum of the Lyα forest in the approximate redshift range 2–4 provides our only estimate of clustering on small scales, down to about one megaparsec, which is not strongly affected by non-linear evolution. This is because at redshifts $z \gtrsim 2$, the distribution of proto-galactic systems has not had time to evolve significantly since they were seeded by *cosmological inflation. Lyα forest clustering measurements thus provide important constraints on *cosmological parameters, particularly on the amplitude and *spectral index of *density perturbations in the early Universe.

Since only neutral hydrogen can absorb photons, the presence of Lyα forest lines at any given redshift tell us that there exists at least some hydrogen in non-ionized form at that redshift. Studies of the Lyα forest can thus constrain the epoch of *reionization in the Universe via the *Gunn–Peterson effect.

http://www.astro.ucla.edu/~wright/
 Lyman-alpha-forest.html

Lyman alpha surveys

The *Lyman alpha (Lyα) line provides an efficient way of finding young *galaxies at high *redshift, because it is strong in systems with young stars and little or no *dust—properties expected in galaxies undergoing their first burst of star-formation.

There are two techniques by which one may search for Lyα-emitting galaxies. The first uses narrow-band imaging to detect flux from a narrow range of wavelengths centred on the redshifted Lyα line. The advantage of this method is that one may explore relatively large areas of sky, but one will only find galaxies in a narrow range of redshift depending on the central wavelength of the imaging filter λ_{obs}. One is most sensitive to the redshift z given by $\lambda_{obs} = (1 + z)\lambda_{em}$, where $\lambda_{em} = 1\,215.67$ Angstroms is the rest wavelength of Lyα. The second technique uses *spectroscopy to search a wider redshift range but over a much smaller area of sky. Both techniques have made use of massive *clusters of galaxies which act as *gravitational lenses in order to amplify the brightness of the extremely faint background galaxies at $z \sim 5$ that one is attempting to detect. Altogether, several hundred Lyα-emitting candidate galaxies at redshifts $z \sim 5$ and greater have been detected.

Lyman break galaxies (LBGs)

Lyman break galaxies are *galaxies at *redshift $z > 2.5$ that are detected in optical passbands but that are not detected in the near *ultra-violet (UV), see Fig. 1. The reason for these 'UV dropouts' is that the 'break' in the spectrum of a galaxy blueward of the **Lyman limit** is redshifted into the near-UV.

The Lyman limit corresponds to the lowest energy *photons that can ionize an *electron from the ground state of neutral hydrogen. These photons have a wavelength of 912Å, not to be confused with *Lyman alpha at 1 216Å, which corresponds to the transition from the lowest $(n = 1)$ to the first $(n = 2)$ excited state of hydrogen. Virtually no flux is detected from galaxies blueward of the Lyman limit. This is because:

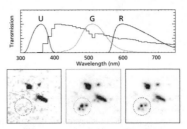

Fig. 1. The lower three panels show the image of a Lyman break galaxy (LBG) (circled) as seen through ultra-violet (*U*), green (*G*), and red (*R*) filters respectively from left to right. The LBG is not detected in the *U* filter due to the lack of flux blueward of the Lyman limit (here redshifted to about 3 500Å or 350nm) as illustrated in the top panel.

- Very few stars are hot enough to produce such energetic photons.
- Any photons from the blue side of the Lyman limit that are emitted are likely to be absorbed by any neutral hydrogen clouds between the galaxy and the observer.

There is thus a sharp break in the spectrum of a galaxy blueward of the Lyman limit. At a redshift $z = 2.5$, the Lyman limit is redshifted to 3192Å, in the near-UV part of the *electromagnetic spectrum.

The Lyman break technique was pioneered in the early 1990s by Charles Steidel and collaborators, and more than 1000 $z > 2.5$ galaxies have been detected. One can push this technique to higher redshifts by searching for dropouts in redder bands. However, at higher redshifts galaxies become fainter and fainter, and spectroscopic confirmation of their redshifts becomes harder to obtain. By redshifts $z \sim 5$, it becomes easier to search for galaxies using *Lyman alpha surveys.

http://www.astro.ku.dk/~jfynbo/LBG.html

M-theory

M-theory is the cutting edge of modern particle theory, having absorbed and supplanted *superstring theory. Yet bizarrely no one knows what the theory actually is. In fact, it isn't even clear what the 'M' might stand for; Edward Witten, inventor of M-theory, suggests 'magic', 'mystery' or 'membrane'. Other authors have suggested it could stand for 'matrix'. We will resist making our own suggestion.

M-theory emerged in response to a crisis in superstring theory. Superstrings had been much vaunted as a candidate to be the unique consistent theory unifying *quantum mechanics and *gravity. Unfortunately, by the mid 1990s it had become apparent that there were five different versions of superstring theory, rather putting paid to the idea of uniqueness. A breakthrough came in 1995 in a famous seminar by Witten in California, where he drew together a number of strands of theoretical reasoning to conclude that the five string theories were in fact different limits of a single all-encompassing theory, to be called M-theory. What precisely this M-theory is remains a mystery; although the limiting cases are the superstring theories, what the main body of the theory looks like has yet to be derived. This situation is schematically shown in Fig. 1.

Like superstring theory, M-theory relies on *extra dimensions beyond the four *spacetime dimensions we perceive. But rather than superstring theory's 10 dimensions, M-theory lives in 11 dimensions (the superstring theories are obtained as limiting cases by removing one of the dimensions by a process called *compactification). Indeed, yet another limiting case of M-theory is a *supergravity theory in 11 dimensions, it having long been known that 11 dimensions is the highest number allowed for a consistent supergravity theory. This is the limiting case believed to be appropriate at low energies, and hence relevant to most of the evolution of the *Universe.

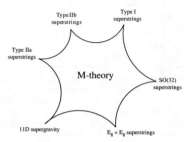

Fig. 1. A schematic illustration of the relations between theories. The structure of M-theory is unknown, but in 6 limiting cases it reproduces the 5 superstring theories and the 11-dimensional supergravity theory.

Since the theory itself has yet to be understood, it is unsurprising that the impact of M-theory on cosmological understanding is relatively unexplored. M-theory does provide the motivation for *braneworld cosmologies, described under that separate topic, which are presently the main area of M-theory motivated cosmology.

MACHOs (massive compact halo objects)

MACHOs are a type of *dark matter postulated to exist in *galactic halos in order to provide *galaxies with sufficient mass to explain their observed dynamics. Examples of MACHOs are small *black holes, brown dwarfs, and Jupiter-sized planets. Thus in contrast with particle-physics dark matter candidates known as WIMPs (weakly interacting massive particles), MACHOs are astrophysical-sized objects.

Why do we think that dark matter should exist in the halos of galaxies? The principal evidence for dark matter in galactic halos comes from the study of *galaxy rotation curves. By comparing the dynamically inferred mass with the mass that can be attributed to visible components of the galaxy, these studies suggest that 50–90% of the mass of most spiral galaxies, including the *Milky Way, is made up of dark matter.

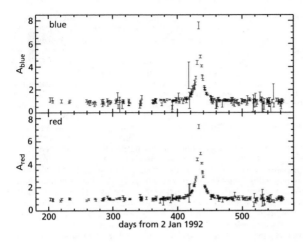

Fig. 1. The first LMC microlensing event observed by the MACHO Project. The two curves show the brightness of a star in the LMC through blue and red filters as a function of time in days. The increase in brightness around day 430 is thought to be due to microlensing by a MACHO in the halo of the Milky Way.

Although MACHOs themselves do not emit any measurable radiation, they do act as *gravitational lenses. If a MACHO happens to pass extremely close to the line of sight from us to a distant star, then the star's light will be amplified by the lensing effect of the MACHO. Since the MACHOs will be moving through the halo, this amplification will be a transient event, and in principle one may detect it by looking for the brightening and fading of a distant star. The amplification can be large, but events are extremely rare: it is necessary to monitor the brightness of several million stars for a period of years in order to obtain a useful detection rate.

In 1992 the MACHO Project began an experiment to detect gravitational lensing from MACHOs. They built a two channel imaging camera that simultaneously imaged a 0.5 square degree field through blue and red filters. This two-colour imaging was vital in order to remove intrinsically variable stars from the sample. Intrinsic variability is a common phenomenon which would numerically swamp MACHO lensing events, but whereas gravitational lensing amplifies all colours by the same amount (**achromatic** in technical parlance), variable stars normally change colour as they change in brightness.

The MACHO camera was mounted on the refurbished Great Melbourne Telescope at Mount Stromlo Observatory near Canberra, Australia. Depending on the time of year,

the observers repeatedly imaged fields in the bulge of the Milky Way, observed through the disk of the Galaxy, and in the Large Magellanic Cloud (LMC), a *dwarf galaxy neighbour close enough to resolve individual stars, observed through the Galactic halo. Purpose-written software then looked for achromatic changes in brightness of any individual stars between repeated observations of the same field.

At the turn of the millennium, after 5.7 years of observing 11.9 millions stars in the LMC, the MACHO team reported 13–17 microlensing events towards the LMC. See Fig. 1 for an example lightcurve. The exact number of events is uncertain, because some types of spectral variability are approximately achromatic and mimic lensing events. In any case, the number of observed events was significantly more than the 2–4 expected from known stellar populations in the halo. It cannot be ruled out that the lensing events were due to lenses within the LMC itself, but it is more likely that the lenses lie within the Galactic halo. Careful analysis showed that between 8 and 50% of the mass of the halo consists of MACHOs (95% confidence interval), with individual MACHO masses in the range 15–90% of the mass of the Sun. One may thus rule out the possibility that the Galactic halo dark matter consists entirely of MACHOs at more than 95% confidence. The likely mass range of the observed MACHOs suggests that they may be white dwarfs: the remnants of intermediate-mass stars.

Sadly, the Great Melbourne Telescope was destroyed along with other telescopes and observatory buildings in a bush fire in 2003. The success of the MACHO Project has however spawned several other microlensing surveys, including the Optical Gravitational Lensing Experiment (OGLE), Expérience pour la Recherche d'Objets Sombres (EROS) and SuperMacho.

http://www.macho.mcmaster.ca/

magnetic monopoles

Magnetic monopoles are an example of a *topological defect, which is a general class of objects that may form during cosmological *phase transitions. The classical theory of electromagnetism, due to James Clerk Maxwell, contains individual electric charges (for instance the *electron has an electric charge) but does not permit isolated magnetic charges—magnetic north poles are always accompanied by magnetic south poles. However fundamental theories of physics such as *grand unification predict that isolated magnetic monopoles exist.

The notion of magnetic monopoles was first suggested by Paul Dirac in 1931, who showed that their existence would explain why electric charge is 'quantized', meaning that it is able to take only integer values. However they were not taken as a serious possibility until it was shown that they were a consequence of grand unified theories. In fact, they appear to be an inevitable consequence of such theories, because the breaking of any large symmetry into the symmetry of the *standard model of particle physics [technically known as $SU(3) \times SU(2) \times U(1)$] always creates them. Such monopoles would be extremely massive, perhaps 10^{16} GeV in *particle physics units, and would form when the *Universe was extremely young.

Observational searches have been undertaken to try to discover magnetic monopoles. Indeed, there was even a claim at Stanford University that one had been found in 1982, but this was never reproduced in subsequent experiments and never pushed very strongly by the discoverer, Blas Cabrera. The most powerful limit on the *density of monopoles in the Universe is known as the **Parker Bound**, and arises because a high density of monopoles would cancel out the galaxy's magnetic field. The Parker bound demands that the total density of monopoles in the present Universe be several orders of magnitude below the total density of matter, and means that the chance of direct discovery in the style of Cabrera should be impossible as events would just be too rare.

The early formation of monopoles represents a significant problem for the hot *big bang cosmology (see 'Overview'), because being so massive they would be non-relativistic early on. This means that their density reduces more slowly with time than that of radiation. If even a tiny density of them were present in what is supposed to be a *radiation-dominated epoch of the Universe, they would rapidly come to dominate the Universe; the present Universe should essentially be entirely comprised of magnetic monopoles. This is in flagrant disagreement with observations, and is known as the monopole problem, an example of a *relic particle abundance problem.

The monopole problem was one of the main motivations that led Alan Guth to propose *cosmological inflation in 1981. He realized that a burst of rapid expansion, taking place after the monopoles formed, would reduce their density to an acceptably low value.

In more exotic grand unification theories, magnetic monopoles can become connected by segments of *cosmic string, with a monopole at one end and an anti-monopole at the other. In such scenarios the string will pull the monopole–anti-monopole pair together, potentially providing another mechanism (known as the Langacker–Pi mechanism after its inventors) which would reduce their density to an acceptable level.

magnitude system

Optical and near-*infrared astronomers tend to describe the apparent and intrinsic brightness of objects using the magnitude system. This dates back to the Greek astronomer Hipparchus in 129 BC. He defined the brightest stars to be of 1st magnitude, slightly fainter stars to be of 2nd magnitude and so on down to the faintest (naked-eye) stars which he defined to be of 6th magnitude. The magnitude system was refined by Claudius Ptolemaeus (now known as Ptolemy) in AD 140 and then remained essentially unchanged until the invention of the telescope in 1608 by Hans Lipperhey. With his own telescope, built in

Fig. 1. Illustration of the visual magnitude system.

1610, Galileo Galilei was able to observe stars of 7th, 8th and fainter magnitudes.

The magnitude system was quantified in 1856 by British astronomer Norman Pogson. He observed that 1st magnitude stars were roughly 100 times brighter than stars of 6th magnitude. He then *defined* a 5 magnitude difference to correspond exactly to a brightness ratio of 100:1. A 1 magnitude difference then corresponds to a brightness ratio equal to the fifth root of 100, roughly a factor of 2.512 (the Pogson ratio). Note that Pogson magnitudes are logarithmic: a difference in magnitudes corresponds to a ratio of brightnesses. Also, due to Hipparchus' original convention, small magnitudes correspond to bright objects. Human visual perception is not quite logarithmic, so in Pogson's system some stars became brighter than 1st magnitude.

Mathematically, the magnitude m of a source with measured flux F is given by

$$m = m_0 - 2.5\log_{10} F,$$

where m_0 defines the **zero-point** of the magnitude system. Rather than using this zero-point directly, it is more common to measure relative magnitudes of stars,

$$m_1 - m_2 = -2.5\log_{10}(F_1/F_2).$$

This equation may also be used to define the colour index of an object, if F_1 and F_2 are the fluxes of the same source measured in two different parts of the *electromagnetic spectrum.

The astronomical source of brightest apparent magnitude is the Sun, which has visual magnitude -26.7. The next brightest star (Sirius) has magnitude -1.5. The deepest *Hubble Space Telescope image reaches a magnitude of $+30$, a factor of $\sim10^{22}$ fainter than the Sun. The visual magnitude system is illustrated in Fig. 1.

The apparent magnitude of a source does not necessarily bear any resemblance to its intrinsic, or absolute, magnitude. The apparent magnitude depends strongly on the *distance of the source; the Sun is the brightest source in the sky merely because it is so much closer to Earth than any other star. By convention, the absolute magnitude of an astronomical object is defined to be the same as its apparent magnitude if it is at a distance of 10 parsecs. Given that observed flux drops inversely with the square of distance (the inverse square law), and the above definition of magnitude, then, one can show that apparent (m) and absolute (M) magnitude are related by the equation,

$$M = m - 5 \log_{10}(d/10),$$

where d is the distance in parsecs.

The absolute visual magnitude of the Sun is $M_V = 4.85$, while a typical galaxy has $M_V \approx -20$, a factor of $\sim 10^{10}$ more luminous.

This formula is accurate within our own Galaxy, where radial velocities are significantly less than the *speed of light, but when estimating the absolute magnitude of other galaxies, one needs to allow for the fact that the light emitted by a distant galaxy has been stretched by the *Hubble expansion of the Universe, and thus that the observed flux was emitted by the galaxy at shorter wavelengths than the light received (*see* DISTANCE MODULUS; K CORRECTION).

matter–anti-matter asymmetry

Our *observable Universe is made from normal matter and contains almost no *anti-matter, even though the laws of physics are practically identical for both. This is known as the matter–anti-matter asymmetry. Theories which attempt to explain this are known as *baryogenesis.

For the observational evidence for a lack of anti-matter, *see* BARYOGENESIS. The evidence basically amounts to the fact that annihilations of matter and anti-matter anywhere within our *Universe would be such a dramatic phenomenon as to be easily detected. One can build a ladder of evidence starting with the Solar System (e.g. Neil Armstrong failed to explode on touching the Moon, thus proving it is not made of anti-matter) and working up to galactic and ultimately cosmological scales (no evidence of catastrophic matter–anti-matter annihilations during the frequent galaxy mergers that take place).

The size of the matter–anti-matter asymmetry is usually described using the *baryon-to-photon ratio, meaning the average number of *baryons in a given volume divided by that of *photons. The baryons in the present Universe are the *protons and *neutrons, and in computing the baryon number one subtracts the number of anti-baryons from that of baryons, so that the answer would be zero if matter and anti-matter were equally prevalent. The photons in the present Universe are predominantly those making up the *cosmic microwave background. From the measured values of these quantities, for instance by the *WMAP satellite, the ratio of baryons to photons is given as about 6×10^{-10}. There are more than one billion photons in the Universe for every baryon! Nevertheless, the typical photon energy is so minuscule compared to the mass–energy of the baryons that in terms of *density it is the baryons which are the dominant material.

The significance of the tiny baryon-to-photon number is apparent when one realizes that in the energetic young Universe baryons and photons would have been in *thermal equilibrium with one another. At such times, the numbers of photons, of baryons and of anti-baryons would have been (almost) equal, with interactions constantly changing one to the other. At such early epochs, the asymmetry between matter and anti-matter would have been a tiny one, just one part in a billion or so. It is only in the present Universe, after the baryons/anti-baryons leave thermal equilibrium and all the anti-baryons annihilate with baryons, that the asymmetry becomes prominent. Theories of baryogenesis aim to create this modest one-in-a-billion asymmetry while the Universe is in this young thermal state.

matter-dominated era

A matter-dominated era of the *Universe is one where the dominant constituent within the Universe is non-relativistic particles. These particles might for instance be *cold dark matter particles or *baryons. According to the *standard cosmological model, the young Universe was *radiation-dominated, and at an age of around fifty thousand years cold dark matter took over and began a long matter-dominated era. The dividing moment is known as *matter–radiation equality. The matter-dominated era then persisted until the Universe was around half its present age, when matter domination gave way to domination by *dark energy.

In assessing which type of material dominates the Universe, the important quantity is the *density, as it is what governs the expansion rate of the Universe via the *Friedmann equation. Matter density and energy density can be used interchangeably, using *Einstein's famous $E = mc^2$ relation. In the phrase 'matter-dominated', the word 'matter' is shorthand for 'non-relativistic matter', meaning that the constituent particles are travelling at much less than the *speed of light. In that case the behaviour of the density is very simple; as the Universe expands, the density reduces in inverse proportion to the volume, since the same amount of material is spread around a larger amount of space. Equivalently, we say that the density reduces as the *scale factor cubed.

The expansion rate of the Universe during a matter-dominated era is that the scale factor grows proportional to time to the two-thirds power. The volume therefore grows proportional to time squared.

The matter-dominated era includes many of the key events in the history of our Universe. Not long after it began, the *cosmic microwave background was created through the processes of *recombination and *decoupling. Later on structures began to form, with the first generations of stars and *galaxies. *Reionization of the Universe also took place during the matter-dominated era.

Although the standard cosmology features only a single matter-dominated era, it is possible that there were other temporary epochs of matter domination in the earlier Universe. This may happen, for instance, if there exists some type of fundamental particle which is both very massive and also long-lived. As the ambient energy reduces, such particles will become non-relativistic and their density will subsequently fall in proportion to the volume, enabling them to overcome any relativistic species whose density is reducing faster due to an additional *redshifting effect. Once their energy comes to dominate, a matter-dominated era begins which will end when the responsible particles decay.

matter–radiation equality

The instant of matter–radiation equality divides the two main epochs of the *Universe's evolution, *radiation domination and *matter domination. It took place when the Universe was approximately 50 000 years old. The *redshift at which it took place can be accurately computed given the present matter and radiation densities, as both have a known scaling with the expansion rate. In the *standard cosmological model, matter–radiation equality was at a redshift $z_{eq} \approx 3100$.

In the historical sequence of events in the Universe, matter–radiation equality comes long after early Universe phenomena such as *cosmological inflation and *nucleosyn-

thesis. It does however happen just before the *cosmic microwave background forms, the proximity of these two events apparently being a coincidence as there seems no fundamental link between them.

One of the most significant aspects of matter–radiation equality is that it marks the epoch when *density perturbations in the *cold dark matter are first able to grow. Indeed, the epoch of matter–radiation equality leads to a characteristic scale in the *power spectrum describing the inhomogeneities in the cold dark matter, and hence ultimately in the *clustering pattern of *galaxies. This characteristic scale can now be measured in galaxy *redshift surveys and is in good agreement with expectations.

Messier catalogue

The French astronomer Charles Messier (1730–1817) was the first person to systematically catalogue *nebulae (the term then used to refer to objects which appear diffuse compared with the point-like stars). He published his catalogue of 110 nebulae in 1784.

The motivation for this work was to help discriminate comets, which also have a diffuse appearance, from the unmoving nebulae. Forty of these nebulae turned out to be *galaxies, separate from our own *Milky Way Galaxy. The remaining objects are mostly clusters of stars within our own Galaxy, too faint to be individually resolved by Messier.

Eleven Messier objects are genuine nebulae as we use the term today, i.e. clouds of gas and dust, rather than collections of unresolved stars.

The Messier objects still bear the designation 'M' today; for example the Andromeda galaxy is also known as M31, being the 31st entry in Messier's catalogue.

http://www.seds.org/messier/

metallicity

Astronomers and cosmologists refer to elements heavier than helium collectively as **metals**, even though many of them, such as oxygen and nitrogen, are gases, and others, such as carbon, are non-metallic solids at standard temperature and pressure. The **metallicity** of an astrophysical object, frequently denoted by the capital letter Z, is the mass fraction of that object comprising metals. It is determined by analysing the object's *spectrum: different elements show spectral features at well-defined wavelengths.

The Sun has metallicity $Z \approx 0.02$. Since metals are synthesized from hydrogen and helium inside stars (with the exception of small quantities of deuterium and lithium that are created by cosmic *nucleosynthesis about one second after the *big bang), and then later released into the interstellar medium by *supernova explosions, metallicity tends to increase as the *Universe ages. The metallicity can thus provide an estimate of the age of an object: recently formed objects have higher metallicities on average than older objects. For this reason, most objects observed at high *redshift, when the Universe was young, have metallicities significantly lower than that of the Sun.

metric

The metric is the mathematical quantity which describes the *geometry of *spacetime. It measures the distance between nearby points in space-time. The usual goal of calculations in *general relativity theory is to determine the metric which describes the space-time generated by a particular distribution of material.

The technical definition of the metric is as follows. The metric is usually written as $g_{\mu\nu}$ where the indices μ and ν take the values 0,1,2,3 corresponding to the time coordinate and the three spatial coordinates. It is defined by the expression

$$ds^2 = \sum_{\mu,\nu=0}^{3} g_{\mu\nu} dx^\mu dx^\nu$$

where ds is the distance between two neighbouring points separated by a distance dx^μ. The metric is symmetric, and hence has ten independent components, being the four diagonal elements g_{00}, g_{11}, g_{22} and g_{33} plus six off-diagonal elements. For instance, g_{00} measures the physical space-time distance corresponding to motion purely in the time direction. In general, each of the ten metric components depends on position in space-time.

In situations where gravity can be ignored, space-time has a flat geometry, corresponding to the Minkowski metric $g_{\mu\nu} = \text{diag}(-1, 1, 1, 1)$. This metric corresponds to the theory of special *relativity, and is used to formulate relativistic physics theories in situations where gravity is negligible, such as the *standard model of particle physics.

In some situations, including non-rotating *black holes and *homogeneous and *isotropic cosmologies, the metric takes on a diagonal form, which simplifies calculations considerably. The corresponding solutions to *Einstein's equation are known as the Schwarzschild metric for black holes, and the *Robertson–Walker metric in cosmology.

*Density perturbations in the *Universe are often described in terms of the perturbation to the metric which they generate, and are sometimes called *curvature perturbations. Provided they are small, a linearized version of Einstein's equations may be used to follow their evolution using *cosmological perturbation theory.

microlensing
See GRAVITATIONAL LENSING.

microwave
The term microwave refers to radiation in that part of the *electromagnetic spectrum which has wavelength in the approximate range 10 cm to 0.1 mm. Microwaves should thus really be called milliwaves or centiwaves (one micron is one-thousandth of a millimetre), although through historical usage, the name microwave has stuck. Microwaves at the long-wavelength end of the range are used to heat food and are also used in radar systems. The most important astrophysical source of microwaves is the *cosmic microwave background (CMB), a direct relic of the *big bang.

Milky Way
The Milky Way is the name given to our own Galaxy (also distinguished from other *galaxies by a capital G). It is thought to be a fairly typical spiral galaxy, but one which we are viewing from within: the name Milky Way derives from the diffuse, milky appearance of a band of light across the sky due to innumerable faint stars in the disk of the Galaxy.

The Milky Way comprises several components, see Fig. 1. Masses and luminosities are expressed relative to that of the Sun, $M_\odot$ and $L_\odot$ respectively.

Bulge—At the centre of the Galaxy is a roughly spherical bulge of stars. At the centre of the bulge is a dense **nucleus** of stars, thought to be orbiting a *supermassive black hole of mass $\sim 10^6 M_\odot$. Away from the nucleus, bulge stars have both circular and random motions of around 100 km/s. The bulge has luminosity $L \approx 5 \times 10^9 L_\odot$ and stellar mass $M \approx 2 \times 10^{10} M_\odot$.

Stellar disk—The Sun is located within the stellar disk of the Galaxy, its most

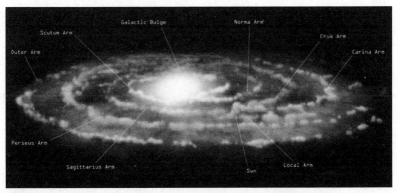

Fig. 1. Schematic illustration of the Milky Way Galaxy, showing the prominent spiral arms, the central galactic bulge, and the location of the Sun. The diffuse light between the spiral arms represents the gas disk. The whole visible galaxy is embedded in a halo of old stars, globular clusters, and dark matter.

prominent feature, at a distance of approximately 8.5 kpc from the Galactic centre. The disk is thin and roughly circular, and appears to us as a band stretching across the night sky. Dark patches in this band mark concentrations of *dust and dense gas. The stellar disk in fact comprises two components: the thin and thick disks. The **thin disk**, roughly 500 pc in height, contains 95% of the disk stars, and all of the young, massive stars. The **thick disk**, in which the thin disk is embedded, is roughly 1 500 pc in height. Stars in the thick disk are older than thin disk stars, and poorer in heavy elements. Disk stars are on nearly circular orbits moving at about 200 km/s; the Sun takes 250 Myr to complete an orbit. Random motions of disk stars are a few tens of km/s. It is worth noting that the rotation speed is approximately independent of distance and thus the Milky Way disk undergoes **differential rotation**. If the disk rotated as a rigid body then rotation speed would increase with distance from the centre. The disk luminosity is around 15–20 $\times 10^9 L_\odot$ and the mass in stars is around $6 \times 10^{10} M_\odot$.

Embedded within the stellar disk are the spiral arms. The spiral arms cannot co-rotate with the disk, otherwise they would be 'wound up' by the differential rotation of the disk, and would be destroyed in $\sim 10^9$ years. The arms are thought to be the result of a **spiral density wave** pattern that flows

through the Galaxy disk. In the same way that waves can move through water without a bulk flow of water molecules in the direction of wave motion, the rotating density wave travels through the material in the disk, and the material within a spiral arm thus changes with time. The density wave causes gas to undergo compression at the peak of the wave thus leading to a local increase in star formation rate, giving rise to the spiral arms. The observational evidence in favour of this spiral density wave model is as follows. High-mass stars are short-lived, and are seen mostly in spiral arms. Medium-mass stars are moderately long-lived, and are seen in the arms and just behind them. Low-mass stars (such as the Sun) are long-lived, and are seen throughout the disk.

Gas disk—Gas and dust lie in a thinner layer than the stars. Near the Sun, most neutral hydrogen gas lies within 100 pc of the mid-plane. The thickness of this gas layer increases with distance from the centre, indeed the gas disk is depleted towards the centre of the Galaxy. Dust in the plane of the Galaxy severely blocks our optical view of the Galactic centre and of the *Universe at low Galactic latitudes (i.e. close to the Galactic plane), leading to the *zone of avoidance in many *galaxy surveys.

Halo—The bulge and disks of the Galaxy are embedded in a roughly spherical halo, home to old stellar populations, including *globular clusters. Globular clusters are compact stellar systems containing $\sim 10^5$ stars. Comparison of the observed colours and luminosities of the constituent stars with stellar evolution models yields estimates for the ages of these clusters, which evidently formed very early in the history of the Galaxy. The most recent estimates of the oldest clusters give ages 13.5 ± 2 Gyr. Galactic globular clusters thus provide useful lower limits to the *age of the Universe.

Halo stars and globular clusters do not have any systematic rotation about the centre, but have random, often eccentric orbits, spending most of the time in the outer Galaxy, but reaching close to the centre at the pericentre of their orbits. Halo stars account for only a small fraction of the Galaxy's luminosity and mass ($M \lesssim 10^9 M_\odot$), but the total mass of the Galaxy is dominated by *dark matter in the halo.

model selection

Model selection (also known as model comparison) is the statistical problem of deciding which theoretical model gives the best description of a dataset. In a cosmological context, the definition of a model amounts to a decision as to which physical processes are thought relevant to the *Universe, quantified by choosing which set of *cosmological parameters will be allowed to vary in a fit to the data. While the technique of *parameter estimation aims to find the best-fit values of a set of parameters which has already been chosen, model selection seeks to use the data to decide the preferred choice of parameter set. Model selection is thus a higher level of statistical inference; the statistics literature sometimes refers to parameter estimation as the first level of inference, and model selection as the second level.

Cosmological model selection is a topic still in its infancy. The usual framework employed is that of *Bayesian inference, where each model (i.e. choice of parameter set) has a probability assigned to it, which is updated each time new data are considered. This updating is achieved by computing a quantity known as the **Bayesian evidence**, which is the average *likelihood over the entire parameter space of a given model. The evidence rewards models for their ability to explain the observed data, but penalizes them if they are unpredictive, setting up a tension

between model complexity and goodness-of-fit to data. Ranking of models according to their evidence values indicates how well a dataset supports or opposes each model.

Computing the evidence is an even harder computational problem than parameter estimation, as it involves determining the likelihood function accurately over all parameter space, rather than mapping out only the best-fitting regions. A numerical code called CosmoNest for achieving this has been made publicly-available, but, as with the parameter estimation code CosmoMC, it requires supercomputer-class facilities in order to achieve the necessary computations.

Model selection techniques are required to rigorously define the *standard cosmological model, with current applications indicating a very simple model with just five fundamental parameters is the best description of existing data. The main application of cosmological model selection in the future will be to give a robust criterion for identifying the need to introduce extra parameters if and when new physical effects become apparent in future observations.

Gregory, P. *Bayesian Logical Data Analysis for the Physical Sciences*, Cambridge University Press, 2005.

CosmoNest Home Page:
http://www.cosmonest.org

modified Newtonian dynamics (MOND)

Modified Newtonian dynamics provides an explanation for the observed dynamical properties of large astronomical systems, such as *galaxy rotation curves and the *Tully-Fisher relation, without the need for any *dark matter. Rather than assuming that the observed dynamics are explained by unseen dark matter, MOND postulates instead that the law of *gravity is modified in the regime of very low accelerations.

Newton's second law states that the acceleration **a** experienced by an object is directly proportional to the external force **F** acting on it, with a constant of proportionality equal to the reciprocal of the object's mass m: $\mathbf{F} = m\mathbf{a}$. Under MOND, Newton's second law becomes $\mathbf{F} = m\mathbf{a}\,\mu(a/a_0)$. The extra factor $\mu(a/a_0)$ is a smoothly varying function of the amplitude of the acceleration, a, relative to a fixed acceleration $a_0 \approx 10^{-10}\mathrm{m/s^2}$, tending towards 1 in the regime of accelerations $a/a_0 \gg 1$, and

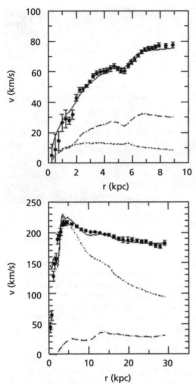

Fig. 1. Galaxy rotation curves for low (top) and high (bottom) surface-brightness galaxies. The points with error bars show the observed rotation curves, and the continuous line shows the MOND prediction, with assumed mass-to-light ratio being the only free parameter. Newtonian predictions from the visible (stellar) and gaseous components of the disk are shown as dotted and dashed lines respectively. These may be added to predict the observed rotation curve, but clearly provide insufficient rotation velocity in the absence of dark matter.

tending towards a/a_0 when the acceleration is very small, $a \ll a_0$.

In the everyday world, and in the case of planets orbiting stars, accelerations are typically much larger than $a_0 \approx 10^{-10} \mathrm{m/s^2}$, and so Newtonian gravity is an adequate description. A relatively small mass in a circular orbit of radius r about a much larger central mass M experiences a centripetal acceleration $a = GM/r^2$, where G is Newton's gravitational constant. The centripetal accelera-

tion of the Earth orbiting the Sun is $a \sim 6 \times 10^{-3} \mathrm{m/s^2}$, well within the Newtonian regime of MOND. The centripetal acceleration of the Sun in its orbit about the Galactic centre is $a \sim 2 \times 10^{-9} \mathrm{m/s^2}$, comparable to the MOND acceleration scale below which non-Newtonian effects become important (see Fig. 1).

While finding little favour with most cosmologists due to its ad hoc nature, MOND, first proposed in 1983 by physicist Mordehai Milgrom, has proved successful in explaining the observed rotation curves of a wide range of galaxies, with the assumed mass to light ratio of the galaxy being the only free parameter in addition to the (fixed) acceleration scale a_0.

A *relativistic version of MOND that agrees with recent observations of *gravitational lensing was proposed in 2004 by Jacob Bekenstein. Tensor-Vector-Scalar gravity (TeVeS) lacks the simplicity of the original MOND proposal, and it is perhaps a matter of taste whether one prefers to accept complications to the law of gravity, or the existence of undetected dark matter. A comparison of the distribution of *X-ray emitting gas and the *gravitational potential, as mapped by gravitational lensing, in the merging galaxy *cluster 1E0657-558, published by Douglas Clowe and collaborators in 2006, appears to provide definitive evidence for the existence of dark matter.

See also ALTERNATIVE COSMOLOGIES.

Sanders, R. H. and McGaugh, S. S., *Annual Reviews of Astronomy and Astrophysics*, 40, 263, 2002.

Mollweide projection

The Mollweide projection is similar to the *Hammer–Aitoff projection, addressing the same problem of how to represent the appearance of the sky around us on a two-dimensional page or screen. The general discussion in that topic is relevant here too. Like the Hammer–Aitoff projection, it maps the sky into an ellipse, and it preserves areas in the sense that equal areas on the sky correspond to equal areas in the projection. Unlike Hammer-Aitoff, lines of constant latitude are represented as horizontal lines (see Fig. 1, over page). It was invented by German astronomer Karl Mollweide in 1805, and is in common use in astronomy. Maps of all-sky *cosmic microwave background anisotropies, for instance from the *WMAP

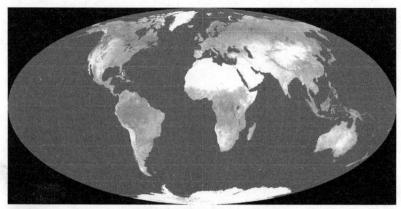

Fig. 1. A map of the surface of the Earth using the Mollweide projection, centred on the Greenwich meridian. Lines of constant latitude and longitude are shown. *Compare* HAMMER–AITOFF PROJECTION.

satellite, are commonly displayed in Mollweide projection.

MOND

See MODIFIED NEWTONIAN DYNAMICS.

monopole problem

See MAGNETIC MONOPOLES; RELIC PARTICLE ABUNDANCES.

Monte Carlo

As well as being a famous casino in the principality of Monaco, Monte Carlo refers to mathematical and computational algorithms which exploit randomness to achieve their aims, rather than the usual deterministic approach typically associated with computer calculations. There are several branches of mathematics where Monte Carlo methods have proved fruitful, particularly in the statistical analysis of observational data and in the computation of complicated multi-dimensional integrals.

Monte Carlo methods typically exploit random numbers. Computers cannot generate genuinely random numbers, but algorithms exist to create very long sequences of almost random numbers, known as pseudo-random sequences, which are good enough for most purposes. For example, a multi-dimensional integral might be evaluated by sampling its integrand at a series of randomly chosen points rather than on a grid. The methods were named after the casino, in recognition

of the key role of randomness which is also essential for a successful casino to operate.

In cosmology, Monte Carlo methods are routinely used for *parameter estimation and are becoming popular for *model selection. They are naturally well adapted to problems in *Bayesian inference which are centred around the manipulation of probability distributions. Monte Carlo methods are also widely used to create simulations of data to be obtained by forthcoming instruments, in order to compare the power of different experiments and to test data analysis tools.

For cosmological parameter estimation, the current technique of choice is Markov Chain Monte Carlo (MCMC). A Markov Chain is a sequence where the next position in the sequence depends only on the previous position (and not on any of the earlier history). The chain explores the parameter space by making a series of random jumps, which are guided by whether the fit to the data at the new point is better or worse than at the old. The jumps are initially trials: if the fit is indeed improved the jump is accepted, but if the fit is worse then the jump may nevertheless be accepted, but only with a probability dependent on how much worse the new point is. Otherwise, the test jump is rejected and the chain remains at the original location from which it makes another trial jump.

If the acceptance criterion is cleverly chosen, in a way known as the **Metropolis–Hastings algorithm**, it turns out that the distribution of points in the chain maps out the probability distribution of the different parameter values (known as the posterior distribution), hence enabling the best-fit values and confidence ranges to be extracted. Typically, the chain starts at some random position in parameter space and wanders, under guidance of the Metropolis–Hastings algorithm, towards the high-probability region or regions. Having arrived there, it then moves about the vicinity probing the probability distribution near the maximum or maxima.

multiverse

In some theories of the *Universe, the laws of physics are different in widely separated regions (separated by distances much larger than our *observable Universe). The term 'multiverse' is sometimes used to describe the collection of different 'Universes', though this is somewhat in contradiction with the usual definition of the Universe as the totality of all that exists.

One way in which a multiverse may arise is via eternal *cosmological inflation, where separate 'Universes' are continually being spawned from an everlasting inflationary domain. Such scenarios are also known as stochastic inflation, or self-reproducing inflationary Universes.

The possible existence of the multiverse is not amenable to direct observational test, since by definition the other regions lie beyond our observable Universe. The existence of a multiverse may however be implied in applications of the *anthropic principle.

N-body simulations

N-body simulations are a type of *numerical simulation that trace the evolution of the distribution of *dark matter in an expanding *Universe via gravitational interactions. The 'particles' used in large-scale simulations typically have masses in the range 10^6–10^{12} Solar masses, and thus are comparable to *galaxy masses. Since the process of *galaxy formation requires considerably more physics than *gravity alone, hydrodynamics must be incorporated to some extent in order to model galaxy formation, rather than just to trace out the distribution of dark matter. Often, simple hydrodynamical physics is incorporated into N-body simulations (*see* SEMI-ANALYTIC GALAXY FORMATION).

Basic principles

The basic idea of N-body simulations is simple. One allows a system of N particles with known masses, initial positions and initial velocities to interact gravitationally in an expanding Universe. With only two particles, the motion can be studied analytically: for a bound system, the two particles will move on elliptical orbits about a common centre of mass. However with three or more particles, the 'n-body problem', there is in general no analytic solution to the particles' motion, which must therefore be solved numerically.

One studies the system at a series of discrete time steps. At each step, one sums up all of the gravitational forces experienced by each particle due to the other $N-1$ particles. Dividing by the mass of the particle, one arrives at its acceleration. Multiplying this by the time step, one finds its change in velocity. Multiplying the average velocity by the time step, one finally arrives at the change in position of the particle. (Acceleration, velocity, and position are of course all vector quantities.) This calculation must be carried out for every particle at each time step, and so there are $N(N-1)$ calculations per time step. To sample a wide dynamic range of structures from individual galaxies to *superclusters, one needs at least many hundreds of thousands of particles, and so a simple direct integration of the forces as outlined above becomes computationally infeasible.

Various techniques have been developed which reduce the number of calculations required at each time step for a system of N particles from of order N^2 to $N \log N$. These essentially rely on dividing the simulation volume up into a hierarchy of smaller and smaller cells, until each cell contains at most a single particle. One then calculates the force on each particle due to single nearby cells, but due to larger and larger groups of cells at larger distances. Such techniques allow much larger systems of particles to be studied.

A brief history

The first N-body simulation of interacting galaxies was performed using an analogue optical computer by Erik Holmberg in 1941. Galaxies were represented by 37 lightbulbs, with photocells to measure the intensity of light, which, like gravity, drops with the square of distance.

The first digital N-body simulations in astronomy were pioneered in the early 1960s by Sebastian von Hoerner in Heidelberg and by Sverre Aarseth in Cambridge. These simulations were limited to systems of less than 100 particles.

The first cosmologically sized N-body simulations were performed by William Press and Paul Schechter in 1974. The early 1980s then saw several theoretical and technical breakthroughs, including realistic models for *dark matter, the theory of *inflation (which provided the initial conditions for the simulations), and the first use of grid-based N-body algorithms, which allowed simulations with more than 10^5 particles to be carried out for the first time. By the mid 1980s, the inflationary *cold dark matter (CDM) model had become the standard framework in which simulations were performed.

Since then, advances in simulations have relied mostly on the increase in available computing power, and in adapting N-body codes to take advantage of the availability of massively parallel computers, which can perform many calculations simultaneously. We next describe two particularly impressive simulations carried out by the Virgo Consortium, a group of astrophysicists from the UK, Germany, Canada and the US.

Hubble Volume Simulations

The Hubble Volume Simulations were the first, and at the time of writing, the only, simulations to model a significant fraction of the entire observable Universe. The two simulations, one a *critical-density model (known as τCDM), the other dominated by a *cosmological constant (ΛCDM), were carried out in 1997 and 1998 using 512 processors of the CRAY T3E supercomputer at the Garching Computer Centre of the Max Planck Society. The simulations traced the distribution of dark matter using one billion particles within a cubic volume up to $3h^{-1}$ Gpc (about 12 billion light years) on a side. Simulations of such a huge volume allowed unprecedented predictions of the evolution of large structures such as *clusters of galaxies, see Fig. 1. The mass of a single particle, however, was 2×10^{12} Solar masses, roughly the mass of a giant *elliptical galaxy, thus limiting the accuracy of small-scale structure in these simulations.

Millennium Simulation

At the time of writing, the Millennium Simulation is the largest cosmological simulation performed in terms of number of particles: over 10 billion particles in a box $500h^{-1}$ Mpc on a side. Thus while the volume surveyed is not as large as the Hubble Volume Simulations, the resolution is significantly higher: the Millennium Simulation is able to predict the positions, velocities, and intrinsic properties of *dwarf galaxies such as the Small Magellanic Cloud, as well as to follow the evolution of giant clusters and superclusters of galaxies (see Fig. 2). In order to track the formation of galaxies and *quasars, the simulation employs semi-analytic models to follow the complex astrophysical processes within *dark matter halos. The Millennium Simulation is able to follow the fate of the first quasars—they end up as central galaxies in rich clusters, and to follow the *clustering

Fig. 1. Light cone output from the Hubble Volume Simulations, also known as the 'cosmic tie'. The long part of the tie extends from the present (at the vertex below the 'knot' in the tie) to a redshift $z = 4.6$ (the tie's end). The comoving length of the image is 12 billion light years, equivalent to looking back to when the universe was 8% of its present age. Growth of large-scale structure is seen as the character of the map turns from smooth at early epochs (the tie's end) to foamy at the present (the knot). The nearby portion of the wedge is widened and displayed reflected about the observer's position. The widened portion is truncated at a redshift $z = 0.2$, roughly the depth of the Sloan Digital Sky Survey.

evolution of both dark matter and galaxies. Comparison with existing and planned *galaxy surveys will provide powerful constraints on both *cosmological models, and on the assumptions that go into semi-analytic galaxy formation models.

The simulation was performed on the 512-processor IBM p690 supercomputer at the Garching Computer Centre of the Max Planck

Fig. 2. Movie frames from a small portion of the Millennium Simulation. Redshift decreases (i.e. time increases) from $z = 127$ (left-most frame) to $z = 0$ (right-most frame). At the same time, the view zooms in on a massive cluster of galaxies. The left-most frame spans several Gpc, via clever slicing of the simulation box, which is in fact only $500h^{-1}$Mpc on a side. Sub-structures on galactic scales (~ 10 kpc) are resolved in the right-most frame, giving some impression of the huge dynamic range encompassed in this simulation. The movie itself is available from the Millennium Simulation website: see link below.

Society, taking 350 000 processor hours of CPU time, 28 days elapsed time. The total number of floating point operations carried out is estimated at 5×10^{17}.

Bertschinger, E. *Annual Reviews of Astronomy and Astrophysics*, *36*, *599*, 1998. A review of N-body simulations.

Springel, V., et al., *Nature*, *435*, *629*, 2005. Describes the Millennium Simulation.

A bibliography of the literature on N-body simulations:

http://www.amara.com/papers/nbody.html

The Millennium Simulation:

http://www.mpa-garching.mpg.de/galform/millennium/

nebulae

The term nebula, the Latin term for 'mist' or 'cloud', originally referred to a diffuse source of light in an astronomical image, including other *galaxies, but is now used to refer to an interstellar cloud of gas, *dust or plasma. Examples of nebulae include **planetary nebulae**, the glowing shells of gas surrounding some stars near the ends of their lives, and the remnants of *supernovae.

The *Messier catalogue compiled in 1784 by Charles Messier contained 110 diffuse sources of light that he referred to as nebulae. Forty of these objects have since been established as galaxies. Most of the remaining objects are actually clusters of stars, with only 11 objects being genuine nebulae as we use the term today.

The term 'galaxy' only gained common usage once Edwin *Hubble had established the extra-Galactic nature of some nebulae. Hubble himself still referred to galaxies as 'nebulae' in his book *The Realm of the Nebulae*.

neutralino

The term neutralino refers to a particular particle proposed to make up the *dark matter in the *Universe, and indeed it is regarded by many as the leading *cold dark matter candidate. It is one of the new particles predicted by *supersymmetry; indeed several neutralino particles are predicted. If one of them proves to be the lightest supersymmetric particle, then at least in the simplest models of supersymmetry that guarantees its stability against decay. It is a *fermionic particle, and the name derives from it having zero electromagnetic charge, an essential property for it to be a dark matter candidate of the class known as *WIMPs.

Supersymmetry associates a new particle with each of the known particles of the *standard model of particle physics. Confusingly, however, the actual physical particles may be made up of fractional combinations; the neutralinos are made of combinations of the supersymmetric partners of the Z *boson, the *photon, and the *Higgs boson. One of these will be the lightest neutralino, and its precise properties would depend on how it is assembled from the possible constituents, which depends on the model of supersymmetry and its parameters. It is also not guaranteed that the neutralino will be the lightest supersymmetric particle, and the partner of the graviton, known as the gravitino and predicted in supergravity theories, is sometimes suggested as an alternative candidate. At present, supersymmetric model-building is too flexible to permit very definite conclusions. That may change in the future either with direct detection of dark matter particles, or through discovery and subsequent examination of supersymmetric particles at the next generation of *particle accelerators.

neutrinos

Neutrinos are fundamental particles and a key part of the *standard model of particle physics. According to the standard model they come in three types, one associated with the *electron and the other two to the electron's heavier cousins the muon and tauon. They are *fermionic particles and part of the grouping of particles known as *leptons. They have no electric charge, and interact only via the weak nuclear force, making them extremely hard to detect.

Neutrinos in particle physics

Neutrinos were first postulated by Wolfgang Pauli in 1931. When a *neutron decayed, the only observed decay products were a *proton and an electron, but the total energy did not add up. Pauli postulated that there must be a third particle carrying away the extra energy, that particle interacting so weakly that it was not seen in any detectors. Famously, he said, 'I have done a terrible thing. I have postulated a particle that cannot be detected.' Fortunately, he was too pessimistic, and around 25 years later the first neutrino was discovered. The second type, the muon neutrino, was discovered in 1962 and the tau neutrino in 2000. The name 'neutrino' was coined by Enrico Fermi and is Italian for 'little neutral one', the name a nice contrast against the much heavier neutral particle, the neutron.

The interaction strength of neutrinos depends on their energy, increasing as the square of their momentum. But even at higher energies interactions are very rare. Neutrinos are produced by nuclear reactions deep in the Sun, and, for their characteristic energy, it would take roughly a light-year-long block of solid lead to make them interact. Accordingly, having been created in the heart of the Sun they fly out of it without further interacting. It is believed that most of the energy of a *supernova explosion is carried away by neutrinos, and indeed in 1987 neutrinos emitted by a supernova in the nearby Large Magellanic Cloud were spotted by detectors on Earth.

Neutrinos provide the most direct evidence that the standard model allows only three generations of particles. The Z *boson (one of the particles mediating the weak nuclear force) can decay into neutrino–anti-neutrino pairs, and the rate of decay depends on how many types of neutrinos there are to decay into. Measurements at the CERN *particle accelerator indicate that the number of neutrinos is 2.994 ± 0.012, perfectly consistent with three types. This result is also supported by cosmological arguments relating to the theory of *nucleosynthesis.

A key question is whether neutrinos have a mass; according to the standard model they are perfectly massless, and if this proves not to be true the standard model requires revision or enhancement. Direct measurements have not revealed a mass, setting upper limits (in *particle physics units) of 3 eV, 0.2 MeV, and 18 MeV for the electron, muon, and tau neutrinos respectively. *Cosmology sets much more stringent limits on neutrino masses, as discussed below.

Nevertheless, there *is* compelling evidence that neutrinos have a mass, coming from a surprising source. It turns out that neutrinos can change type (the usual jargon for the different types is **flavour**) while in flight. What started out as an electron neutrino, generated for instance in nuclear reactions deep within the Sun, may turn into a muon or a tau neutrino before arriving at our detectors. This phenomenon is known as **neutrino oscillation**.

The possibility of neutrino oscillations was first raised in 1968, when Ray Davis examined neutrinos from the Sun. To see such weakly interacting particles, he procured 600 000 litres of chlorine (in the form of cleaning fluid) and sited it deep underground where only neutrinos could reach. Neutrinos interacting with chlorine generate argon, and counting those argon atoms yields the intensity of neutrinos. He discovered only about 30% as many electron neutrinos as expected. While the community was initially sceptical that such a measurement was real, Davis was absolutely vindicated by subsequent experiments and shared the *Nobel Prize in 2002 for this remarkable achievement.

Modern neutrino detectors are essentially scaled-up versions of Davis's original experiment, but with more sophisticated detector techniques. Fig. 1 shows the interior of the massive Japanese detector Super-Kamiokande, which when operating is filled with water. Neutrinos interact with electrons in the atoms, forcing them to travel faster than the *speed of light in water (though not of course faster than the speed of light in vacuum). Such electrons slow down by emitting *photons known as Cherenkov radiation (the optical equivalent of a sonic boom), which

Fig. 1. The interior of the Super-Kamiokande experiment during construction, roughly half-filled with water. For scale, on the right-hand side is a rowing boat containing two scientists. The vast array of what look like light-bulbs are photo-multiplier tubes which amplify any light generated within the detector by neutrino interactions.

are amplified by detectors mapping out the trajectory of the electron, from which the neutrino properties can be inferred. A similar experiment in Canada, the Sudbury Neutrino Observatory (SNO), is smaller scale but achieves enhanced sensitivity by using only heavy water, where the hydrogen atoms in the water molecules are replaced by deuterium atoms. In combination, these experiments all confirm neutrino oscillations.

Neutrino oscillations require that neutrinos have a mass, and are currently the strongest evidence for physics beyond the standard model. However the oscillations don't probe the mass itself, but instead the *difference* between the squares of the masses of the two neutrino types in question. Here it all gets a little confusing, because the neutrinos that have a well-defined mass are not the same as the different types. If we label the lowest-mass neutrino as ν_1, it is made up of a mixture of fractions of the electron, muon, and tau neutrinos (the particle physics jargon is that the mass eigenstates are superpositions of the flavour eigenstates). The mass difference between the first two neutrino types is determined by observations to be

$$m_{\nu_2}^2 - m_{\nu_1}^2 \approx (9 \times 10^{-3} \text{ eV})^2 ,$$

but these experiments do not measure the two masses separately. The simplest interpretation is that the first neutrino is much lighter than the second neutrino, and then the second will have mass 9×10^{-3} eV, but an alter-

native interpretation is that the two could have much bigger masses with just a small difference between them.

The mass-squared difference between the second and third neutrinos is also now quite accurately measured, as

$$m_{\nu_3}^2 - m_{\nu_2}^2 \approx (0.05 \text{ eV})^2 .$$

Again, the simplest interpretation is that the second one is much lighter than the third, implying $m_{\nu_3} \approx 0.05$ eV, but that need not be the case. This number is however a lower limit on the mass of the third neutrino type. Determining the absolute scale of neutrino masses is a key goal of neutrino physics.

Although the standard model contains only three types of neutrino, many extensions of the standard model, particularly those based on *grand unification, predict additional neutrinos. Because the Z boson decay sees only three types, these neutrinos must not even have weak interactions, making gravity their only interaction. Such neutrinos are known as **sterile neutrinos** in recognition of their lack of interaction. Their existence is hard to verify in particle experiments, but presuming the sterile neutrinos have mass they can still have gravitational interactions, with cosmological implications.

Neutrinos in cosmology

According to the *standard cosmological model, nearly half the particles in the *Universe are neutrinos. Specifically, it predicts that the most numerous species are photons and neutrinos, and that for every 11 photons there are three neutrinos or anti-neutrinos of each type, the numbers of neutrinos and anti-neutrinos being essentially identical. In numerical terms all other particles are utterly negligible; there is only about one proton or electron per billion photons or neutrinos, and probably even fewer *dark matter particles though the number density of those is not known. Despite this vast numerical disadvantage, the dark matter and baryons are nevertheless the dominant particles in the present Universe in terms of their total mass–energy; each proton carries around a trillion times the mass–energy of a typical photon.

The Universe is believed to be filled with neutrinos—the neutrino equivalent of the *cosmic microwave background (CMB) of photons—because during the early stages of the Universe's evolution the neutrino energies and the ambient density of material would have been sufficient for neutrino interactions to be frequent. They would be created and destroyed by a variety of reactions, and be in *thermal equilibrium with all the other particles present, including the photons. This would guarantee a similar number of photons and neutrinos.

This equilibrium phase would persevere until the Universe was around one second old. The strength of neutrino interactions decreases as their energy gets less, which happens due to *redshifting as the Universe expands. The same expansion is also reducing the density of the Universe, leaving less to interact with. In combination, these lead to the neutrino equivalent of *decoupling; there simply becomes a negligible chance of a given neutrino experiencing any more interactions. After that the neutrinos travel freely through the Universe, known as *free streaming, and are sometimes known as the **cosmic neutrino background**.

If photons and neutrinos had identical properties, this process would lead to there being one of each type of neutrino per photon. This is a good first approximation but in detail needs some correcting. Firstly, neutrinos are fermionic particles while photons are bosons, which slightly modifies their relative density. Secondly, after neutrino decoupling, the photon energy gets a boost when electrons and positrons drop out of thermal equilibrium and annihilate, which

they do predominantly into photons rather than neutrinos. In combination, these two corrections give the factor 11/3 mentioned above, which leads to there being slightly more photons than neutrinos.

The neutrino background may not interact directly with other particles, but nevertheless it is an important component of the Universe and has observable consequences through its gravitational effect. During cosmic *nucleosynthesis, when the light elements are first created, following the detailed pattern of chemical reactions requires that one includes the contribution of the neutrinos to the expansion rate of the Universe. According to the *Friedmann equation the expansion rate is proportional to the square root of the density; during the *radiation-dominated era the neutrinos contribute nearly half the total density, and so the expansion rate at a given temperature would be quite different were they absent. Indeed, nucleosynthesis calculations only match observations provided there are three neutrino species, in agreement with what is found in particle accelerator experiments.

Neutrinos also have important effects on the development of cosmic structures. For instance, calculations of *CMB anisotropies must also include the neutrinos to make predictions accurate enough to compare with observations, and in principle need even to include perturbations in the neutrino density from point to point (the neutrino equivalent of the CMB anisotropies). The neutrinos influence the epoch of *matter–radiation equality, which imprints a characteristic scale on the clustering pattern of dark matter and galaxies.

The above discussion is valid for massless neutrinos. A neutrino with mass–energy less than about 5×10^{-4} eV will still be relativistic today, and its mass will never have had any effect on its properties. If its mass is higher than this, and we have seen that neutrino oscillations indicate it is for at least two of the neutrino species, we should consider the cosmological effects of neutrino mass. It is convenient to divide the discussion into two separate possibilities, known as **light** and **heavy neutrinos**.

Light neutrinos—This refers to neutrinos of mass–energy up to about 1 MeV. Such neutrinos may be non-relativistic now, but at neutrino decoupling they would have been relativistic with their mass playing a negligible role. The production of neutrinos would therefore be just the same as in the massless case, meaning that the same number would have been left after neutrino decoupling. However, their present contribution to the density is dominated by their mass, and so the total density of neutrinos is proportional to the mass. Expressed as a fraction of the *critical density using the *density parameter Ω_ν, their density is predicted to be

$$\Omega_\nu = \frac{\sum m_\nu c^2}{94 h^2 \, \text{eV}}$$

where h is the *Hubble constant, believed to equal about 0.7, and the summation is over all types of neutrinos whose mass is less than 1 MeV.

We see that a light neutrino species could readily provide the observed dark matter density of $\Omega \approx 0.3$, which would require a neutrino of mass–energy around 14 eV. Particle experiments exclude this possibility for the electron neutrino but not the other two. Unfortunately though, as discussed below, such a neutrino would be hot dark matter which is not favoured.

Heavy neutrinos—This refers to neutrinos of mass–energy greater than 1 MeV. In this case the neutrinos are non-relativistic at neutrino decoupling, which means that their final numbers do depend on the mass and the calculation becomes rather more involved. A detailed calculation shows that the present predicted density then begins to fall, and if the neutrino mass is about 1 GeV the observed dark matter density is once again reproduced. Such neutrinos would be a *cold dark matter candidate, but unfortunately particle accelerator experiments already exclude such a large mass for the three standard neutrino types.

If neutrinos have a mass, then they have gravitational attraction and hence act as dark matter. From now on we consider only the more interesting case of light neutrinos. As we have already seen, the highest permitted mass, corresponding to neutrinos providing all the dark matter, is about 14 eV. On the other hand, the *minimum* neutrino mass inferred from the oscillation experiments is 0.05 eV, which would contribute a density of about one-thousandth the critical density. As

it happens, this is about the total density of all the stars in the Universe so, as a minimum, neutrinos contribute as much to the density of the Universe as stars do.

Unfortunately, light neutrinos are not a good candidate for the dark matter responsible for structure formation. Although non-relativistic today, they will have been travelling at relativistic speeds for a significant part of the history of the Universe, and this free streaming opposes gravitational collapse. It does so because there are more neutrinos in the higher-density regions, and hence their random motions lead to a higher net flux from high to low density than vice versa. Only once they slow down to non-relativistic speeds can neutrinos begin to cluster and provide gravitational attraction to bring galaxies together.

In the extreme case where a neutrino is all the dark matter, we have the hot dark matter scenario. Initially popular in the 1980s, this scenario is hopelessly excluded today; if the theories are to provide the right magnitude of CMB anisotropies it turns out that no galaxies form at all, obviously in contradiction to the true Universe. Indeed, even a fairly small level of neutrino dark matter, combined with cold dark matter from another source, is strongly constrained by this suppression effect. Detailed analyses indicate a cosmological constraint that the sum of the three neutrino masses cannot exceed 1 eV, and this limit is substantially stronger than any non-cosmological bounds. Even if neutrinos have the minimum mass suggested by neutrino oscillations, it should be possible to spot their dynamical effect on structure formation with upcoming experiments.

Super-Kamiokande WWW site:
http://www-sk.icrr.u-tokyo.ac.jp/index_e.html

Sudbury Neutrino Observatory WWW site:
http://www.sno.phy.queensu.ca

neutrons

Neutrons are particles which, along with *protons, are the constituents of atomic nuclei. They were discovered by James Chadwick in 1932. At a more fundamental level neutrons are made up of three *quarks, and are members of the classes of particles known as *baryons and *hadrons.

The mass of an individual neutron is 1.7×10^{-24} grams, more usefully expressed in *particle physics units by its mass–energy of 939.6 MeV. This is just fractionally larger than the proton mass–energy of 938.3 MeV. They possess no electric charge and thus are said to be neutral, from which their name derives. Isolated neutrons are unstable, decaying into a proton, an *electron and an anti-*neutrino, a process known as beta decay. The half life for this process is 614 seconds, which is extremely long by the usual standards of particle physics. Neutrons which have formed atomic nuclei may be stable, for instance the helium-4 nucleus is a stable configuration of two protons and two neutrons. Atomic nuclei are not always stable; the conversion of a neutron into a proton, electron and anti-neutrino by beta decay is one of the mechanisms of radioactive decay, the escaping electron being the measured radioactivity and the anti-neutrino being too weakly interacting to detect directly.

The initial abundance of neutrons in the *Universe is set by the process of cosmic *nucleosynthesis. This theory indicates that roughly 10% of baryons are neutrons, the rest being protons. Almost all of these neutrons end up in the form of helium-4, with trace amounts of other light nuclei present. Much of this helium remains in the *intergalactic medium, having not yet had the opportunity to form stars. Neutrons that do end up in stars may find themselves in the heavier elements created by the nuclear fusion that provides the star's energy.

A remarkable phenomenon is the **neutron star**, which is the evolutionary endpoint of stars whose mass lies in the range around two to ten times the mass of the Sun. Such stars end their life in a *supernova explosion, with the stellar core collapsing as the star runs out of fuel and hence its ability to support itself against *gravity. In this dense environment, protons and electrons are forced together to form neutrons (in the process releasing a considerable number of neutrinos), and the resulting object, of radius perhaps 20 kilometres, is a dense ball of neutrons held together by gravity and resembling a giant atomic nucleus. Such neutron stars have intense magnetic fields and some, known as **pulsars**, are observed via repetitive bursts of radio waves sent out as they spin. Neutrons are able to form such objects only because they have no electrical charge; if one attempted the same trick with protons it would immediately be blown apart by the repulsive effect of all the positive charges.

Nobel Prize

The Nobel Prize in Physics is the most prestigious award for achievement in physics research. It is awarded annually by the Royal Swedish Academy of Sciences, one of several Nobel Prizes in different areas of knowledge. The prizes were set up according to the wishes of Alfred Nobel, who invented dynamite and made his fortune from establishing companies in many countries. According to his will, referring to the interest on his bequest and the five subjects he defined, the prizes will be awarded '… to those who, during the preceding year, shall have conferred the greatest benefit on mankind. The said interest shall be divided into five equal parts, which shall be apportioned as follows: one part to the person who shall have made the most important discovery or invention within the field of physics; …'. The original five subjects were physics, chemistry, physiology or medicine, literature, and peace. In 1968 the Bank of Sweden instituted a sixth prize in the area of economic science.

The physics prize was first awarded in 1901, to Wilhelm Röntgen for the discovery of *X-rays. It is awarded each year to between one and three people, who receive a medal and a share of the prize fund which is presently ten million Swedish Kronor (approximately one million euros/dollars). The decision process is a carefully guarded secret with an extensive nomination and assessment procedure, ending with the Nobel Committee making a recommendation to the Academy whose vote, usually in mid October, finalizes the award. The announcement is made with a short citation describing the achievements being recognized by the award, and the award ceremony takes place in December.

A complete list of winners and their achievements can be found at the Nobel Prize website listed below. A brief list of physics Nobel Prize awards directly relevant to cosmology appears below with excerpts from their citations. In the case of Albert *Einstein, his 1921 Nobel Prize was awarded principally for his explanation of the photoelectric effect and not for his work on *relativity.

1921 Albert Einstein: for his services to theoretical physics, and especially for his discovery of the law of the photoelectric effect.

1922 **Niels Bohr:** structure of atoms and radiation emanating from them.

1932 **Werner Heisenberg:** creation of *quantum mechanics.

1933 **Erwin Schrödinger and Paul Dirac:** new forms of atomic theory.

1935 **James Chadwick:** discovery of the *neutron.

1936 **Victor Hess:** discovery of cosmic radiation [in modern terminology, *cosmic rays] (share of prize).

1965 **Sin-Itiro Tomonaga, Julian Schwinger, and Richard Feynman:** physics of elementary particles.

1967 **Hans Bethe:** energy production in stars.

1969 **Murray Gell-Mann:** classification of elementary particles.

1974 **Martin Ryle:** pioneering research in radio astrophysics.
      Antony Hewish: discovery of pulsars.

1978 **Arno Penzias and Robert Wilson:** discovery of the *cosmic microwave background radiation (share of prize).

1979 **Sheldon Glashow, Abdus Salam, and Steven Weinberg:** unified weak and electromagnetic interaction.

1983 **Subrahmanyan Chandrasekhar:** structure and evolution of stars.
      William Fowler: formation of chemical elements in the Universe.

1993 **Russell Hulse and Joseph Taylor:** discovery of a binary pulsar.

1999 **Gerardus 't Hooft and Martinus Veltman:** electro-weak interactions in physics.

2002 **Raymond Davis Jr and Masatoshi Koshiba:** detection of cosmic neutrinos.
      Riccardo Giacconi: discovery of cosmic X-ray sources.

2004 **David Gross, David Politzer, and Frank Wilczek:** theory of the strong interaction.

2006 **John Mather and George Smoot:** discovery of the black-body form and anisotropy of the cosmic microwave background radiation.

Home Page: http://Nobelprize.org

non-Gaussianity

The simplest type of random process, discovered by Carl Friedrich Gauss in the 1800s, is a *Gaussian one, where the probability of obtaining a value follows the bell-shaped Gaussian distribution. A powerful mathematical theorem—the central limit theorem—

indicates that many random processes result in a Gaussian distribution. In *cosmology, the most important random process is the mechanism which generated the initial *density perturbations from which *cosmic structure grew. A key question, then, is whether this process follows Gaussian statistics. If it does not, the initial perturbations are said to be non-Gaussian.

So far, all measurements have proved consistent with the initial cosmic perturbations being Gaussian. The best tests come from the *cosmic microwave background (CMB) anisotropies, and in particular the measurements by the *WMAP satellite, as those observations probe the perturbations when they are in a relatively pristine state. By contrast, testing for non-Gaussianity in the galaxy distribution is difficult because the *gravitational instability process generates non-Gaussianities even if the initial perturbations are Gaussian, which must then be untangled from any primordial effect.

While Gaussianity is a well-defined and predictive scenario, non-Gaussianity is rather slippery as there are many (indeed, infinitely many) different types of non-Gaussian randomness, many of which become very nearly Gaussian in some limit. Moreover, none are particularly well motivated from theoretical considerations. Because of this a large battery of cosmological tests for non-Gaussianity have been developed, which have sensitivities to different types of non-Gaussianity. Some of these methods are computationally very intensive. Over the years there have been several reports of deviations from non-Gaussianity, but these have so far been ultimately traced to instrumental effects rather than a true primordial origin.

As far as theories for the origin of structure are concerned, the simplest models of *cosmological inflation, presently the leading candidate, predict very nearly Gaussian perturbations. In this case the deviations are too small to ever be measured. However, more complex models may be able to produce measurable non-Gaussianity, the *curvaton variant being a case in point. Perturbations generated by *topological defects such as *cosmic strings are expected to be significantly non-Gaussian, at least on small scales, but WMAP satellite observations require such a contribution, if present at all, to be subdominant to the inflationary one. If there is a non-trivial *topology of the Universe, this will necessarily lead to non-Gaussianities in the CMB anisotropies, which would be the most promising way of discovering it.

A detection of non-Gaussianity would provide powerful indications as to how the initial perturbations were generated, and is an important target of upcoming projects including the *Planck satellite.

non-singular cosmologies

Conventional *cosmological models begin with a *big bang, which corresponds to a singularity in the *space-time where the laws of physics cease to apply. There have been many attempts to create cosmological models which do not contain an initial singularity, and these are collectively known as non-singular cosmologies. It is not known whether or not there was an actual big bang.

In non-singular cosmological models, the big bang is typically replaced by an epoch where the *Universe is collapsing. It then passes through a minimum size and begins expanding again. The collapsing phase might extend into the infinite past, or alternatively be the latest collapse in an endless cycle of collapse and re-expansion which may continue into the future. The simplest model of the former type is to have a particularly strong *cosmological constant, which can generate a sufficient repulsive force for an initially collapsing Universe to halt and re-expand. Such a large cosmological constant is however ruled out by observations such as those of distant *supernovae.

The cosmological constant model relies solely on Albert *Einstein's theory of *general relativity to describe *gravity, but typically, attempts to avoid the initial singularity assume that during the Universe's earliest stages this no longer applies, perhaps because *quantum mechanical effects relating to gravity become important. Possible alternative theories applying at this early stage include *superstring theory and *M-theory, and there is also the possible effect of *extra dimensions of space as might be invoked in the *braneworld scenario. These mechanisms for avoiding the singularity would operate when the Universe was extremely hot and dense.

A particular example is a class of cosmological models known as **pre big bang** models, introduced by Maurizio Gasperini and Gabriele Veneziano in the early 1990s. They realized that superstring theory predicted

that for every expanding Universe model, there was an equivalent collapsing Universe model related to it by a property known as string duality. If those two models could be connected, it would provide a way to extend the usual cosmological model to an epoch before the big bang. This model is echoed in the later development of the *ekpyrotic Universe scenario. However in neither case is the transition from collapsing to expanding phase understood, and so both proposals remain controversial.

See also STEADY-STATE COSMOLOGY.

nucleosynthesis

Cosmic nucleosynthesis is the process by which atomic nuclei first formed in the *Universe. It took place when the Universe was approximately one second old. After nucleosynthesis, the Universe contained mainly nuclei of hydrogen and helium-4, with additional trace amounts of deuterium, helium-3, and lithium-7. The success of nucleosynthesis theory is one of the strongest pieces of evidence in support of the hot *big bang cosmology (*see* 'Overview').

History

The subject of nucleosynthesis arose out of a desire to explain the observed abundances of chemical elements in the Universe. By the 1940s, it was already known that stars generated their energy from nuclear fusion of elements, particularly hydrogen to helium as proposed by Arthur *Eddington in 1920. This theory was refined by Hans Bethe and Carl Friedrich von Weizsäcker, who detailed the main reactions—the *proton–proton chain in low-mass stars and the carbon–nitrogen–oxygen cycle in high-mass stars. These theories were later to be refined and unified in a landmark 1957 paper by Margaret and Geoffrey Burbidge, William Fowler, and Fred *Hoyle, which demonstrated that stellar nucleosynthesis could explain the observed abundances of all heavy elements from lithium up to iron. Yet heavier elements are believed to be produced in *supernovae.

A simple and elegant hypothesis was that the Universe began consisting entirely of hydrogen, with all heavier elements created in stars. Unfortunately, while this worked for all elements heavier than lithium, it had become apparent that the observed primordial abundance of helium was, at nearly 25%

by mass, far in excess of anything that could be produced from the observed rate of stellar fuel consumption. During the 1940s, George *Gamow turned his attention to the problem, and came to believe that the helium could have been produced in the early stages of the Universe's evolution.

His ideas first found a proper formulation in a 1946 paper 'Expanding Universe and the Origin of Elements', and were refined in subsequent work with Ralph Alpher and Bethe (Bethe's role in the paper was actually not on the physics at all, as explained in the biographical entry on Gamow), and later also Robert Herman. At around this time, Alpher and Herman would also predict the existence of the *cosmic microwave background radiation. Important steps were also taken by Chushiro Hayashi, who used statistical physics arguments to predict the relative abundance of protons and *neutrons at the onset of nucleosynthesis, and by Alpher and Herman in collaboration with James Follin. The work indicated that the observed abundance of helium could indeed have been generated during the hot early stages of the Universe.

Cosmic cookery: the basic picture

The process of nucleosynthesis is very similar to that of *recombination and *decoupling, which leads to formation of the cosmic microwave background. However in that context the physics is atomic physics, concerning the attachment of electrons to atomic nuclei. In nucleosynthesis, we are concerned with the formation of the nuclei themselves. The characteristic energy scale of nuclear processes is roughly one million times greater than that of atomic processes, and so they are relevant at a much younger and hotter stage of the Universe's evolution. [Note that the phrase 'atomic bomb' is a misnomer; the drastic power of such weapons arises from nuclear reactions, either fission of very heavy nuclei such as uranium or fusion of light nuclei.]

The raw ingredients for nucleosynthesis are the *baryons, namely the protons and neutrons in the Universe. Essentially, the procedure follows a simple recipe: take a collection of baryons, heat to a trillion degrees, and leave to cool for a few minutes. You'll be serving a meal consisting mostly of hydrogen, a reasonable portion of helium-4, and with a light seasoning of deuterium, helium-3, and lithium-7.

Let's follow it from the beginning. We chose a trillion degrees, a temperature attained when the Universe was about one ten-thousandth of a second old (*see* TEMPERATURE–TIME RELATION), because this was well before the main events take place. The Universe is in *thermal equilibrium, with frequent interactions amongst all the particles establishing an equilibrium distribution at that temperature. The typical energy per particle at such temperatures is 100 MeV in *particle physics units. This is less than the proton and neutron masses, (938.3 MeV and 939.6 MeV respectively), and so they are already non-relativistic (moving much slower than the *speed of light). The frequent interactions ensure that there are essentially the same numbers of protons and neutrons at this time.

At this stage nuclei cannot form, because the ambient medium is simply too energetic. The binding energy of typical light nuclei is approximately 1 MeV per particle, and as soon as two particles try to join they are immediately blasted apart by the high-energy photons pervading the Universe.

As the Universe continues to cool, approaching a temperature of ten billion degrees and an ambient energy of 1 MeV, the difference in mass between the proton and neutron starts to become important. Being slightly heavier, it is somewhat harder to make neutrons than protons in nuclear reactions, and without much spare energy around this starts to tip the balance in favour of the protons. At around this time, interactions changing protons to neutrons and vice versa begin to become inefficient, and the proton-to-neutron ratio settles at around five protons per neutron. The Universe is about one second old.

Still nuclei cannot form due to the ambient energy, and a new phenomenon comes into play. Free neutrons are unstable, undergoing beta-decay to become a proton, electron, and an electron anti-*neutrino. Mediated by the puny weak nuclear force, the half-life for this process is astonishingly long by particle physics standards, around ten minutes. The neutrons now urgently need to combine into a nucleus, because there they can be stabilized by the nuclear binding energy.

Eventually, with the Universe around 400 seconds old, it becomes cool enough that nuclei can start to form and survive. Due to decays, by that time there is only around one neutron for every eight protons. The neutrons prefer to live in the most tightly-bound light nucleus, which is helium-4, made up of two protons and two neutrons. Almost all the neutrons combine this way; just a small number fail to find a partner neutron and make the best of an unfortunate situation by joining one or two protons to form deuterium or helium-3. And even more occasionally, some helium-4 undergoes further reactions to make lithium-7. Bound into the light nuclei, the neutrons are safe from decay.

In order to form helium-4, an equal number of protons and neutrons are required. But we started with far more protons, and the remainder have no neutrons to merge with. They are destined to remain single, but at least at this time we should recognize that a proton is the same thing as a hydrogen nucleus, and begin to call it that. Eventually, much later in the Universe's evolution, the solitary protons will undergo recombination with electrons to become fully fledged hydrogen atoms.

Totalling it up, we find that there is one helium-4 nucleus for every fourteen or so hydrogen nuclei. The helium nucleus has four times the mass, however, so the fractions by mass of hydrogen and helium-4 are approximately 78% and 22% respectively. This is in good agreement with the observed compositions of slow-burning stars such as our Sun.

The present Universe contains many heavier elements than those discussed above; for instance as well as hydrogen you are made of substantial amounts of carbon, oxygen, and other elements. Heavier elements were not produced in the big bang, but were made subsequently by nuclear reactions within stars and released into the interstellar medium by *supernovae.

Cosmic cookery: the full story

The above description indicates some of the key physics but is oversimplistic. A full calculation has to follow a complicated network of nuclear reactions corresponding to creation and decay of the many species involved, and needs to be done on a computer. Results are shown in Fig. 1. There are two particularly important constraints which can be derived from nucleosynthesis, one on the *density of

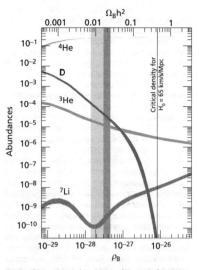

Fig. 1. The predicted abundances of the main four light elements, as a function of the baryon density in the Universe. The present baryon density is given in kilogrammes per cubic metre along the bottom axis, or as a density parameter (multiplied by the Hubble constant squared) along the top axis. It is assumed that there are three neutrino families, and the width of the bands indicates the size of calculational uncertainties. The vertical bands indicate the observationally allowed regions, the darker band being a strong interpretation of the observational data, and the combined light and dark bands being a more conservative interpretation. Finally, the critical density is shown for a particular choice of the Hubble constant, showing that the permitted values of the baryon density are only around 0.04 of the critical density.

baryons and the other on the number of types of neutrino.

The initial density of baryons is the main input into a nucleosynthesis calculation. Because the early Universe was in a state of thermal equilibrium, the total density is all that need be specified and the relative abundances of protons and neutrons follow from statistical physics arguments. Because the Universe is *radiation-dominated throughout nucleosynthesis, the density of baryons has no effect on the expansion rate, but it does feed into the network of chemical reactions governing how the nuclei build up. Fig. 1 shows how the predicted abundances vary with the baryon density. Agreement with observations holds only for a narrow range of densities around $\rho_{\text{baryon}} = 4 \times 10^{-28}\,\text{kg}\,\text{m}^{-3}$. Deuterium is a particularly powerful constraint in this density range, as its predicted abundance falls rapidly with baryon number, and much recent effort has gone into measuring its abundance, particularly through *quasar absorption systems.

These observations show that the baryon density is much less than the *critical density, indeed being only about 4% of it. This is also sufficient to indicate that the *dark matter in the Universe, whose measured density is about 25% of the critical density, cannot be made entirely from baryons.

For most of the history of modern cosmology nucleosynthesis gave the best bound on the baryon density, but it has recently been overtaken by measurements using *cosmic microwave background anisotropies.

The second input is the number of types of neutrino. This affects the total energy density corresponding to a given temperature during the radiation-dominated era; more neutrinos imply a faster expansion rate as nucleosynthesis occurs. This alters the balance between the nuclear reaction rates and the rate of expansion, and hence changes the final yields of the elements in a characteristic way. It was already apparent in the mid 1980s that a good fit to the observed element abundances was only possible with three neutrino families, rather than two or four. This was subsequently verified by analysis of Z *boson decays at the CERN *particle accelerator.

Weinberg, S. *The First Three Minutes*, Basic Books, 1994.

numerical simulations

Numerical simulations are a vital tool for the modern cosmologist. They provide a crucial link between *cosmological perturbation theory, which provides a description of small *density perturbations on large scales, to observations of large density perturbations on small scales. Simulations thus enable one to predict how observed structures such as *galaxies and *clusters of galaxies form in an assumed *cosmological model.

There are two main types of cosmological simulations: *N-body simulations and *smoothed particle hydrodynamics. The former consider a system of N discrete particles which interact solely via their mutual gravitational attraction. The 'particles' used in large-scale simulations typically have masses in the range 10^6–10^{12} solar masses, and thus are comparable to galaxy masses. Hydrodynamical simulations treat the contents of the Universe as a fluid, and incorporate the flow, compression, expansion, heating, and cooling of gas. Since the process of *galaxy formation requires considerably more physics than *gravity alone, hydrodynamics must be incorporated to some extent in order to model galaxy formation, rather than just to trace out the distribution of *dark matter. Often, simple hydrodynamical physics is incorporated into N-body simulations (*see* SEMI-ANALYTIC GALAXY FORMATION).

Bertschinger, E. *Annual Reviews of Astronomy and Astrophysics*, 36, 599, 1998.

Springel, V. et al., *Nature*, 435, 629, 2005.

http://www.amara.com/papers/nbody.html

observable Universe

The observable *Universe is that part of the Universe that we can in principle make observations of. It takes the form of a large sphere centred on our location, with radius such that if a ray of light set out towards us at the instant of the big bang, then it would just reach us today. As nothing can travel faster than light, we cannot make any observations telling us what the Universe looks like beyond that distance, known as the *particle horizon.

If the distance to an object is a significant fraction of the horizon distance, then any light we receive from it will have been travelling for a sizeable fraction of the Universe's age, and so the Universe will have expanded as the light travels. The light from such objects is therefore *redshifted. The closer to the edge of the observable Universe the object is, the larger the redshifting will be.

Almost all the information we receive from distant objects is in the form of *photons of light. Photons could not in fact reach us from the horizon distance, because the Universe was opaque to photons during its early stages. Only at the epoch of *decoupling, when the *cosmic microwave background came into being, does the Universe become transparent. Hence in practice light can only reach us from within a sphere known as the *last-scattering surface, which is located slightly inside the observable Universe. It is quite common therefore for the term observable Universe to be used to refer to the region within the last-scattering surface, rather than the true horizon distance. However more exotic forms of radiation, such as *gravitational waves, or very weakly interacting particles, such as *neutrinos, could in principle travel to us from distances beyond the last-scattering surface.

The distribution of matter, such as *galaxies, within the observable Universe is found to be very even, demonstrated most clearly by the near uniformity of the temperature of the cosmic microwave background in different directions. One explanation for this uniformity comes from *cosmological inflation, which postulates that our entire observable Universe lies within what was a tiny region of the young Universe, whose size was greatly increased by a period of rapid cosmic expansion.

observational selection effects

Whenever one analyses observational data, one needs to bear in mind that, taken at face value, the data will not be a truly representative sample of the *Universe.

Several observational selection effects can bias our data.

Flux-limited samples—In an astronomical image, the apparent brightness of each source depends on the observed flux from that source. Since flux decreases with the square of distance, only intrinsically luminous sources at large distance will make it into a flux-limited sample. This is particularly apparent in a *redshift survey in which the *galaxies have been selected by apparent flux. First, the overall number density of galaxies will decrease with *redshift, and second, the objects with higher redshift will tend to be more intrinsically luminous. It is important to account for both of these effects when analysing such a sample.

Surface brightness selection—This selection effect applies to galaxies, and arises because galaxies are extended sources with low *surface brightness. Bright spiral galaxy disks have a typical central surface brightness in the B band of $I_B(0) \approx 21.7$ mag arcsec^{-2}; the surface brightness drops off exponentially with radius. The moonless night sky has a surface brightness around 22.5 mag arcsec^{-2} in the B band and so is typically brighter than all but the inner core of a galaxy. Some *dwarf galaxies, which tend to be of low surface brightness, are entirely fainter than the night sky.

A shallow image (taken with a small telescope and with a short exposure) can thus only detect galaxies of high surface brightness. In 1970, Ken Freeman pointed out that the majority of spiral galaxies have a B-band central surface brightness within 0.3 *magnitudes of $\mu_B = 21.65$ mag arcsec^{-2}. This result, known as Freeman's Law, came about because the sample used was biased against galaxies of low surface brightness, and the majority of galaxies are almost certainly fainter than this.

Since surface brightness scales with redshift as $(1 + z)^{-4}$, distant galaxy samples are particularly biased against low surface brightness objects. Even those galaxies which are of high enough surface brightness to be detected are likely to have their total fluxes underestimated, unless an accurate model is fitted to the galaxy's profile.

Observational passband—A galaxy classified as an elliptical in a red passband may be classified as a spiral in a blue passband, since the bulges of galaxies are redder than the spiral arms (*see* GALAXY CLASSIFICA-TION).

Aperture effects—When obtaining *spectroscopy through a fixed angular-size aperture, such as an optical fibre, one frequently does not sample all of the light from an extended source such as a galaxy. For nearby galaxies, one may only sample light from the central bulge, and so star-forming regions in the disk of a galaxy may not be sampled. One thus needs to take care in comparing the spectra of galaxies of different angular size and, implicitly, at different distances.

One might also regard the *anthropic principle as a selection effect. While not biasing our observations, it does require that we live in a Universe that is capable of supporting human life, thus providing limits to the allowed values of the *cosmological parameters.

Olbers' paradox

Before the expansion of the *Universe was demonstrated by Edwin *Hubble in 1929, it had been commonly assumed that the Universe was infinite, eternal, and static. If these assumptions were true, however, then in whatever direction you looked, your line of sight would eventually intersect a star, and so the entire sky would be approximately as bright as the Sun. (This is because the *surface brightness of a star is independent of distance: while the flux received decreases

with the square of distance, so does the apparent size of the star, and so the flux per unit area is constant.) This paradox, 'Why is the night sky dark?', was written about as early as 1576 by Thomas Digges, but was stated clearly in 1823 by the Prussian astronomer Heinrich Olbers, and so is known today as Olbers' paradox.

The paradox cannot be resolved by postulating a large amount of interstellar or intergalactic dust blocking out the distant starlight, since the dust would eventually heat up and re-radiate the energy at longer wavelengths, leading to a much higher *cosmic infrared background than is observed.

The resolution of the paradox is in recognizing that two of the assumptions made are false, at least according to *big bang cosmology (*see* 'Overview'). First, the Universe is of finite age, thus limiting the total amount of light emitted by stars and *galaxies. Second, the Universe is expanding, which reduces the intensity of the observed light. It turns out that it is the finite age of stars and galaxies that is the most important factor in resolving Olbers' paradox. Expansion of the Universe has only a factor of two effect on the background light.

Overduin, J. M. and Wesson, P. S. *Dark Sky, Dark Matter*, IoP Publishing, 2003.

open Universe

An open *Universe is one which possesses the hyperbolic *geometry, meaning that the *density of material within it is less than the *critical density. Such a Universe has an infinite extent and volume. In the simplest *cosmological models introduced by Alexander *Friedmann, which did not possess a *cosmological constant, an open Universe would expand forever, eventually coasting at a constant rate.

Modern cosmological models do include a cosmological constant, or more generally *dark energy, which widens the possibilities. If the cosmological constant is negative, it will always eventually come to dominate the density of the Universe and will cause the open Universe to undergo recollapse into a *big crunch. Such models are however disfavoured by modern observations. Another possibility that should be borne in mind is that the Universe may have a non-trivial *topology, and even if our local patch were found to have the hyperbolic geometry, the Universe might still be finite.

Up until the late 1990s, it was considered quite possible that our Universe could be distinctly open, with density around one-third of the critical density. Observations of *cosmic microwave background anisotropies have now provided convincing evidence that the Universe either has precisely the flat geometry or is very close to it, within a few percent, though if it is really flat it will be impossible to ever definitively rule out that it might be marginally open or closed. Whether or not the Universe is truly infinite in extent is something we will never know, because the finite *age of the Universe and the finite *speed of light restrict our knowledge to the *observable Universe.

optical depth

The optical depth of an object measures how transparent it is to radiation. As radiation from a distant object traverses, for instance, a cloud of gas, some fraction of the *photons may be absorbed or scattered. The optical depth τ (Greek tau) is the probability of a photon scattering. If this is much less than one, the intervening material is said to be optically thin, meaning that it is close to transparent. In the opposite limit it is said to be optically thick. If gas clouds are optically thick that means that photons cannot penetrate to their inside. The optical depth may be different depending on the energy of the photons under consideration, as the photon energy determines the type of interactions it can have with atoms and *electrons.

[It is tempting to think that if τ is the scattering probability, then $1 - \tau$ must be the fraction that arrives without scattering. However this ignores the possibility of multiple scatterings, which if accounted for properly modify the unscattered fraction to $\exp(-\tau)$. In the limit where τ is small this can be approximated by $1 - \tau$.]

Of particular importance in cosmology is the optical depth of the *Universe to low-energy photons after *decoupling, as this determines the fraction of *cosmic microwave background (CMB) photons that travel to us uninterrupted since that epoch. CMB photons are only able to interact with free electrons, as they have insufficient energy to interact with atoms. The optical depth therefore receives contributions both from *residual ionization left over by the decoupling process, and from electrons liberated by the *reionization process due to early *structure formation.

The optical depth is one of the *cosmological parameters that cosmologists aim to determine from observational datasets. According to the 2006 analysis by the *WMAP satellite team, its current best-fit value is approximately $\tau = 0.10$, indicating that about 90% of CMB photons we receive are pristine photons which have not interacted since the Universe was about 400 000 years old.

A quantity closely related to the optical depth is the *visibility function.

pairwise velocity dispersion

The relative velocity between a gravitationally bound pair of *galaxies provides an estimate of the galaxy masses. Measuring the relative velocity of an individual galaxy pair is extremely difficult, however, requiring a measurement of the *distance to each of the galaxies (to confirm that they form a genuine physical pair) as well as their recession velocities. It is easier to measure the dispersion of pairwise velocities statistically from *redshift surveys.

Redshift surveys provide the angular coordinates and the redshift for each galaxy. Galaxy *peculiar velocities distort the apparent spatial distribution: galaxies within bound systems have large random motions and lead to an apparent elongation of structures along the line-of-sight. Conversely, the systematic collapse of newly forming structures make them appeared flattened along the line of sight. (*See* PECULIAR VELOCITIES, Fig. 1 for an example of both types of *redshift-space distortion.) These distortions can be exploited statistically in order to probe the dynamics of galaxies without the need to measure redshift-independent distances.

To do this, the *correlation function $\xi(\sigma, \pi)$ is measured as a function of the two variables σ and π: respectively the separations perpendicular and parallel to the line of sight. Peculiar velocities, including pairwise velocities, affect only the line-of-sight component of separation π between two galaxies: the projected separation on the sky σ is unaffected. According to the *cosmological principle, the real-space correlation function $\xi(r)$ should be isotropic. Any anisotropy in the two-dimensional redshift-space correlation function $\xi(\sigma, \pi)$ is then due to peculiar velocities. One assumes a functional form for the pairwise velocity distribution, $f(v)$, and fits this function to the measured redshift-space correlation function $\xi(\sigma, \pi)$, which is given by a convolution between $\xi(r)$ and $f(v)$. An exponential distribution of the form

$$f(v) = \frac{1}{a\sqrt{2}} \exp\left(-\frac{\sqrt{2}|v|}{a}\right),$$

where $a \approx 400$ km/s is the pairwise velocity dispersion, is found to give a good fit.

One can show that the *density parameter Ω_m is proportional to the square of the velocity dispersion. Numerically, one finds $\Omega_\mathrm{m} \approx 0.3$, consistent with other estimates.

parameter estimation

A *cosmological model contains various *cosmological parameters describing the relative importance of different ingredients and effects, and an important goal of observational *cosmology is to determine suitable values of these parameters. This process is known as parameter estimation.

The standard tool of parameter estimation is *likelihood analysis. The likelihood function of a dataset measures how probable the observed dataset was given a particular set of parameter values. The likelihood is therefore a function of the parameter values, with the best models being those yielding a high likelihood. If you have decided how probable you thought the different parameter values were before the data came along, known as the prior probability distribution of the parameters, then the product of the prior distribution and the likelihood gives the probability distribution of parameters after the data are incorporated, known as the posterior distribution.

In the above process, known as *Bayesian inference, the process of comparing theory to observations is an ongoing process, with the posterior distribution from one dataset becoming the prior distribution for the next one. Originally, the prior distribution may be primarily motivated by theoretical considerations, but eventually once the data becomes powerful enough, the influence of the prior should become negligible.

The standard technique for determining the posterior distribution is known as Markov Chain *Monte Carlo, or MCMC for short. We start with a point randomly chosen within the prior probability distribution of the parameters, and then make random exploratory jumps to new locations. If the likelihood of the new point is higher than the original, then the jump is made, but if it is lower then a random choice is made whether or not to accept the jump anyway, or to remain at the starting point and try a different jump. The general trend is therefore to move to the highest likelihood region, and then to move around exploring it by occasional jumps to lower likelihood. Provided the acceptance condition is cleverly chosen (the Metropolis–Hastings algorithm), the distribution of points explored matches the posterior probability distribution, and hence measures the probability of the different parameter values after comparison to the data.

The complete data analysis process is therefore quite complicated, with three key ingredients:

- The ability to make accurate predictions of observables, for instance *cosmic microwave background anisotropies, from theoretical models.
- Possession of an accurate likelihood function describing the different datasets to be included in the analysis.
- An MCMC algorithm for efficiently exploring the parameter space.

Luckily, a computer program which does all this, CosmoMC written by Antony Lewis and Sarah Bridle, has been made publicly available (see below). Before trying to use it yourself however, be aware that the calculations are so intensive that in practice they can only be done using multi-processor supercomputers. A typical computation will require of order tens or hundreds of thousands of jumps, the points thus generated being known as (Markov) chain elements.

Distinct from, but related to, parameter estimation is the statistical problem of *model selection. While parameter estimation seeks to determine the values of a set of parameters which have already been chosen, model selection seeks to use the data to determine the best set of cosmological parameters to be varied. Model selection is the appropriate approach if one wishes to determine whether new data justify the introduction

of extra cosmological parameters to describe new physical processes.

CosmoMC Home Page:
  http://www.cosmologist.info

particle accelerators

To understand the *Universe, we need to understand the properties of its constituent particles. Important tools for doing so are particle accelerators, which smash fundamental particles together in order to elucidate how they behave in a range of physical circumstances.

All particle accelerators function by using electric or magnetic fields to accelerate charged particles such as *electrons and *protons. They fall into two main classes, circular or linear.

The first circular particle accelerator, known as a cyclotron, was built in the late 1920s by Ernest Lawrence and was a table-top experiment. By contrast, modern particle accelerators are the largest pieces of experimental physics apparatus in existence, the biggest being the CERN accelerator in Switzerland with a ring length of 27 kilometres, shown in Fig. 1. [Incidentally, CERN is where the World Wide Web was invented in 1989 by Tim Berners-Lee, who wished to devise a more user-friendly way of accessing documents both at CERN and at participating universities around the globe.] An even more ambitious project, the Superducting SuperCollider, began construction in the 1980s in Texas, but eventually spiralling costs led to its cancellation. Relying on powerful magnets to steer and accelerate the particle beams, these devices are known as synchrotrons.

In a circular accelerator, particles are accelerated by electromagnets. The advantage of the circular design is that the particles can loop around more or less indefinitely, gaining more and more energy from kicks from the electromagnets. The downside however is that charged particles accelerated on circular orbits also lose energy, by emitting *electromagnetic radiation known as synchrotron radiation, and this sets a practical limit to the energies that can be achieved in a given accelerator. Minimizing this energy loss is what forces the experiments to ever larger scales.

Circular accelerators may direct their beams into fixed targets, or more ambitiously two separate sets of particles can be

Fig. 1. Aerial view of the CERN accelerator ring on the Swiss–French border, with its position highlighted.

accelerated round the ring in opposite directions and made to collide, thus increasing the collision energy. The cleanest experiments come from colliding electrons and their *anti-particles positrons, but these light particles are particularly vulnerable to synchrotron losses. *Protons and anti-protons can be accelerated to much higher energies, but as they are each composed of three *quarks/anti-quarks, the collision events tend to be much messier and harder to interpret. CERN's original successes were as an electron–positron collider, but it will reopen as a proton collider around 2008 having had its magnets upgraded to superconducting ones. In this incarnation it will be known as the Large Hadron Collider (LHC).

An alternative to a circular collider is a linear one, where typically the particles are accelerated in rapidly varying electric fields, rather in the manner of a surfer riding a wave. This largely removes the problem of synchrotron losses, but the particles cannot be recirculated through the apparatus for repeated acceleration. The largest linear accelerator in the world is at Stanford in California and is around three kilometres long. It is widely expected that the next large particle accelerator after the LHC will be a linear accelerator colliding electrons and positrons, almost certainly as a worldwide collaboration.

It's one thing to collide the particles together, and quite another to figure out what actually happened in those collisions. The range of types of particle detectors is immense, and a proper discussion would need a book devoted to this topic alone. An example image of a reconstructed particle decay is shown in Fig. 2, in this case being the decay of a Z_0 *boson (*see* STANDARD MODEL OF PARTICLE PHYSICS) which would have been

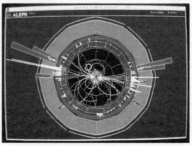

Fig. 2. The decay of a Z_0 boson as witnessed by the ALEPH detector at CERN.

created shortly before in an electron–positron collision.

Particles in an accelerator can be travelling very close to the speed of light, perhaps 99.99% of it. *Relativity theory is therefore required and is routinely tested in accelerator operation.

CERN Home Page: http://www.cern.ch

particle astrophysics

'Particle astrophysics', also referred to as astroparticle physics, is a term which covers quite a disparate range of topics, the general theme being that astrophysical objects are explored by any means other than the traditional astronomical tool of the *electromagnetic spectrum. It is sometimes taken to include *particle cosmology, though strictly these are really quite separate disciplines. The aim of particle astrophysics studies can be both to understand the physics responsible for astronomical phenomena, and to obtain information about the fundamental particles observed that might not be achievable through Earth-bound experiments.

The four main branches of particle astrophysics are *cosmic rays, *neutrinos, *dark matter detection, and *gravitational waves, each being described in more detail at their own entries. Some also consider gamma-rays, the most energetic form of *photons, as a particle astrophysics topic, although they are a type of electromagnetic radiation.

'Cosmic rays' is a term generically referring to any high-energy particles which arrive at the Earth from distant sources. Charged particles are accelerated by the magnetic field of our own Galaxy, and perhaps in the *galaxy from which they originated, and can reach extremely high energies well beyond anything achievable at terrestrial *particle accelerators. On impact with particles in the Earth's atmosphere, they generate a series of interactions (known as a cosmic ray air shower) enabling their properties to be studied, though especially at the highest energies their astrophysical origin is unclear.

Neutrinos are very weakly interacting particles, and vast numbers are passing through the Earth, and indeed your own body, each second, only very occasionally interacting. Neutrinos from the Sun, known as Solar neutrinos, are now routinely detected and played a key role in the discovery of neutrino mass.

Neutrinos are ... *supernovae, though ... supernova in 1987 in the ... Cloud (a satellite galaxy of the *M... has it yet been possible to detect these ne... trinos. Nineteen neutrinos from that supernova were detected, in two different neutrino experiments in Japan and the US. (This number is consistent with the theory of supernova explosions. In total, a supernova is believed to produce about 10^{58} neutrinos, meaning that even at the distance of the Large Magellanic Cloud around 10^{28} of them will have passed through the Earth. Roughly 10^{17} of these will have passed through a neutrino detection experiment, but the interactions are so feeble that this resulted only in the handful of detected particles.)

Neutrino experiments also detect what are called atmospheric neutrinos, which are produced in the sequence of interactions when cosmic rays hit particles in the Earth's atmosphere.

Dark matter is believed to pervade the Universe and is observed indirectly through its gravitational effects, which are responsible for *structure formation. The favourite hypothesis is that it is in the form of fundamental particles, whose properties must be such that existing particle accelerators have been unable to produce them. Dark matter detection attempts to identify the presence of these particles through interactions, and thence to constrain their properties. At the time of writing there are no generally accepted direct detections of dark matter.

Gravitational waves are not fundamental particles, but rather waves in *space-time predicted by *general relativity. *Gravity is an extraordinarily weak force, making gravitational waves hard both to create and to detect. They are thought to be produced in dramatic events, such as *black hole mergers and during *cosmological inflation in the *early Universe. At time of writing direct detection of gravitational waves has not been achieved, but large-scale experiments seeking them are entering operations and may prove able to initiate a new field in astrophysics—gravitational wave astronomy.

particle cosmology

Particle cosmology refers to research which takes ideas from fundamental particle physics

...tages, perhaps up to one hundred mil-
..n years old, all that the Universe contained
was fundamental particles, with stars and
*galaxies yet to form.

The units of measurement of metres, kilo-
grams and seconds (known as SI units)
are well suited to our everyday experience,
but rather inconvenient otherwise. Cosmologists deal with phenomena on a much
larger scale, and tend to prefer such units
as megaparsecs for length, solar masses for
mass, and gigayears (i.e. billions of years)
for time. By contrast, particle physicists
require units designed to conveniently measure very small masses, separated by very
short distances, and evolving on very short
timescales.

For the most part, the interesting quantity is the mass of the particles. According
to Albert *Einstein's famous $E = mc^2$ equation, mass and energy are interchangeable
simply by multiplying/dividing by the *speed
of light squared, and traditionally it is the
mass–energy that tends to be quoted. Rather
than using the SI energy unit joules, a useful
energy unit is the electron-volt, denoted eV
and defined as the energy an *electron gains
on moving through a potential difference of
one volt. It is equal to 1.6×10^{-19} joules, and
happens to be quite convenient for describ-
ing particles (for instance, the separation of
electron energy levels in atoms is of order of
an electron-volt). Its multiples are also widely
used, for instance a kilo-electron-volt (keV) is
a thousand eV, and a giga-electron-volt (GeV)
is a billion eV. If you plan only to remember one thing from this topic, a useful one is
that the mass–energy of a *proton is about
1 GeV.

Cosmologists often talk about the energy
scales at which different things happen. For
instance, the *electro-weak phase transition
is a process which takes place at an energy
of around 100 GeV; this is much higher than
the proton mass and so should be thought
of as a high energy (in fact it is around
the limiting energy of current *particle accelerators). *Grand unification, on the other
hand, has the much higher characteristic
energy of 10^{16} GeV, and occurs at energies way beyond those we can reproduce on
Earth.

It is useful to know when the Universe was
at different energy scales; for its early history
the Universe was in a thermal state with its
constituent particles having a corresponding

...only in one instance, a
...ly produced copiously in
particle cosmology... in the Large Magellanic
Milky Way), also

...ring
...s con-
...ns, rather
...f large num-

...cludes all the top-
i... ...verse topic, and addi-
tio.. ...y is taken to encompass
*nuc.. ...s, and perhaps all the events
up un.. ...ecombination and *decoupling.
Working particle cosmologists tend to come
from a particle physics background rather
than an astronomical one.

particle horizon

The particle horizon marks the limit of how
far, in principle, we can see in the *Universe.
It is a finite distance, because both the *age of
the Universe and the *speed of light are finite.

In reality our view is limited to within
the particle horizon, because the Universe is
opaque at its early stages. For electromagnetic radiation, the practical limit is the *last-
scattering surface corresponding to the epoch
of *decoupling. The *cosmic microwave background brings us an accurate view of conditions at that location. In the distant future
it may be possible to do cosmology with
*neutrinos or *gravitational waves which
decoupled earlier, thus seeing slightly further
away.

In the simplest *Friedmann cosmologies
the difference between the particle horizon
and the last-scattering surface is quite modest. However an early period of *cosmological
inflation radically changes this, as the particle
horizon becomes enormous during the inflationary expansion. The distance to the last-
scattering surface, by contrast, is unchanged
by inflation since decoupling took place long
after inflation was over.

The particle horizon is not to be confused
with the concept of the *event horizon, which
in *cosmology marks the furthest distance we
will ever be able to see, not just what we can
see by the present.

particle physics units

While this book is about the *Universe, which
is of course a big thing, in order to understand some of its properties we have to recognize that much of its content is in the form
of fundamental particles. Indeed, in its ear-

typical energy. During the hot *radiation-dominated era, the relation between time t and energy E is approximately

$$\left(\frac{E}{10^6 \, \text{eV}}\right)^2 \approx \frac{1 \, \text{sec}}{t}.$$

For instance, we can substitute on the right-hand side to say that when the Universe was one second old, the typical energy was 10^6 eV, while at 100 seconds it had fallen to 10^5 eV. Alternatively we can substitute on the left-hand side; if we are interested in the time of the electro-weak phase transition, we substitute in $E = 10^{11}$ eV (= 100 GeV) to find it took place when the Universe was around 10^{-10} seconds old, i.e. after a tenth of a nanosecond. (For further discussion of these useful relations, *see* TEMPERATURE–TIME RELATION.)

To simplify mathematical expressions, particle physicists often use dimensionless units. The archetypal example is speed. At school you are told that a quantity should always have units otherwise it is meaningless; speed might for instance be in metres per second. Particle physicists, however, take advantage of there being a characteristic speed in the Universe, the *speed of light c which gives the universal limit. What they do is specify speed as a fraction of the speed of light; this is expressed succinctly by saying they set $c = 1$. A speed of one-half (no units needed), means half the speed of light.

The speed of light is one fundamental constant of Nature that can be set to one. Two other such units are Newton's constant G, expressing the strength of gravity, and Planck's (reduced) constant $\hbar$ measuring the strength of *quantum mechanical effects. If these are also both set equal to one, then all dynamical quantities become dimensionless.

It turns out that making absolutely everything dimensionless is rather confusing, so often just c and $\hbar$ are set to one. The remaining constant G can then be expressed as an equivalent mass. Its value is rather large; in terms of the units defined above it is about 10^{19} GeV, and known as the Planck mass or Planck energy (more generally, the *Planck scale). This is the energy scale at which quantum gravitational effects will become important; for example in *superstring theory this would be the energy at which the stringiness of the fundamental particles would be evident. However it is a very high energy indeed,

and lasted in the Universe only until it was 10^{-44} seconds old, known as the Planck time.

peculiar velocities

In an idealized picture of the *Universe, the whole of space is expanding and objects such as *galaxies embedded within it are carried passively along by the expansion. These objects would then be stationary in *comoving coordinates. In practice, galaxies are moving relative to the comoving coordinate system. They may have both a random component, reflecting the initial random motion of the material from which they formed, and a *bulk motion towards regions of higher than average density due to the subsequent actions of *gravity. Together, these deviations from uniform *Hubble expansion are known as 'peculiar velocities'.

Peculiar velocities must be taken into account when estimating *distances to nearby galaxies from their *redshifts. For these nearby galaxies, distance is normally assumed to be given by recession velocity divided by the *Hubble parameter, $d = v/H_0$. If the peculiar velocity v_{pec} of the galaxy is comparable to the expansion velocity $v_{\text{exp}} = cz$ given by the *speed of light c times the redshift z, then the estimated distance would be seriously in error. In these cases, one needs a redshift-independent estimate of distance (*see* DISTANCE LADDER). By comparing this independent distance estimate with that predicted by the Hubble expansion, the line-of-sight component of the galaxy's peculiar velocity may be inferred.

Typical peculiar velocities are of order a few hundred kilometres per second. Since the speed of light $c \approx 3 \times 10^5$ km/s, peculiar velocities are significant for redshifts z of order 0.001 or less.

In practice, redshift-independent distance estimates are difficult and time-consuming to obtain. The peculiar velocity distribution function may be studied statistically by investigating the *anisotropy of the *correlation function obtained by using redshift-derived distances. The correlation function gives the excess probability above random of finding two galaxies separated by a distance π along the line of sight to the galaxy pair and with a separation σ perpendicular to the line of sight. Peculiar velocities affect only the line-of-sight component of separation π between two galaxies: the projected separation on the sky σ is unaffected. On small scales one finds

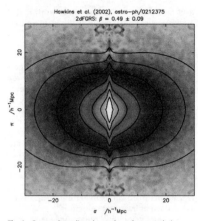

Hawkins et al. (2002), astro-ph/0212375
2dFGRS: $\beta = 0.49 \pm 0.09$

σ /h^{-1}Mpc

Fig. 1. Grey-scale coding shows the galaxy correlation function $\xi(\sigma, \pi)$ measured from the two degree field galaxy redshift survey (2dfGRS). The best-fit model is plotted as contour lines at $\xi = 4.0, 2.0, 1.0, 0.5, 0.2, 0.1$ and 0.05 from the centre outwards. At small projected separations ($\sigma \lesssim 5h^{-1}$Mpc) the correlation function is elongated along the line-of-sight direction π by random peculiar velocities. At larger separations the correlation function is squashed along the line of sight by coherent bulk motions.

that the correlation function is elongated along the line of sight, the so-called 'finger of God' effect. This effect is dominated by galaxies in *clusters whose peculiar velocities are essentially random, thus weakening the line-of-sight clustering signal (see Fig. 1). On larger scales, galaxies will have dominant peculiar velocities towards large mass overdensities: positive peculiar velocities for galaxies on the near side, and negative peculiar velocities for galaxies on the far side of the overdensity. This results in a flattening of the correlation function on large scales along the line of sight.

perfect fluid

The phrase perfect fluid refers to materials which possess neither viscosity nor heat conduction. Such a fluid is completely described by its *density and *pressure. A more technical definition is that the fluid looks isotropic in a reference frame moving with the fluid, meaning that the pressure must be the same in each direction. Many of the material components of the observed *Universe can be described by perfect fluids, in particular *cold dark matter, *baryons and *radiation.

Additionally, for many fluids the pressure and density are related, the formula connecting them being known as the *equation of state.

perturbations

A perturbation is an irregularity, usually small, superimposed on something uniform. In *cosmology, to a first approximation the *Universe is taken to be *homogeneous and *isotropic, meaning that all locations are equivalent. This is obviously not completely true, however; the surface of the Earth is rather different from the centre of the Sun, and the regions within *galaxies much more dense than the *voids in between. There are therefore perturbations to the distribution of material in the Universe.

According to the standard picture of cosmology, these irregularities started as small perturbations, corresponding to genuinely small deviations from a perfectly homogeneous Universe described by the *Robertson–Walker metric. These are sometimes known as linear *density perturbations. According to *cosmological perturbation theory the perturbations evolve, with the density perturbations growing according to *gravitational instability. Ultimately, the irregularities become large deviations from homogeneity, and *structure formation commences leading to the formation of galaxies.

In describing the evolution of the Universe, several different types of perturbation can be considered. Describing the material in the Universe as a single fluid, one would be interested in following the perturbations in the *density and also in the velocity. Perturbations in the *pressure might also be important. Bearing in mind that *gravity is interpreted in Albert *Einstein's theory of *general relativity as being due to *curvature of *space-time, one needs also to consider the *curvature perturbations which are induced by the perturbations in the matter.

It is often important to separately follow perturbations in each of the different types of material in the Universe. According to the *standard cosmological model, the present Universe has five ingredients: *dark energy, *dark matter, *baryons, *photons and *neutrinos. As structure formation develops, the different physical properties of each material may lead to their perturbations behaving differently. The distribution of dark matter is the most important in governing where galaxies form in the Universe, as it provides the dominant gravitational attraction. The perturbations in the photons, at our position in the Universe, are the *cosmic microwave background anisotropies.

Finally, during its early stages the Universe may have undergone a period of *cosmological inflation, during which its density would have been dominated by one or more *scalar fields. Perturbations in those scalar fields, known as *inflationary perturbations, are the leading candidate to be the origin of structure formation in the Universe.

phantom dark energy

*Dark energy is believed to be responsible for the observed present *acceleration of the Universe. While the fundamental nature of this substance remains an enigma, it is normally assumed that its *density either stays constant or decreases as the *Universe expands. A more radical possibility, introduced by Robert Caldwell, is that the dark energy density may increase with time. This leads to an unstable feedback, whereby the increased density forces the Universe to expand faster, which in turn increases the phantom energy density further. The Universe ends in a *big rip. Caldwell named the possibility 'phantom dark energy', to represent the idea of something that comes as if from nowhere and overwhelms the Universe.

At present it is not known whether phantom dark energy is a possibility consistent with all physical laws, and so its study is controversial. But we can say that it is permitted by current observations. The simplest model of dark energy is Albert *Einstein's *cosmological constant, whose density remains constant as the Universe expands. This gives predictions consistent with observations. Since observations can only ever have a finite accuracy, that implies that models where the dark energy density is either increasing or decreasing can also be compatible with observations, provided the variation is sufficiently slow.

Future observations may be able to definitively select whether the dark energy density is increasing or decreasing, thus confirming or refuting phantom dark energy. However if observations remain consistent with a cosmological constant then they will not be able to definitively exclude the possibility of phantom dark energy. In that case improved

understanding of fundamental physical laws may be needed to determine whether or not phantom dark energy can play a role in the Universe's evolution.

phase transitions

A physical system undergoing a dramatic change in its properties is said to be undergoing a phase transition. Usually it is a change in temperature that induces the transition. A classic everyday example is the freezing of water to ice, or its evaporation to steam. Phase transitions are ubiquitous in Nature, examples being the change in state between solid–liquid–gas, changes in electric or magnetic properties (for instance the onset of superconductivity), or changes in the nature of *fundamental forces.

In particle physics, phase transitions are often known as symmetry breaking, as they correspond to some reorganization of the fundamental symmetries relating particle properties. An example is the *electro-weak phase transition at which two of the fundamental forces of Nature, electromagnetism and the weak nuclear force, develop separate identities.

As the *Universe cooled from the extremely high temperatures of the early hot *big bang, for instance following *reheating after *cosmological inflation, it is believed to have undergone a series of phase transitions. A partial list of possible transitions is as follows, corresponding roughly to the chronological order in which they would be expected to occur.

- A transition in the dimensionality of *space-time giving the present three space dimensions.
- A transition bringing a period of inflation to its end.
- The *grand unification phase transition where the strong nuclear force first develops a separate identity.
- The *supersymmetry phase transition, at which particles and their corresponding supersymmetric partners develop different properties.
- The electro-weak phase transition.
- The *quark–hadron phase transition, where individual *quarks first combine to form *protons and *neutrons.

Whether these all occurred is quite uncertain. The mechanism for ending inflation might or might not involve a phase transition, and grand unification might never be achieved thus avoiding the need for a phase transition ending it. How the dimensionality of space-time was determined is essentially unknown.

Phase transitions commonly lead to the formation of defects in the physical system. An example is faults in ice crystals where the crystal lattices have grown together but don't match properly. These defects correspond to trapped energy which can no longer be released after the phase transition has completed. In fundamental physics these are known as *topological defects, an example in cosmology being (so far hypothetical) *cosmic strings. They are also observed in laboratory systems such as superfluid liquid helium.

photometric redshift

Determination of an accurate *redshift for an astronomical source requires observation of the object's *spectrum, so that the wavelengths of spectral features may be measured. Such observations are time-consuming and are still not feasible for objects with an apparent *magnitude of around 25 or fainter.

An alternative, less accurate estimate of an object's redshift may be obtained from photometric observations obtained through several medium- or broad-band filters. In this case, rather than identifying particular spectral features, one uses the overall shape of an object's spectrum to estimate a photometric redshift, see Fig. 1. This technique is less accurate than *spectroscopy, since the overall spectral shape of a galaxy is affected by its stellar population as well as redshift. One thus needs to determine the object's type concurrently with its redshift: there is often a *degeneracy between the two.

There are two classes of techniques currently in use for estimating photometric redshifts from broad-band photometric observations. The first class, template fitting methods, assume that the distribution of flux with wavelength (the *spectral energy distribution, or SED) is known for a set of template objects. One then finds the type and redshift of the template which agrees best with the observed magnitudes. The second class, empirical fitting methods, make no assumption about the SEDs of the sources to

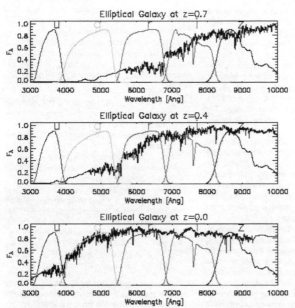

Fig. 1. A model elliptical galaxy spectrum shown at three redshifts along with the five-filter response function of the Sloan Digital Sky Survey. Note how the distinctive 'break' in the spectrum at 4000Å shifts through the filter system, changing the observed colours, as redshift increases.

be redshifted, but do require a training set of galaxies with known spectroscopic redshifts in order for the methods to be calibrated. Both methods can give estimated redshifts with accuracy $\Delta z \sim 0.05$.

Probably the most ambitious survey to date to use photometric redshifts is the COMBO-17 survey, see weblink below, which has imaged 1 square degree of sky through 17 filters, obtaining photometric redshifts accurate to about 2%.

See also REDSHIFT SURVEYS.

http://www.mpia-hd.mpg.de/COMBO/

photons

Photons are the individual particles of light. While it is often convenient to think of light as a wave, at the microscopic level *quantum mechanics indicates that it is made up of 'packets' of energy, behaving as individual fundamental particles. In a wide range of astrophysical situations, one has to consider the particle behaviour of light.

The first indications that light is made up of particles came with the work of Max Planck at the onset of the 20th century in understanding the nature of the *black-body spectrum. At the time classical physics was in crisis, because it predicted that the spectrum should contain infinite energy at high frequencies, in contradiction to observation. Planck introduced the assumption that radiation could not be emitted in arbitrary amounts, but rather had to be emitted in discrete bundles known as quanta. This assumption was sufficient to explain the form of the black-body spectrum and ensure that the total energy remained finite. It was Albert *Einstein, in one of the key papers of his miraculous 1905 output, who first showed that the particle view was necessary to explain the interaction of radiation and matter; it was his explanation of the photoelectric effect, not of *relativity, that was cited for his *Nobel Prize.

Photons were identified as belonging to a class of particles known as *bosons (possessing integer spin), and in the quantum theory of atoms photons are emitted or absorbed as *electrons move between energy levels. Photons have zero mass, which means that they always travel at the *speed of light. [Since photons *are* light this statement appears rather tautologous, but the term 'speed of light' is used here in the sense of the universal speed limit set by relativity theory, with all massless particles travelling at that speed.] In

modern particle physics, the photon is seen as one of the particles of the *standard model of particle physics, its role being to supply the electromagnetic force between charged particles. As photons have no electric charge, they themselves do not feel the electromagnetic force. They are also an example of a particle which is its own *anti-particle.

The main distinguishing feature between different photons is their energy, which is entirely determined by their kinetic energy or, equivalently, their momentum or frequency. The energy E is related to their frequency f via the expression

$$E = hf,$$

where h is Planck's constant, one of the fundamental constants of Nature. The higher the energy, the higher the frequency. This relation provides the link to the *electromagnetic spectrum, which is usually expressed in terms of the frequency or wavelength of light, and it allows us to work out the corresponding energy of photons given their frequency. The relation of that energy to properties of matter (e.g. the energy levels of electrons in atoms, or the binding energy of atomic nuclei) helps indicate how that particular radiation will interact.

Photons can also be distinguished by their *polarization, having two possible polarization states (also known as spins or helicities).

That light is made up of photons is important to a wide range of astronomical and cosmological applications. For example, the spectral lines seen in the electromagnetic spectrum, used for instance to measure the *redshift of objects, originate from the absorption or emission of photons matched to the energy level differences in atoms. At very high energies, corresponding to *X-rays and gamma-rays, modern detectors are actually sensitive to individual photons; the faintest X-ray objects yet imaged, in a survey by the *Chandra satellite, correspond to a single X-ray photon hitting the detectors every four days! In a cosmological context, the process of *decoupling, leading to formation of the *cosmic microwave background, is best understood by considering how photons interact with free electrons and atoms.

pixelization

Astronomical data comes in various types. Point sources such as stars can be specified by giving their location and other properties such as brightness and *spectrum. Some types of data however are continuous functions of position on the sky, such as the intensity of *cosmic microwave background radiation coming from different directions. In the latter case, one cannot specify the intensity coming from every one of the infinite directions. Rather, the data must be **pixelized**, with the region of the sky broken up into a finite number of small units in which properties such as intensity can be specified. A good choice of pixelization must be high enough resolution to accurately represent the data which has been obtained, but should not be too large, so as to avoid storing redundant information. An example pixelization is the image on a computer display, which from a distance can give high-quality images but which, close up, reveals itself as a set of mono-chrome dots.

For data covering only a small part of the sky, the geometry of that patch can be treated as flat and there is no reason not to use a square grid of pixels. The main issues arise when data covers a substantial fraction of the sky, so that the spherical nature of the sky must be taken into account. A square grid no longer works as it won't join up properly once it completely encircles the globe. Quite a number of datasets are now available that cover most or all of the sky, and a consistent use of pixelization is important to allow the different sets to be compared.

Especially for *cosmic microwave background anisotropy studies, a popular choice of pixelization is HEALPix (Hierarchical Equal Area isoLatitude Pixelization), invented by Krzysztof Górski in 1997, which combines a number of desirable properties encoded in its name. 'Hierarchical' means that the resolution can be reduced by averaging four small pixels into one bigger one, giving a new pixelization which also respects the HEALPix scheme. Conversely, pixels can be split in four to enhance resolution. 'Equal area' refers to the pixels all having the same size. Finally, 'isolatitude' indicates that the pixel centres follow lines of latitude, a technical requirement which greatly speeds up certain key analysis computations. Fig. 1. shows an example HEALPix pixelization of the sphere.

Occasionally, for instance in computer *numerical simulations, there is a need to pixelize in three dimensions, by breaking space up into little cubes or cuboids. These go by the rather ugly name of 'voxels'.

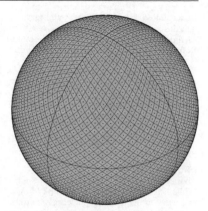

Fig. 1. A low-resolution HEALPix pixelization of the sphere, in this case into 768 pixels giving approximately 7 degree resolution. Typical actual applications may involve many millions of pixels.

HEALPix Home Page:
   http://healpix.jpl.nasa.gov

Planck length
See PLANCK SCALE.

Planck mass
See PLANCK SCALE.

Planck satellite
The Planck satellite (Fig. 1, over page) is a European Space Agency project scheduled to launch during 2008, with the aim of providing high-resolution all-sky maps of the *cosmic microwave background (CMB) and hence making precision determinations of *cosmological parameters. It is the natural successor to the *WMAP satellite, offering better angular resolution, higher sensitivity detectors, and observations across a wider range of frequencies.

The project was conceived in the mid 1990s following the successful first detection of *CMB anisotropies by the *COBE satellite. It arose as a merger of two separate proposals for a satellite mission, known as COBRAS and SAMBA. This history is partly preserved as the mission features two separate detector instruments, the French-led High-Frequency Instrument (HFI, principal investigator Jean-Loup Puget) and the Italian-led Low-Frequency Instrument (LFI, principal investigator Reno Mandolesi). Contributions

Fig. 1. A computer generated image of the Planck satellite. Planck will use a single mirror to measure the absolute temperature as it scans the microwave sky, unlike the WMAP satellite which used two mirrors to compare the temperatures from two different directions.

to the project come from a wide range of European institutions, and also from the US and Canada.

Planck will operate at the Earth–Sun Lagrange 2 point, orbiting the Sun at a distance of around one and a half million kilometres outside the Earth's orbit. After taking a few months to reach this orbit, its baseline mission involves two complete surveys of the sky over a period of fourteen months, which may then be extended to additional surveys.

The LFI instrument is quite similar to WMAP in terms of the observational resolution and microwave frequencies used, but with significantly more sensitive detectors. HFI, as its name suggests, operates at a higher frequency range, using sets of a different type of detector known as bolometers. The HFI instrument benefits from being able to survey at a higher angular resolution due to its range of observing frequencies, with the highest-resolution channels achieving 5 arcminute resolution. Both instruments will measure CMB *polarization as well as temperature.

While the CMB anisotropies are its principal target, its wide frequency coverage means that Planck is able to carry out a wide range of other science projects, including an all-

sky survey for the *Sunyaev–Zel'dovich effect in galaxy *clusters, surveys of emission from other *galaxies, and surveys of the interstellar medium of the *Milky Way.

Planck scale

The physical properties of the *Universe are governed by the fundamental constants of Nature. Such constants, an example being the *speed of light c, are not predicted by theoretical models, but rather have to be determined by measurement. There would be nothing mathematically inconsistent about a Universe where the speed of light was one metre per second, but it happens that its value is much larger than that, around 300 000 kilometres per second.

There are in fact three key dynamical constants of Nature. They are the speed of light, Newton's constant G which expresses the strength of gravity, and Planck's reduced constant $\hbar$ measuring the strength of quantum effects. [Planck's actual constant, denoted h, is the constant of proportionality relating the frequency of a *photon to its energy. However this quantity usually appears in formulae divided by 2π, and to avoid writing all those factors the reduced constant is defined by $\hbar = h/2\pi$, and pronounced *aitch-bar*.]

Each of these has units made up from combinations of the fundamental dynamical units of mass, length, and time. For reference, c has dimensions LT^{-1}, G has dimensions $L^3 M^{-1} T^{-2}$ and $\hbar$ has dimensions $L^2 MT^{-1}$, where L, M, and T indicate length, mass, and time. Accordingly they can be uniquely combined to give quantities which have the units of time, length, and mass; for example the combination $\sqrt{\hbar G/c^5}$ has units of time. These are known as Planck units after the German physicist Max Planck whose work initiated *quantum mechanics.

By substituting in the measured values of the fundamental constants we can obtain values for the Planck units. They are

Planck time	5×10^{-44} seconds
Planck length	2×10^{-36} metres
Planck mass/energy	2×10^{-8} kilograms /
	1×10^{19} GeV

Notice that the Planck time and length are extraordinarily small. By contrast the Planck mass is quite a substantial number, bigger than a microgram, and when translated into energy units is very large in terms of known particle properties (e.g. the proton mass is about 1 GeV).

In the system of *particle physics units often used to describe fundamental particles, the three constants of Nature are often set equal to be 1. All dynamical quantities then become dimensionless, and are measured as fractions/multiples of the corresponding Planck units such as Planck length or time.

The Planck energy density (being the Planck energy divided by the Planck length cubed) is significant because it indicates the density at which quantum effects and gravitational effects are of comparable importance, meaning that a theory of *quantum gravity would be needed to describe the physics. However this corresponds to a very high energy scale, around 10^{19} GeV, and was achieved in the Universe only up until it was a Planck time old. This very early era has so far defied attempts to model it, and is often called the Planck era.

Planck time

See PLANCK SCALE.

polarization

The most important property of light of a given wavelength arriving from distant objects is its intensity, but it also has a second property, known as polarization. The polarization measures the relative intensity of light in the two directions perpendicular to the direction the light is travelling.

At the classical level, light waves correspond to oscillations in the electromagnetic field, where the oscillations take place at right angles to the direction of propagation of the wave. As there are two such directions, there are two independent modes of oscillation, whose amplitude and phase can be specified independently. Depending on the relative phases, the combined oscillation may be in a particular plane (known as linear polarization), or trace out a circle (known as circular polarization), or more generally may trace out an ellipse. A general plane wave can be decomposed into a sum of two orthogonal linear polarizations, or alternatively into a sum of two oppositely directed circular polarizations. We will henceforth assume the decomposition is into linear polarization modes.

Typically, light we receive is not a perfect plane wave, but rather is incoherent, meaning that it is made of light from many separate emitters, that need not share the same polarization. Nevertheless, the physical situation may dictate that there be some level of polarization; this is common for instance in situations where the light is scattered into a new direction. Incoming light can be split into a sum of unpolarized light and perfectly polarized light, the degree of polarization then being given by the ratio of polarized to total intensity.

The standard description of polarization used by astronomers is the Stokes parameters, invented by George Stokes in the mid 1800s. There are four Stokes parameters, the first being the total intensity and the other three providing a full description of the polarization state of incoming light (the degree of polarization and the magnitude of the two polarization states).

Study of polarization adds an important new dimension to several areas of astronomy. Within *cosmology, the most important include the polarization of the *cosmic microwave background radiation, which was first detected in 2001, and polarization in the lobes of *radio galaxies.

If polarized light crosses a region containing a magnetic field, the plane of polarization undergoes a rotation, known as Faraday rotation after its discoverer Michael Faraday. This effect can be used to estimate the strength of interstellar magnetic fields, both within our Galaxy and in distant *galaxies.

Polarization is also an important property viewed at the quantum level, where light is described as made up of individual *photons. Photons have two possible polarization states, corresponding to the possible spin directions relative to the direction of motion. The superposition of a large number of individual photons recovers the classical description of polarization using, for instance, the Stokes parameters.

Chandrasekhar, S. *Radiative Transfer*, Dover Publications, 1960 [Technical].

potential

The term potential is used in various contexts in *cosmology. Typically, the potential is a measure of the energy of a given physical configuration. Forces normally act on objects in the direction that would reduce their potential.

A simple example is *gravity, where the Newtonian potential provides a description of the gravitational forces at work. The potential at any point in space indicates how much gravitational potential energy an object of a particular standard mass (for instance a kilogram) would have if placed at that point. The Earth's orbit is caused by the gravitational potential due to the Sun, which provides a force towards the Sun as that is the direction in which gravitational potential energy reduces the fastest. This acceleration causes the Earth to follow its nearly circular orbit.

In many cosmological contexts, the Newtonian potential gives a sufficiently accurate description of gravity. For instance, *N-body simulations of *structure formation use it. However on length scales comparable to or greater than the *Hubble length, it is essential to instead use *Einstein's theory of *general relativity, which features two separate potentials to describe the gravitational effects caused by *density perturbations.

A different, though related, use of the word potential comes from fundamental particle physics, where the potential describes the energy of a configuration of particles. The form of the potential determines how particles interact with each other. In cosmology the classic example is the potential of the *scalar field presumed to cause *cosmological inflation, known as the *inflaton. Observations of *inflationary perturbations are employed to try to determine the form of this potential, which governs how inflation would have taken place.

power spectrum

The term 'power spectrum' originates from the field of electrical engineering, where it refers to the power per unit frequency of a time-varying signal. In physics, the signal can be any type of wave. For example, if the surface of a lake is smooth apart from the occasional large wave reaching the shore, the power in the wave will be concentrated at large wavelengths and low frequencies. Conversely, if there are no large waves, but the surface of the lake is covered with small, closely spaced ripples, then the power will be mostly at small wavelengths and high frequencies.

In *cosmology, the power spectrum $P(k)$ quantifies the scale-dependence of *density perturbations, in terms of a superposition of waves with wavenumber $k \equiv 2\pi/\lambda$, λ being the wavelength, see Fig. 1. The power spec-

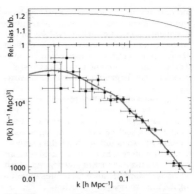

Fig. 1. The power spectrum of galaxies measured from the Sloan Digital Sky Survey (points with error bars) along with the prediction from the standard cosmological model with parameters determined from the first year WMAP satellite data. The top panel shows the relative galaxy bias as a function of scale.

trum and the *correlation function $\xi(r)$ form a *Fourier transform pair, and contain, in principle, equivalent information. One advantage of using the power spectrum rather than the correlation function is that, as the Universe evolves, modes of the power spectrum with a particular wavenumber k evolve independently of all other modes. Conversely, two regions of the Universe at fixed *comoving separation r will become less and less independent as *gravity has time to act on larger and larger scales.

The power spectrum is also used to quantify the scale-dependence of *cosmic microwave background (CMB) anisotropies. Since we measure the CMB temperature integrated through the (relatively narrow) surface of *last-scattering, we use a two-dimensional form of the power spectrum known as the *radiation angular power spectrum C_ℓ, where ℓ is the multipole.

preheating

The word 'preheating' is used in *cosmology in two unconnected settings.

In the first, it refers to part of the *reheating process which takes place at the end of *cosmological inflation, whereby the energy trapped in the form of the inflating material is released to form a thermalized background of conventional particles such as *baryons and *photons. Preheating is distinguished from the rest of the reheating process by happening at a catastrophic rate, corresponding to coherent decays of many particles simultaneously rather than by individual particle decays. Whether or not it takes place depends on the details of the inflation model, but if it does it is likely to give way to standard reheating.

In the second, it refers to the heating of intergalactic gas before it collapses to form galaxy *clusters, for instance due to *feedback mechanisms such as *supernova explosions. The infall process further heats the material, then known as the *intra-cluster medium, but it may retain a memory of any heating which occurred before the formation, thus measuring the importance of feedback.

Press–Schechter theory

Press–Schechter theory, invented by William Press and Paul Schechter in 1974, is a way of estimating how many objects of a given mass have formed as a function of time. The objects in question are usually *galaxies or galaxy *clusters. Press–Schechter theory can be used to provide estimates of formation rates that would otherwise have to be determined by time-consuming *N-body simulations.

The basic idea of Press–Schechter theory is to say that once *density perturbations on a given scale have grown beyond a certain threshold, the material within them can be said to have collapsed and formed objects of masses at least of that scale. By considering different scales, one can then extract the fractions of objects in different mass ranges. For example, one might wish to calculate how many galaxies of *Milky Way size have formed in the present *Universe, and how that compares to the numbers at earlier epochs.

Given that the actual amount of physics going into the Press–Schechter approximation is rather limited, it has proved surprisingly successful in matching results obtained from N-body simulations. That is particularly true when one uses such simulations to calibrate the parameters of the theory, such as the formation threshold, rather than attempting to predict them from first principles.

While Press–Schechter theory has been widely used ever since its invention, with the recent advent of very large N-body simulations such as the Virgo consortium's **Hubble Volume** simulation, it has become apparent that it does underestimate the numbers of very rare, very massive objects. Several refine-

ments are now in use which correct this, generally known as **mass functions**.

pressure

The pressure of a fluid governs how its density changes in response to changes in volume. In the usual case of a positive pressure, the fluid loses energy as it expands, and accordingly the *density falls more rapidly the larger the pressure is. Note that in a *Universe which is *homogeneous there are no pressure forces, since the density is everywhere the same, but nevertheless the pressure does affect the evolution of the density via the energy-loss rate (sometimes expressed as 'work done equals pressure times change in volume').

A fully relativistic fluid, in which the particles move at the speed of light (*photons of light being the obvious example), has pressure equal to one third of the energy density, while *cold dark matter and *baryons presently behave as pressureless fluids. Other examples sometimes encountered are the case of a 'stiff fluid', which has pressure equal to the energy density, and a *cosmological constant which behaves as a fluid with pressure equal to minus the energy density. More generally, the relation between the pressure and energy density is known as the *equation of state.

One of the key roles of the pressure is that, along with the density, it determines the rate of *acceleration of the Universe. Negative pressure is an essential condition if the Universe is to be accelerating rather than decelerating, and hence is a required property of the *dark energy. To be precise, acceleration requires that the pressure is less than (i.e. more negative than) minus one-third the energy density.

primordial black holes (PBHs)

Primordial black holes are *black holes which form during the *early Universe, as opposed to ones which form in the later Universe, for instance from the collapse of supermassive stars. There is no direct evidence for the existence of PBHs. Some theories predict excessive production of PBHs and are ruled out, so even if they are never detected their apparent non-existence is a significant constraint on the types of physics allowed in the early Universe.

Several mechanisms by which PBHs might form have been proposed, including (in rough order of plausibility):

- Via the collapse of short-scale high-amplitude *density perturbations, which may for instance have been created during the late stages of *inflation.
- During the coalescence of bubbles in a cosmological *phase transition.
- Through the collapse of nearly circular loops of *cosmic string.

Early production of PBHs, during the *radiation-dominated era of the hot *big bang, is strongly constrained. Once formed, the *density of PBHs reduces with the expansion as non-relativistic matter, which is more slowly than the dominant radiation. Accordingly, PBHs can quickly come to be the dominant objects in the Universe, giving a *relic particle abundance problem (the term 'particle' being used rather loosely here).

One intriguing possibility is that PBHs might be light enough to experience Hawking evaporation. In 1975 Stephen Hawking showed that black holes are not in fact perfectly black, but rather evaporate due to quantum processes. The evaporation is however extremely slow, and utterly negligible for the sorts of black holes forming in the present Universe. However it can be important for very light black holes. If a PBH were formed with a mass around 10^{15} grams (about the mass of a decent-sized mountain), they would take about the *age of the Universe to evaporate, and any which were lighter would already have evaporated by the present.

PBHs have a number of potential observational consequences, which strongly constrain their formation rate. For those with masses above 10^{15} grams, evaporation is negligible but their present 'relic' abundance must be low enough to be compatible with the established inventory of the Universe's contents. For those with lower masses, they will have evaporated to nothing by the present and hence not pose a relic abundance problem, but the evaporation products (for instance gamma-ray *photons, *protons, and positrons) may be detectable as *cosmic rays, or may have disturbed cosmological processes such as *nucleosynthesis. These constraints demand that, at the time of formation, PBHs can comprise no more than a tiny fraction, typically 10^{-20}, of the density of the Universe—see Fig. 1. The strongest constraints are on PBHs that would be evaporating to nothing at the present epoch.

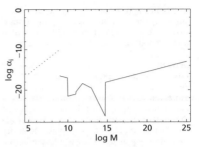

Fig. 1. Limits on the fractional density of the Universe that can be in the form of PBHs at the time of their formation (here denoted α_i). The masses M are given in grams. Note that both axes use base-10 logarithms. For masses above 10^{15} grams the constraint comes from the total mass density of the PBHs. At lower masses various constraints due to the Hawking evaporation come into play, giving a complicated shape to the constraint curve. The dashed line shows the constraint if PBH evaporation leaves behind a relic particle, but it is not known whether it does.

It has also been speculated that PBH evaporation might not completely destroy the black hole, but might instead leave behind a stable relic particle. If true, this introduces further constraints as those relic particles must also have a low enough density to be compatible with observations.

PBHs are a candidate to be the *dark matter in the Universe, and provided they are small enough are astrophysically indistinguishable from other *cold dark matter candidates such as *WIMPs. They are an attractive possibility because they avoid the need to postulate new types of fundamental particle. However such PBHs would have to have formed before the epoch of cosmic nucleosynthesis to avoid spoiling the predictions of that theory, and in the absence of any plausible way of producing the right density of PBHs the possibility of them being the dark matter receives surprisingly little attention from cosmologists.

proper coordinates

A proper coordinate system is fixed in time, in contrast to a *comoving coordinate system that expands in tandem with the expansion of the *Universe. Light will always take the same time to travel between two fixed points (in the absence of deflection of the light by *gravitational lensing) in proper coordinates, whereas light travel time will increase with the expansion *scale factor between two fixed points in comoving coordinates.

It is the appropriate system to use when studying the evolution of a tightly bound gravitational system, such as a *galaxy, which does not partake in the *Hubble expansion.

protons

Protons are particles; protons and *neutrons are the constituents of atomic nuclei. At a more fundamental level protons are made up of three *quarks, and are members of the classes of particles known as *baryons and *hadrons. They make up the majority of your body mass, slightly outnumbering neutrons.

The mass of an individual proton is 1.7×10^{-24} grams, more usefully expressed in *particle physics units by its mass–energy of 938.3 MeV. They possess an electric charge of plus one, opposite and equal to that of an *electron. According to the *standard model of particle physics protons are absolutely stable, though *grand unification theories typically predict that they do decay on a very long timescale (such decays have not however been observed).

In the present *Universe protons are found in a variety of forms. A key *cosmological parameter is the total density of baryons, now measured to be approximately 4% of the *critical density; according to the theory of cosmic *nucleosynthesis approximately 90% of those baryons are protons. The majority exist as isolated ionized hydrogen nuclei in the *intergalactic medium (a proton and a hydrogen nucleus are one and the same thing). Others are in the form of atomic or molecular hydrogen, in clouds of cold atomic gas. Hydrogen, either ionized into a plasma or in atomic form, is the predominant constituent of stars and giant planets, where helium is also common. Only a small fraction of the protons are in heavier atomic nuclei, such as the carbon and oxygen from which you are primarily made.

quantum cosmology

The subject of quantum cosmology concerns the creation of the *Universe during an epoch where both *quantum mechanics and *gravity are believed to be important, known as the *quantum gravity regime. As such, it is a highly speculative and controversial corner of *cosmology, particularly as it is unclear whether any of the ideas will ever be testable observationally.

The main idea in quantum cosmology is that the Universe came into existence via a quantum tunnelling event. Quantum tunnelling—the process whereby objects can penetrate through barriers which in classical physics would be impassable—is a well-known feature of quantum mechanics, responsible for instance for radioactive decay. The novelty in quantum cosmology is that the object in question is the entire Universe, and moreover that the initial state from which it tunnels is absolutely nothing. Not just empty space, but a non-existence of space and time themselves. Such a process is also called 'tunnelling from nothing'.

Such tunnelling is described in quantum mechanics by what's called a wavefunction, in this case the **wavefunction of the Universe**, which is rather more grandiose than corresponding examples such as the Schrödinger wavefunction for an *electron in a hydrogen atom. There are two rival proposals for this wavefunction. One, motivated by Euclidean quantum gravity, is known as the Hartle–Hawking wavefunction after its founders James Hartle and Stephen Hawking. The other, known as the tunnelling wavefunction, was studied independently by Andrei Linde and Alex Vilenkin. These two proposals make quite different predictions, and it is unclear which, if either, might be the correct one. Analysis of this type has typically been carried out in the traditional four-dimensional *space-time of *general relativity, which is currently rather unfashionable as compared to the ideas of *superstring and *M-theory.

quantum gravity

The two great theoretical physics achievements of the 20th century, *quantum mechanics and *general relativity as a description of *gravity, are unfortunately inconsistent with one another; quantum mechanics as applied to gravity simply does not make sense and computations lead to irretrievably infinite answers. 'Quantum gravity' refers to any theories which propose to repair this inconsistency. There are several candidate quantum gravity theories but no established consensus on which is likely to be correct.

Quantum mechanics and gravity ordinarily operate in two very different physical regimes, the former applying to the very small and the latter to the very large. However, on a sufficiently small length or timescale, known as the *Planck scale, both theories should be taken into account and that is where the problem lies. Applying the normal rules of quantum mechanics to general relativity produces a theory of a type known as **unrenormalizable**, which means that it yields nonsensical infinite answers to what should be well-posed problems.

[By contrast, the theories of the other forces comprising the *standard model of particle physics are renormalizable. This means that they too often give alarmingly infinite answers to simple questions such as masses and interaction rates, but in this case there is a well-defined procedure—renormalization—which makes sense of those answers by carefully subtracting off the troublesome infinities to leave a finite answer. This sounds like witchcraft, but is devastatingly successful in bringing theory in line with observation, in many cases to several significant digits.]

Quantum gravity effects would be important both at the initial *singularity of the hot *big bang cosmology (*see* 'Overview'), and at the central singularities of *black holes. One manifestation of the incompatibility is that quantum mechanics is formulated in a fixed *space-time with a well-defined time variable, whereas in general relativity space-time is dynamic and there is no preferred definition of time.

There are a variety of approaches to quantum gravity, either by modifying the behaviour of space-time, and thus changing the structure of general relativity near the Planck scale, or by modifying the laws of quantum mechanics to enable a formulation without an explicit definition of time. The current leading candidates are known as loop quantum gravity and *M-theory, the latter being the successor to *superstring theory. M-theory has the more solid heritage in fundamental particle physics theory, and as such would be regarded by the majority as the currently preferred option. Loop quantum gravity, by contrast, addresses only gravity and not other *fundamental forces. In addition to these two there are a number of other suggestions including Euclidean quantum gravity and Roger Penrose's twistor theory, which are currently less popular.

It is rather unclear how one might go about carrying out observational tests to distinguish between different candidate theories, since the Planck scale is vastly out of reach of any conceivable experimental probe. It may well be that astrophysics and cosmology offer better hopes for testing quantum gravity theories than lab-based experiments. In particular, quantum gravity is presumably the appropriate theory to deal with the very first micro-instants of the Universe's existence, and some ideas in this vein are considered under the heading *quantum cosmology.

Smolin, L. *Three Roads to Quantum Gravity*, Perseus Books, 2002.

quantum mechanics

The theory of quantum mechanics is one of the crowning glories of 20th century physics. It seeks to describe phenomena on microscopic scales, normally at the level of individual fundamental particles. The first hints towards the theory came with Max Planck's theory of radiation as discrete packets of energy in 1900, and Albert *Einstein's explanation of the photoelectric effect in 1905. The full development of theory came via the work of Danish physicist Niels Bohr on the structure of the atom, and the complementary mathematical descriptions of quantum theory of Werner Heisenberg and Erwin Schrödinger in the early 1930s.

The basic precept of quantum mechanics is that physics on microscopic scales becomes probabilistic in nature. Rather than having a definite location, an *electron exists in a state where it may be at any number of different positions, each with its own probability. Only the act of measurement can determine where it actually is. Further, objects have complementary properties which cannot be known simultaneously, an example being their position and velocity. This is known as Heisenberg's Uncertainty Principle.

Quantum mechanics has the public perception of a mysterious and incomprehensible theory, and indeed Bohr said 'Anyone who is not shocked by the quantum theory has not understood it'. It is true that at a conceptual level the theory is very troublesome, with competing (and ultimately equivalent) interpretations as to what is really going on. But this shouldn't detract from the fact that mathematically the theory is unambiguous and enables calculations in striking accord with observations, in some cases to many decimal places. Quantum mechanics is routinely tested and verified in physics laboratories every day, and is now embedded in the way our modern world operates. Indeed, without quantum mechanics the laptop computer on which this article was typed just simply wouldn't work.

Quantum mechanics plays a wide role in *cosmology. Physical processes which rely on quantum mechanics include the generation of primordial *density perturbations in the *inflationary cosmology, and the Hawking evaporation of *black holes. In developing our understanding of the *Universe, crucial knowledge comes from studying features in the *electromagnetic spectrum of radiation emitted and absorbed by objects, the main features of which are determined by the quantum mechanics of atoms.

Modern theories of particle physics, such as the *standard model of particle physics, *supersymmetry, and *superstring theory, are all constructed to automatically incorporate the laws of quantum mechanics. The last of these is viewed as the leading candidate theory to provide a unification of quantum mechanics and *gravity into a theory of *quantum gravity, seen by many as the major challenge currently facing theoretical physics.

quark–hadron phase transition

*Quarks are one of the fundamental building blocks of Nature, and the particles that those building blocks are assembled into are known as *hadrons ('hadron' literally meaning any particle made up of quarks). While there are only six types of quark, of which only two occur naturally in the present *Universe, there are a vast array of particles they can be built into. The most important by far are *protons and *neutrons.

A peculiarity of quarks is that they cannot be isolated—they always have to be near either two other quarks or one anti-quark. In the young hot Universe, it is believed that the Universe was sufficiently dense that the quarks would behave as individual particles, in a state known as a quark–gluon plasma (*gluons are the particles which mediate interactions between quarks). In this state all the quarks are so close that they have suitable close companions and can switch them at will. By contrast, in the present Universe all quarks are bound into hadrons, either bundles of three quarks or in quark–anti-quark pairs. The transition is believed to have happened around ten microseconds after the *big bang, and is known as the quark–hadron *phase transition. Although potentially of some cosmological significance, not much is known about the nature of this transition. In particular it is not known whether it happens smoothly everywhere in the Universe at once (called a second-order phase transition) or via the appearance of bubbles containing hadrons within a sea of quarks, which then expand and coalesce (a first-order phase transition). Numerical simulations of the transition are being employed to try to address this issue.

Experimentalists try to recreate the quark–hadron phase transition in the laboratory by colliding heavy ions together, aiming to create high enough densities to create the free quark state. The most powerful such experiment is RHIC (relativistic heavy ion collider) at the Brookhaven National Laboratory in the US, which fires gold ions together. It remains unclear to what extent such experiments can actually reproduce conditions pertinent to the young Universe.

quarks

Quarks are elementary particles. They are one of the fundamental building blocks of Nature, and a key ingredient of the *standard model of particle physics. Quarks come in six types, known as **flavours**; historical development has led to the mystifying names of up, down, strange, charm, bottom, and top, each of which has been detected in *particle accelerator experiments. However only the up and down quarks are stable under normal circumstances, and so those are most probably the only types present in significant quantity in the *Universe, though it has been speculated that strange quarks might exist within very dense objects such as *neutron stars.

The idea of quarks emerged from independent work in the early 1960s by American physicists Murray Gell-Mann and George Zweig and Japanese physicist Kazuhiko Nishijima. They provided a way of simplifying the classification of known particles, which by that time had become very complex due to frequent discoveries of new particles. Postulating that those particles (specifically the sub-class of particles known as *hadrons) were made up from more fundamental constituents brought much greater order to the classification of particles. Gell-Mann invented the name 'quarks', inspired by the line, 'Three quarks for Muster Mark' from James Joyce's rather inpenetrable masterpiece *Finnegan's Wake*, and went on to win the *Nobel Prize for Physics in 1969. Originally it was thought that quarks were merely a convenient mathematical fiction for classifying particles, but it later became apparent from particle accelerator experiments that they were real particles.

Each quark has a corresponding *antiparticle known as an anti-quark. However the Universe is dominated by matter rather than *anti-matter, so quarks dominate greatly over anti-quarks.

Quarks possess a property known as 'colour', which is analogous to electric charge, and there are three possible colours. That's not to say that they are actually coloured, but it can be quite convenient to think of them as if they are. Anti-quarks have anti-colours (rather harder to visualize!). According to the Standard Model, isolated free quarks cannot exist. Rather, they must be in groups of three, each with a different colour, or alternately in groups of a quark and an anti-quark with matching colour/anti-colour. Only if the colours cancel out in those ways is the energy of the configuration finite, and so no experiment is able to separate out a single quark.

The groups of three quarks are known collectively as *baryons, and the only two stable ones are *protons (two up quarks and one down quark) and neutrons (one up quark and two downs). Those are, of course, the constituents of atomic nuclei, and hence the main components out of which we ourselves are made; human beings are predominantly made from quarks. Objects made from a quark and anti-quark are known as **mesons**, and hadron is the collective word for all particles made up entirely of quarks.

Interactions between quarks are mediated by particles known as gluons, the name being given as they are responsible for 'gluing' the quarks together. The interaction, one of Nature's *fundamental forces, is known as the strong nuclear force or strong interaction, and it is what prevents atomic nuclei from disintegrating. Quarks also interact via the electromagnetic and weak nuclear forces.

Although isolated quarks cannot exist, if the medium is dense enough they can be so close together that they lose track of which particle they are actually in, forming a sea of quarks and gluons (this is similar to the behaviour of some metals in which the *electrons are free to travel around rather than being associated with one particular nucleus). It is believed that the *early Universe was sufficiently dense to have been a quark–gluon plasma of this kind. Only after about ten microseconds since the *big bang did it became cool and diffuse enough for

hadrons to form, the transition being known as the *quark–hadron phase transition. Particle physicists attempt to recreate the conditions of a quark–gluon plasma by crashing together heavy atomic nuclei at high speed, but experiments so far are inconclusive.

Not all particles are made from quarks. There is a second class of fundamental particles known as *leptons, examples of which are electrons and *neutrinos. *See also* STANDARD MODEL OF PARTICLE PHYSICS.

quasar absorption systems

The name 'quasar absorption system' (sometimes called QSO absorption line system, or QSOALS for short) is a bit of a misnomer, since these are clouds of gas that lie between us and distant *quasars, and are not physically associated with the quasars themselves. The quasars act merely, albeit valuably, as distant light sources, by which the absorption systems may be detected.

An example of a quasar absorption system is shown in Fig. 1, a *Keck telescope spectrum of a quasar at *redshift $z = 3.18$. Blueward of the *Lyman alpha (Lyα) emission line seen at an observed wavelength $\lambda_{obs} \approx$ 5100Å, one sees the *Lyman alpha forest of hydrogen lines produced by intervening gas clouds. The particularly strong line at $\lambda_{obs} \approx$ 4650Å is a *damped Lyman alpha absorption feature produced by a cloud at a redshift of $z = 2.82$ which contains enough neutral hydrogen to absorb almost all Lyα photons

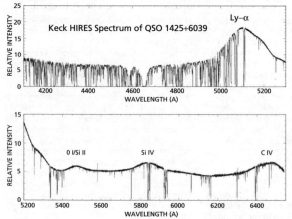

Fig. 1. Spectrum of the quasar 1425+6039 obtained with the High Resolution Echelle Spectrograph (HIRES) at the Keck Observatory.

passing through it. Many of the weak absorption features seen redward of the Lyα emission line are low ionization lines of heavier elements (OI, SiII, CII) associated with this system. Other features are CIV and MgII lines produced in the halos of intervening galaxies.

Absorption lines from these heavier elements are generally found in denser gas clouds, and since they have longer wavelengths than Lyα, they are easier to detect at relatively low redshifts. Studies of quasar absorption systems thus allow us to trace out the distribution of possible proto-galaxies and to trace the evolution of the chemical composition of the *intergalactic medium. As an example, measurements of the deuterium abundance in quasar absorption systems have provided important constraints on models of big bang *nucleosynthesis (BBN).

quasar surveys

The first *quasars were discovered in the 1950s and 1960s as powerful sources detected in the radio part of the *electromagnetic spectrum (at wavelengths longer than about 10 cm). The third and fourth radio surveys carried out at Cambridge, and denoted 3C and 4C respectively, found large numbers of quasars in the northern hemisphere. The Parkes radio telescope in Australia made a comparable survey in the south. However, optical *spectroscopy was needed in order to confirm the *redshifts of quasar candidates, and it soon became apparent that most quasars are in fact radio-quiet, and so do not appear in radio surveys. Modern surveys for radio-loud quasars include the National Radio Astronomy Observatory (NRAO) Very Large Array (VLA) Sky Survey and the Faint Images of the Radio Sky at Twenty-centimetres (first) survey also made with the VLA.

Surveys for radio-quiet quasars are possible via multi-colour surveys: surveys made through more than one passband. Quasars may then be distinguished from stars by their characteristic colours. In particular, quasars at redshifts $z \lesssim 2.3$ appear very bright in the *ultra-violet (UV) , due to their strong *Lyman alpha emission. These quasars appear as 'UV excess' objects in surveys comprising UV and optical imaging. At higher redshifts, Lyman alpha is redshifted into the optical and near-*infrared passbands,

and so high-redshift quasars may be distinguished from stars by their colours in these bands.

Modern optical surveys for high-redshift quasars include the *two degree field quasar redshift survey, the *Sloan Digital Sky Survey, and the COMBO-17 survey, which has imaged 1 square degree of sky through 17 optical filters. The highest-known redshift quasar, at a redshift $z = 6.4$, was discovered by the Sloan Digital Sky Survey. It is hoped that the near-infrared United Kingdom InfraRed Telescope Infrared Deep Sky Survey (UKIDSS) will break the redshift seven barrier for quasars.

Finally, quasars are also strong *X-ray emitters. The early X-ray satellites lacked the sensitivity and resolution to carry out quasar surveys, but now the *Chandra and *XMM-Newton satellites are yielding larger numbers of quasars per square degree than any other surveying technique.

quasars

Quasars are a sub-category of *active galactic nuclei (AGN), and are amongst the most powerful known sources of energy in the *Universe. Although similar in appearance to ordinary stars in optical images, *spectroscopy reveals most quasars to be at vast distances and hence extremely luminous. They are thought to be powered by the accretion of matter onto a *supermassive black hole.

The first quasars (a contraction of the term 'quasi-stellar radio source') were discovered in the 1950s and 1960s in radio surveys. These powerful radio sources appeared star-like in optical images. Only their enormous *redshifts indicated they were not Galactic stars but extremely distant and luminous sources. **Radio-quiet quasars** (also known as **quasi-stellar objects** or QSOs) were subsequently found by searching for objects that appeared stellar on optical photographs but emitted too strongly at IR or UV wavelengths. Radio-quiet quasars outnumber radio-loud quasars by a factor 10–30. Both radio-loud and radio-quiet QSOs are now thought to be variants of the same type of object, and the term 'quasar' now includes QSOs. Deep imaging in the 1980s revealed that quasars are bright nuclei of *galaxies outshining the surrounding stars, and quasars are now regarded as more powerful versions of Seyfert nuclei (*see* ACTIVE GALACTIC NUCLEI).

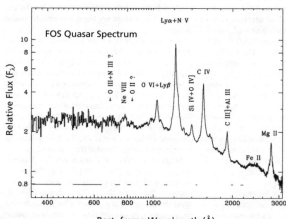

FOS Quasar Spectrum

Fig. 1. A composite quasar spectrum obtained from 101 quasars observed with the faint object spectrograph (FOS) on the Hubble Space Telescope.

Rest–frame Wavelength (Å)

A composite quasar spectrum is shown in Fig. 1. This was obtained from observations of 101 quasars at redshift $z > 0.33$ from the faint object spectrograph (FOS) on the *Hubble Space Telescope. Each spectrum was shifted to its restframe by dividing observed wavelength by a factor of $(1 + z)$, where z is the redshift of the quasar. The restframe spectra were then co-added. The prominent lines which are used in identifying quasars and in measuring their redshifts are indicated in the figure. Due to the high energies and hence large velocities within quasars, these emission lines are widened by Doppler broadening, and several consist of blends of more than one ion.

Quasars frequently vary in brightness on a timescale of order months down to a few hours. Such variability sets an upper limit on the size of the quasar, since the most rapidly varying cannot be more than a few light-hours in diameter (light travels about 10^{12} metres in one hour, less than the distance between the Sun and the planet Saturn in our Solar System). From this relatively small volume is emitted radiation equivalent to about 10^{12} Suns. The only plausible explanation for generating such large amounts of energy within such a compact volume is from the accretion of material onto a supermassive black hole, at the rate of order 10–1000 Solar masses per year. This model also explains why quasars are only seen at relatively high redshifts (quasar activity appears to peak at a redshift $z \sim 2$): by today the supply of mate-rial to 'feed' the central black hole has been exhausted, and the quasar becomes an ordinary galaxy. It appears that most galaxies have a massive black hole at their centre, and it is conjectured that the quasar phenomenon is closely linked to the process of *galaxy formation and evolution.

http://www.astronomynotes.com

quintessence

Quintessence is a proposal for the nature of the *dark energy responsible for the observed *acceleration of the Universe. It supposes that the dark energy takes the same form—a fundamental *scalar field—as that usually assumed to cause *cosmological inflation in the *early Universe. Quintessence models typically predict a slow evolution of the *density of dark energy, which may be observable in future experiments. However at present the *cosmological constant is the simplest model for the dark energy and remains observationally viable.

Etymology

The word 'quintessence', literally 'fifth essence', was introduced to *cosmology in a 1998 paper by Paul Steinhardt and Robert Caldwell, though the idea, then unnamed, can be traced to work a decade earlier by Bharat Ratra and James Peebles, and independently by Christof Wetterich. The name is appropriate as dark energy is the fifth major component of the Universe's composition (the others being *dark matter, *baryons, *photons, and *neutrinos). Their original use of the word was to describe any material capable of driving cosmic acceleration, but usage has evolved so that quintessence is nowadays almost universally used to indicate models of dark energy using scalar fields, with the term 'dark energy' used for the general description.

Quintessence models

Quintessence models postulate that the dark energy is driven by the energy density of a scalar field. Specifically, if the potential energy of the scalar field dominates over its kinetic energy, it will behave as if it has a negative *pressure, as is necessary to drive an accelerated expansion.

There are many different possible quintessence models, corresponding to different choices of the potential energy of the scalar field. Each model will in detail make different predictions for observables, as the evolution of the dark energy density will be different. The aim of observations is to test whether the quintessence idea works at all, and if so then to constrain the allowed forms of the potential energy, hopefully allowing a link to be made to fundamental physics models.

quintessence and the coincidence problem

The *coincidence problem asks why the dark energy is just coming to dominate the Universe at the epoch that we are able to make observations. Quintessence has not so far offered any resolutions, but does offer some hints. Because the energy density of quintessence is able to vary, one could imagine that the dominance of dark energy is triggered by some other phenomenon happening in the Universe. A popular suggestion is that the transition from a *radiation-dominated era to a *matter-dominated era, which happened not too far back in the Universe's history, might have provided such a trigger, though as yet it is unclear how.

Perhaps related is that some quintessence models have a property known as **scaling solutions** or **tracking solutions**, where their evolution is determined by the dominant type of material in the Universe. In a scaling solution, the quintessence field mimics that dominant form, so that for instance it behaves like radiation in a radiation-dominated Universe and like matter in a matter-dominated one. This however is not that useful as dark energy behaves as neither. In tracking solutions, the quintessence field's evolution is determined by the dominant material but is not identical to it, which allows it to come to dominate as dark energy by the present epoch.

Observational signatures

Quintessence is a special case of dark energy, and as such makes the same general predictions for observational signatures that are found in many other dark energy models. However quintessence scenarios are based on well-defined physical models, and hence their predictions can be unambiguously calculated, which is not true of all dark energy models.

Amongst the signatures of quintessence models, most of which will be sought in upcoming experiments, are the following. The slow variation of the energy density means that the *scale factor has a different evolutionary history as compared to a cosmological constant model, which may be observable via the luminosity–*redshift relation of distant type Ia *supernovae, and also affects the angular size of the observed peaks in the *cosmic microwave background (CMB) anisotropy spectrum. Its presence also affects the growth rate of *density perturbations at recent epochs, which might for instance be probed by weak *gravitational lensing. Finally, the quintessence field itself is predicted to have initial perturbations which will dissipate, in doing so leaving potentially observable signatures in the CMB anisotropies via the integrated *Sachs–Wolfe effect.

radiation angular power spectrum

The radiation angular *power spectrum is the main statistical descriptor of *cosmic microwave background (CMB) anisotropies (and indeed gives a complete description of their properties if they are *Gaussian). It measures the typical size of irregularities on a given angular scale. Given a cosmological model, it can be calculated using *linear perturbation theory.

The CMB radiation can be thought of as originating on the surface of a large sphere surrounding us, the *last-scattering surface. The angular size of the perturbations is indicated by an index ℓ (technically, the index of a spherical harmonic expansion of the temperature anisotropy), and the radiation angular power spectrum is denoted C_ℓ (pronounced see ell or see sub ell). The index ℓ is an integer, whose required range is determined by the angular resolution of an experiment. A rough correspondence is that the angular scale θ corresponding to a given ℓ is about $180°/\ell$. For example, the *WMAP satellite measures C_ℓ in the range from $\ell = 2$ up to approximately 1000.

Calculation of the C_ℓ predicted by a particular *cosmological model is challenging, and can only be accurately achieved by numerical calculation. *Cmbfast is an example computer program able to do these calculations. See COSMIC MICROWAVE BACKGROUND ANISOTROPIES, Fig. 2, for an example calculation, and that entry for observational determinations.

The same formalism can be extended to describe CMB *polarization.

radiation-dominated era

A radiation-dominated era of the *Universe is one where the dominant constituent within the Universe is relativistic particles. These particles might for instance be *photons of light or *neutrinos. According to the *standard cosmological model, the young Universe was radiation-dominated from some very early epoch, up until an age of around 50 000 years when it gave way to a *matter-dominated era.

In assessing which type of material dominates the Universe, the important quantity is the *density, as it is what governs the expansion rate of the Universe via the *Friedmann equation. Matter density and energy density can be used interchangeably, using Albert *Einstein's famous $E = mc^2$ relation. In the phrase 'radiation-dominated', the word 'radiation' can mean any form of relativistic matter, i.e. that the constituent particles are travelling at or near to the *speed of light. As the Universe expands, the energy density of such particles reduces in proportion to the *scale factor of the Universe to the fourth power; three of these powers correspond to the volume increasing (as the particles are spread more thinly), and the fourth to *redshifting reducing the energy of individual particles.

The expansion rate of the Universe during a radiation-dominated era is that the scale factor grows proportional to time to the power one-half. The volume therefore grows proportional to time to the power three-halves.

In the present Universe, photons and neutrinos are the only relativistic constituents, but their total energy density is subdominant to that of *baryons, *dark matter, and *dark energy. However, as the Universe expands their density has been dropping more quickly than that of the non-relativistic particles (which reduces as the scale factor cubed rather than the fourth power), implying that at a sufficiently early epoch the relativistic particles must have dominated.

In the late stages of radiation domination only photons and neutrinos were relativistic, with photons contributing slightly greater total density. However in the earlier, hotter, stages of the Universe the ambient energy would have been higher and other particles could also have been relativistic. For instance, electrons would have been relativistic when the Universe was less than about one second old.

It is not known when radiation domination might have started. As far as observations are concerned, the earliest clear evidence that the Universe was radiation dominated came at the time of *nucleosynthesis, when the Universe was about one second old. However there is no reason why it shouldn't have started long before that. Indeed, early *cosmological models took the radiation era to begin at the *big bang itself, but it is now widely assumed that there was a period of *cosmological inflation during the Universe's earliest stages, with the radiation era only becoming established after inflation came to an end. It is also possible that the radiation-dominated era was briefly punctuated by eras of matter domination, or even brief extra periods of inflation.

radio galaxies
Radio galaxies are a class of *active galactic nuclei that emit strongly in the radio part of the *electromagnetic spectrum. They were first identified as strong sources in radio surveys carried out in the 1950s and were later identified with luminous *elliptical galaxies in optical surveys. Many radio galaxies feature twin radio-bright lobes up to 3 Mpc across centred on the galaxy; see, for example, Fig. 1. The radio emission is produced by energetic particles moving through magnetic fields. Radio lobes are thought to result from jets of energy emitted by the galaxy, sometimes visible in the optical as well as the radio.

Fig. 1. Composite optical and radio image of the radio galaxy 3C296. Note that the radio jets extend significantly further than the visible light of the host galaxy (NGC5532).

Galaxies with the largest lobes are giant ellipticals. Since nothing can travel faster than the *speed of light, the size of these lobes implies that the nucleus has been active for at least 10–50 Myr.

The large radio luminosity of these objects makes them useful probes of the *cosmic structure at high *redshift and in directions of large optical *extinction, such as behind the plane of the *Milky Way.

http://www.astronomynotes.com

recombination
Recombination is part of the sequence of events leading to the formation of the *cosmic microwave background, taking place when the *Universe was around one thousandth of its present size and aged a few hundred thousand years. It refers to the epoch when atoms first form in the Universe, meaning that *electrons bind with atomic nuclei. As the young Universe is comprised primarily of hydrogen and helium nuclei (see *nucleosynthesis), those are the atoms that form. Rather embarrassingly, the term 'recombination' is in fact a misnomer, since the electrons were never previously combined into atoms. Technically it should therefore just be 'combination', which however doesn't seem to have the same ring to it.

A nucleus which has lost its electrons is referred to as *ionized, and the fraction of nuclei in that state is known as the ionization fraction. During recombination the ionization fraction falls from one, meaning full ionization, to practically zero.

The young Universe is highly ionized because it is a dense energetic environment. In particular, it is filled with radiation whose constituent *photons are sufficiently energetic to disrupt any atoms as soon as they form. However as the Universe expands and cools, the photons lose energy due to *redshifting and become less able to destroy atoms, which eventually begin to form.

Since most nuclei in the Universe are hydrogen, the key determining factor is the binding energy of electrons in hydrogen atoms; recombination will take place once the photon energy is less than this binding energy. In fact a naive calculation based on equating the typical photon energy with the hydrogen binding energy predicts too high a temperature for recombination (about 50 000 kelvin); to get a good estimate one has to take into account the fact that there are around a billion times more photons than electrons in the Universe. With such a numerical imbalance, the rare photons in a black-body distribution with energy much higher than the average are able to keep the Universe ionized until the temperature is much lower, around 3 000 kelvin. While interactions are common, the ionization fraction can be calculated using a simple balance equation known as the Saha equation, but the accuracy of this equation fails during the recombination process since, by definition, interactions become rare.

To specify a definite epoch of recombination requires a precise criterion. We cannot define it as the epoch when all electrons are in atoms, as not all electrons do manage to form atoms; the last few electrons have no good way of finding the last few nuclei in order to pair up. This *residual ionization is predicted to be at the level of one thousandth. The standard definition of recombination is the epoch at which 90% of electrons are in atoms. The recombination redshift so defined has some dependence on cosmological parameters such as the *Hubble constant, but is at about redshift 1300.

By this definition, recombination does not precisely correspond to the time the microwave background forms, because photons are very good at interacting with free electrons and even a 10% population is easily enough to keep the Universe opaque. The last interactions of the photons, giving the epoch of *decoupling, therefore happens somewhat later.

Recombination is not in fact the last word as far as the ionization state of the Universe is concerned, since the Universe is believed to have undergone *reionization in the relatively recent past, with electrons once more evicted from atoms by energy released by early star formation. The entire evolution of the ionization fraction is known as the *ionization history of the Universe. In general there are separate ionization histories for each type of nucleus, though usually only hydrogen and helium need be considered.

redshift

Redshift is the term given to the apparent reddening of distant sources due to the expansion of the *Universe. In *cosmology the phrase is used with several different meanings.

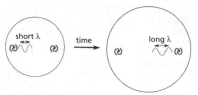

Fig. 1. Illustration of redshift as due to expansion of the Universe. In the time it takes light to travel from one galaxy to the other, the separation between the galaxies and the wavelength of light have both been stretched by the same multiplicative factor of $(1 + z)$, where z is the redshift at which the light was emitted.

Redshift of light

As light emitted from distant objects comes towards us, the Universe is expanding according to the *Hubble law. This causes the wavelength of the light, observed perhaps via a *spectrum of the object, to shift towards the red end of the *electromagnetic spectrum.

There are two ways to think about redshift. The first is to think of it as due to the radial velocity of a distant object moving away from us. Light received by us from this object will be stretched to redder wavelengths, just as the pitch of an ambulance siren is lowered as it races away from you, an effect known as the *Doppler shift. Small radial velocities v, much less than the *speed of light c, are related to redshift z by the formula $z \approx v/c$. The exact relation between redshift and velocity, valid for any measured redshift, may be derived from the special theory of *relativity and is given by

$$1 + z = \left(\frac{1 + v/c}{1 - v/c} \right)^{1/2} ,$$

though even this expression ignores the expansion of the Universe as the light propagates across it.

The second, and much better, way to think of redshift is as due to the stretching of space as the Universe expands (Fig. 1). Since light travels at a finite speed (close to 300 000 km/s), the Universe will have expanded by a certain amount in the time it takes light to travel from a distant object to us, the observer. The wavelength of light expands at the same rate as the Universe, and so by the time the light has reached us, it will have increased its wavelength by this amount.

Consider light which is emitted with a certain wavelength, λ_0. By the time the light reaches us, its wavelength has increased to an observed value $\lambda = \lambda_0 + \Delta\lambda$. Redshift, denoted by the letter z, is defined to be the ratio of change in wavelength over initial wavelength,

$$z = \frac{\Delta\lambda}{\lambda_0}.$$

The quantity $(1 + z)$ is then the factor by which the Universe expanded whilst the light was travelling to us,

$$1 + z = \frac{\lambda_0 + \Delta\lambda}{\lambda_0} = \frac{\lambda}{\lambda_0}.$$

Thus by measuring the observed wavelength λ of a spectral feature of known rest wavelength λ_0 in a distant object's spectrum, the redshift of that object can be determined.

The first *galaxy radial velocity was obtained in 1912 by Vesto Slipher using the 24-inch refracting telescope at Lowell Observatory. A spectrum of the galaxy M31 showed that the characteristic absorption lines were shifted to the blue end of the spectrum by an amount corresponding to a radial velocity of 300 km/s towards the Sun. Thus the first galaxy radial velocity was negative, corresponding to a blueshift in the spectrum. (M31 is a companion of ours in the *Local Group. The *peculiar velocity of this galaxy towards us is larger than the *Hubble expansion on this scale.)

By 1925, Slipher had obtained radial velocities for 41 galaxies, the majority of which were positive and surprisingly large, ranging from -300 to $1\,800$ km/s. These velocities were much larger than any others then known, thus providing the first evidence that the *nebulae, as they were then known, were independent bodies from our own Galaxy. This 'island universes' interpretation of nebulae was definitively proved by Edwin *Hubble in 1924 when he published distances to some of these galaxies based on measurements of *Cepheid variables.

Redshift as distance or time

The word 'redshift' is also used to refer to the distance of objects, and to the time at which they were observed. If a *quasar, for instance, is said to be at redshift $z = 4$, that means that its distance is such that any light received from it is redshifted by a factor of $(1 + z) = 5$ before it is received by us. That implies that the light set out when the Universe was one fifth of its present size. When used to refer to

a time, the phrase 'at redshift z' means the epoch at which any light emitted and later received by us would be redshifted by a factor $(1 + z)$.

For nearby objects at redshift $z \ll 1$, redshift is often quoted in units of km/s, i.e. as a recession velocity. In this case, the velocity quoted is the product of redshift multiplied by the speed of light c: $v = cz$. Redshift proper, being a ratio of two wavelengths, is a dimensionless quantity, and so has no units.

Redshift as evolution of density

Yet another use of the word 'redshift' is to describe how the *density of material varies as the Universe expands. For instance, one may see the density of non-relativistic matter described as redshifting as a^3 as the Universe expands (a being the *scale factor), and that of radiation as a^4. This description is technically inaccurate, in that most (all in the case of non-relativistic matter) of the decrease is due to the increasing volume of the Universe, rather than to actual redshifting of the energy of the constituent particles. Nevertheless, it is useful to refer to the different 'redshifting' laws in order to compare how different types of material evolve as the Universe expands.

See also REDSHIFT SURVEYS.

redshift space

Redshift space is the name given to the coordinate system in which the *distance to an extra-Galactic object, such as a *galaxy, is determined from its *redshift. Relative to real space, which uses true distances, redshift space is distorted by *peculiar velocities. These are velocities relative to the *Hubble expansion. Since only the radial coordinate is affected, structures which are intrinsically isotropic appear to be anisotropic in redshift space. Redshift space is distorted on both small and large scales. On small scales, galaxies within bound systems such as *clusters of galaxies have large random motions, leading to an apparent elongation of structures along the line of sight. Conversely, on large scales, the systematic collapse of newly forming structures makes them appeared flattened along the line of sight. (*See* PECULIAR VELOCITIES, Fig. 1 for an example of both types of distortion.) These distortions can be exploited statistically in order to probe the dynamics of galaxies without the need to measure redshift-independent distances (*see* PAIRWISE VELOCITY DISPERSION).

redshift surveys

Redshift surveys provide a way of mapping out the three-dimensional distribution of *galaxies and *quasars in the Universe.

Measuring redshifts

Much of the radiation we detect from astronomical sources is due to transitions in the energy state of an *electron in the *quantum mechanical model of the atom due to Niels Bohr. In this model, an electron transition either absorbs or emits radiation of a certain energy. This corresponds to a *photon of a certain wavelength, which can be accurately measured in the laboratory. These transitions often occur in groups or series, such as the Balmer series of hydrogen, which involves the transition of an electron in the second-lowest energy level to or from higher levels, accompanied by the emission or absorption of a photon in the optical part of the spectrum. Astronomers can thus recognize these same transitions in distant astronomical sources, whose light has been *redshifted to longer wavelengths.

Redshift surveys are based on the technique of *spectroscopy, in which light is broken down into its component wavelengths, or *spectrum, just as a prism will turn sunlight into a rainbow. Atomic transitions then show themselves as a series of features either in emission (above the continuum) or absorption (below the continuum). The observed wavelengths of these features can then be compared with the known laboratory rest wavelength, and hence the redshift can be calculated (Fig. 1). In practice, since absorption features tend to be broadened by the differing velocities of stars or gas clouds within a *galaxy or *quasar, the redshift is often determined from cross-correlation with a template spectrum, rather than from the wavelengths of individual spectral features.

Once the redshift z of an object has been determined from its spectrum, its recession velocity may be calculated. For objects at low redshift, $z \ll 1$, velocity v is given approximately by $v = cz$, where $c \approx 300\,000$ km/s is the *speed of light. At higher redshift, the velocity must be determined using a *relativistic formula, since the velocity v cannot exceed the speed of light c.

The distance to the object may then be found via the *Hubble law, $d \propto v$, which again is only valid at low redshifts. Relativistic expressions relating distance to redshift

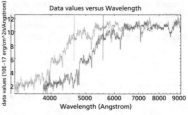

Fig. 1. Spectrum of a galaxy at redshift $z = 0.2038$ as observed (on the right) and as it would be seen at redshift zero (on the left). Note that the distinctive absorption features (dips) in the spectrum are shifted to higher wavelengths by a factor of $(1 + z)$.

have been derived which are valid for any measured redshift and which depend on the assumed *cosmological parameters.

History of redshift surveys

Three techniques have been used to compile redshift surveys.

Targeted spectroscopy—This is the most frequently used technique, and consists of two steps. The first step involves identifying a suitable set of targets (galaxies or quasars), or taking a pre-existing galaxy or *quasar survey. The second step involves measuring redshifts for these targets, most frequently by obtaining optical spectra, or alternatively (for spiral, irregular, and dwarf galaxies) by measuring the redshift of the 21-cm line of neutral hydrogen.

Objective prism survey—Here one places a prism in front of the objective lens of a telescope. The resulting image consists of a series of low-dispersion spectra of bright objects, from which their redshifts may be estimated. This technique only works for bright objects in uncrowded fields; in crowded fields the spectra overlap and are difficult to associate with individual objects.

Photometric redshifts—Here one images the same area of sky through several passbands. One then simultaneously fits a *spectral energy distribution (SED) and a redshift to the observed colours of each object. The advantages of this technique are that many thousands of redshifts may be estimated using just a few hours of telescope time, and one may estimate redshifts for objects too faint to yield a high-quality spectrum. The disadvantage is that there is frequently a *degeneracy in simultaneously fitting an SED and a redshift, and so the resulting redshift may be quite uncertain. Typical uncertainties in photometric redshifts are $\Delta z \approx 0.05$.

Some of the more important redshift surveys are listed below, in chronological order within each category. Unless otherwise stated, they utilize the technique of targeted spectroscopy. We have subdivided them into wide-area and narrow surveys.

Wide-area surveys

All of the following surveys cover at least 100 square degrees of sky. The primary goals of these surveys are to understand the properties of the local galaxy population and to measure the *clustering of galaxies on large scales.

Revised Shapley–Ames Catalogue—In the 1930s, Milton Humason and Nicholas Mayall began a programme to obtain redshifts for the Shapley–Ames galaxies (*see* GALAXY SURVEYS). They published a catalogue in 1956 that included redshifts for 63% of Shapley–Ames galaxies north of declination $\delta = -30°$. Allan Sandage and Gustav Tammann compiled redshifts from 430 separate sources for all but six of the original Shapley–Ames galaxies and included this redshift information in *A Revised Shapley–Ames Catalogue* published in 1981.

This catalogue proved too shallow to make reliable measurements of galaxy clustering, but it did provide some of the first estimates of the galaxy *luminosity function and of the galaxy *pairwise velocity dispersion.

*CfA Redshift Survey**—The Center for Astrophysics (CfA) Redshift Survey was started in 1977 and the first part was completed in 1982. It contained redshifts for 2 401 galaxies brighter than *magnitude 14.5 at high galactic latitude in the merged catalogues of *Zwicky and Nilson and produced the first large area and moderately deep three-dimensional maps of *cosmic structure in the nearby Universe.

The second CfA survey (CfA2), carried out between 1985 and 1995, extended the original CfA survey one magnitude fainter and contains redshifts for about 18 000 bright galaxies in the northern sky. Although it was already known from statistical analyses that galaxies are clustered, a plot of redshift against right ascension for the first strip of galaxies in the CfA2 survey

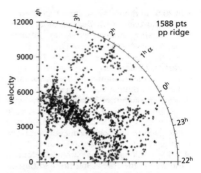

Fig. 2. Distribution of galaxies in the Pisces–Perseus supercluster as a function of recession velocity (km/s) and right ascension (hours).

(*see* CfA REDSHIFT SURVEY, Fig. 1) gave the striking impression that galaxies appear to be distributed on surfaces, almost bubble like, surrounding large empty regions, or *voids.

These surveys allowed the first large-scale, quantitative analyses of the clustering of galaxies in three dimensions.

Southern Sky Redshift Survey (SSRS)—The SSRS provides a southern-hemisphere complement to the CfA2 survey. Originally based on a diameter-limited sample of galaxies from the ESO Survey (*see* GALAXY SURVEYS), the final version of the SSRS, published in 1998 and known as SSRS2, contains redshifts for 5 369 galaxies with blue magnitude $m_B \leq 15.5$.

Pisces–Perseus supercluster—Riccardo Giovanelli and Martha Haynes used the Arecibo radio telescope to obtain 21-cm line redshifts for about 2 800 galaxies in the Pisces–Perseus supercluster between 1983 and 1993. A cone plot of the Pisces–Perseus supercluster is shown in Fig. 2. The main ridge of the supercluster can be traced over 90 degrees across the sky. It lies at a mean redshift of about 5 500 km/s and is an approximately linear filament inclined by less than 12 degrees to the plane of the sky. The supercluster extends over 45 Mpc in length before it disappears into the *zone of avoidance around $\alpha \approx 4^h$.

Stromlo–APM Redshift Survey—The Stromlo–APM Redshift Survey consists of 1 797 galaxies brighter than blue magnitude $b_J = 17.15$, selected randomly at a rate of 1 in 20 from the *APM galaxy survey. Despite the relatively small number of redshifts, which were obtained using the 2.3-metre telescope of Mount Stromlo and Siding Spring Observatories between 1987 and 1991, the sparse-sampling strategy ensured that the survey sampled a volume of more than one million cubic megaparsecs, an order of magnitude larger than any other redshift survey then in existence. This large survey volume proved ideal for measuring the galaxy luminosity function and the clustering of galaxies on large scales.

IRAS PSCz—Based on the *IRAS Point Source Catalogue, the IRAS PSCz Redshift Survey, completed in 2000, mapped out the three-dimensional distribution of 15 000 galaxies over 83% of the sky. Important results from this redshift survey include measurements of the *cosmic dipole and the *power spectrum of galaxies.

A predecessor of the PSCz survey, known as the QDOT survey, provided independent confirmation in 1991 of the result published a year earlier from the APM galaxy survey, that the large-scale clustering of galaxies was inconsistent with the prediction of the then popular *critical density (or 'standard') *cold dark matter (SCDM) model. It was at about this time that cosmologists began to seriously doubt the validity of the SCDM model. Suggested changes to make the CDM model viable included allowing a significant contribution to the mass density from hot *dark matter, and allowing the *density parameter Ω to be less than one, the latter now being the established favourite.

Las Campanas Redshift Survey—The Las Campanas Redshift Survey (LCRS) was the first large-area galaxy redshift survey to be selected from a CCD-based catalogue, as opposed to a photographic catalogue. It was also the first large-area survey to use a fibre-optic system so that 112 spectra could be simultaneously observed in the 1.5 degree diameter field of the Las Campanas 2.5-m DuPont telescope.

The survey, published in 1996, includes 26 418 redshifts for galaxies selected in the R-band. It covers over 700 square degrees in six strips, each 1.5×80 degrees, three each in the North and South galactic caps. The typical redshift in the survey is 30 000 km/s. Two-dimensional representations of the redshift distribution (see Fig. 3) reveal

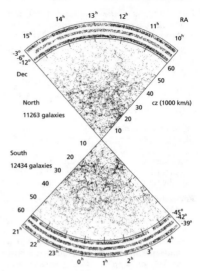

Fig. 3. Distribution of galaxies in the Las Campanas Redshift Survey as a function of recession velocity (1000 km/s) and right ascension (hours).

many repetitions of voids, on the scale of about 5 000 km/s, sharply bounded by large walls of galaxies.

Muenster Redshift Project—The Muenster Redshift Project (MRSP) is the only large galaxy survey to have measured galaxy redshifts using an objective prism. Thirty pairs of United Kingdom Schmidt Telescope survey plates (one of each pair being a direct image, the other being a low-dispersion objective prism plate) have been scanned with a Photometrics Data Systems (PDS) machine at the Muenster Astronomical Institute in Germany. The survey includes 900 000 galaxies with redshifts accurate to $\sigma_z = 0.03$ over an area of 750 square degrees in the southern sky, making this the largest existing redshift survey. Although individual galaxy redshifts are highly uncertain, the large number of redshifts and large survey volume make this a powerful sample for studying the large-scale properties of the galaxy distribution. In particular, the redshift–volume *cosmological test was applied to a subset of the MRSP data in 1998 and used to place an upper limit on the *deceleration parameter of $q_0 < 0.1$ (95% confidence).

Durham/UKST—Quite similar to the Stromlo–APM Survey, the Durham/UKST survey consists of 2 500 galaxy redshifts to a limiting magnitude of $b_j = 17$. The survey covers an area of approximately 1 500 square degrees around the Southern Galactic pole. The galaxies in the survey were selected randomly at a rate of 1 in 3 from the Edinburgh/Durham Southern Galaxy Catalogue (see GALAXY SURVEYS), and thus the resulting survey provides a denser sampling of a smaller volume than the Stromlo–APM survey. Redshifts were obtained on the 1.2m UK Schmidt Telescope at Siding Spring Observatory in Australia using the FLAIR multi-object spectrograph.

Two degree field galaxy redshift survey—
See TWO DEGREE FIELD GALAXY REDSHIFT SURVEY.

6dF Galaxy Survey—The 6dF Galaxy Survey aims to measure redshifts for 150 000 galaxies, and *peculiar velocities for 10% of these, using the six-degree field (6dF) multi-fibre spectrograph on the UK Schmidt Telescope. Galaxy targets are objects brighter than $K = 12.75$ in the *Two Micron All Sky Survey (2MASS) extended source catalogue. The second data release (6dFGS DR2), issued in May 2005, included 89 211 spectra. When complete, the 6dFGS will cover the entire southern sky at galactic latitude $|b| > 10°$, more than eight times the area of the 2dF galaxy redshift survey and twice that of the Sloan Digital Sky Survey.

Sloan Digital Sky Survey—See SLOAN DIGITAL SKY SURVEY.

Narrow surveys

The following surveys cover fields from of order 10 square degrees down to as small as a few square arcminutes. In the early days, the primary motivation of these surveys was to constrain the galaxy distribution using deeper samples than could feasibly be observed over wide areas. With the advent of multi-object spectrographs in the late 1980s, the primary goal of these deeper surveys has generally been to understand *galaxy evolution.

KOS/KOSS—The first deep (for their time) redshift surveys were carried out by Robert Kirshner and collaborators. The KOS survey, published in 1979, included 164 galaxies brighter than 15th magnitude in eight widely separated fields of 15 square degrees

each. The deeper KOSS survey (1983) covered six smaller fields but included 280 galaxies as faint as 17th magnitude. All three northern fields in this survey, which were separated on the sky by about 35 degrees, showed an underdensity of galaxies at a redshift corresponding to 15 000 km/s recession velocity. The same observers confirmed in 1987 the existence of a roughly spherical void with radius of 62 Mpc, known as the Bootes void. The existence of such a large void presented challenges for contemporary models of *structure formation.

AARS—The Anglo-Australian Redshift Survey (AARS, 1986) contained 329 galaxies down to 17th magnitude in five fields of 15 square degrees each. The AARS provided useful estimates of the clustering and peculiar velocities of galaxies and, in particular, provided one of the first truly reliable estimates of the galaxy luminosity function.

Durham/AAT—A significant leap in survey depth was made with the Durham/AAT redshift survey, published in 1988. By measuring redshifts for 200 galaxies to blue magnitude $b_J = 21.5$ in five 20 × 20 arcminute fields, this was the first redshift survey deep enough to constrain models of galaxy evolution. In particular, these observations suggested that strong, short-lived bursts of star formation were required. An extension to this survey, published by Tom Broadhurst and collaborators in *Nature* in 1990, indicated an apparent periodicity in the redshift distribution of $128h^{-1}$Mpc. It is now widely believed that this apparent periodicity was simply a result of undersampling the true galaxy distribution with a series of very narrow, and quite closely spaced, 'pencil-beam' surveys.

ESO Slice Project—The ESO Slice Project (ESP) was a key project of the European Southern Observatory. This survey includes 3 342 galaxy redshifts in a slice of sky 22 × 1 degrees near the South Galactic Pole. At redshift $z \sim 0.1$, which corresponds to the peak of the *selection function of the survey, this corresponds to physical dimensions 110 × 5 Mpc.

Key science results include a measurement of the galaxy luminosity function, a study of the statistics of emission line galaxies, and a measurement of the size distribution of inhomogeneities in the galaxy distribution over a large volume.

The two degree field galaxy redshift survey and Sloan Digital Sky Survey have largely superseded the ESP since they survey much larger areas to a comparable depth.

CNOC Surveys—The Canadian Network for Observational Cosmology (CNOC) consortium has carried out two redshift surveys using the 3.6m Canada–France–Hawaii Telescope on Mauna Kea, Hawaii. The first, known as CNOC1, surveyed 16 rich galaxy *clusters with redshifts in the range $0.17 < z < 0.55$ from January 1993 to March 1995. The main goal of CNOC1 was to measure the density parameter Ω. Galaxy cluster *virial masses and total luminosities were calculated from about 1 200 cluster redshifts, and the *luminosity density was estimated from 1 000 field galaxies from the same dataset. The CNOC1 survey determined a density parameter $\Omega = 0.19 \pm 0.1$, in good agreement with the currently-preferred value (*see* STANDARD COSMOLOGICAL MODEL).

The CNOC2 survey observed 6 200 field galaxies in the redshift range $0.1 < z < 0.7$ from February 1995 to May 1998 over four widely separated areas totalling 1.5 square degrees of sky. The main goals of CNOC2 were to measure galaxy evolution. The survey found that galaxy clustering evolves weakly with redshift from $z = 0$ to $z = 0.65$, in the sense that clustering was slightly weaker at higher redshift, consistent with predictions from the standard cosmological model. It was found that luminosity function evolution depends on galaxy type: primarily density evolution for the late-SED-type galaxy population and primarily luminosity evolution for early/intermediate SED types.

DEEP Surveys—The Deep Extragalactic Evolutionary Probe (DEEP) is a long-term programme which is using the twin 10 m *Keck Telescopes and the *Hubble Space Telescope (HST) to conduct a truly large-scale survey of distant, faint, field galaxies. The broad scientific goals include: the formation and evolution of galaxies, the origin of large-scale structure, the nature of the dark matter, and the *geometry of the Universe.

Phase 1 (DEEP1) used the Keck LRIS spectrograph to obtain 1 100 redshifts in the range $0.3 < z < 1.3$ in two fields and to $z \sim 3$ in the *Hubble Deep Field. Phase 2 (DEEP2) used the improved DEIMOS spectrograph to observe galaxies that have been pre-selected by *photometric redshift to lie in the range $0.7 < z < 1.55$ in four 0.54 square degree fields. The improved resolution of the DEIMOS spectrograph enabled line-widths and *rotation curves to be obtained for about 50% of the galaxies observed. These dynamical observations enable galaxy masses to be estimated, providing powerful new constraints on evolution models. Already more than 100 publications have resulted from the DEEP surveys.

VIRMOS-VLT—The VIRMOS-VLT Deep Survey (VVDS) will provide a complete picture of galaxy and structure formation over a very broad redshift range $(0 < z < 5)$ over 16 square degrees of the sky in four separate fields. This ambitious survey is possible thanks to the impressive multiplex gains of the VIRMOS instruments (VIMOS and NIRMOS) built by the Franco-Italian VIRMOS consortium for the ESO *Very Large Telescopes. The survey will be carried out over the consortium's 120 nights of guaranteed time, and will contain in total a sample of 150 000 redshifts. This unique database will enable the team to trace back the evolution of galaxies, *active galactic nuclei, and clusters to epochs where the Universe was a fraction (about 20%) of its current age. The VVDS will be comparable in size to the largest redshift surveys currently underway, but will probe to much higher redshifts. It will provide an unparalleled description of how structures and galaxy populations evolved in the universe from high redshift to the present day.

The VVDS will consist of three subsamples: 100 000 galaxies to $I = 22.5$, 50 000 galaxies to $I = 24$, and 1 000 galaxies to $I = 26$. The goals of the survey include: mapping the distribution of galaxies at various epochs, establishing the evolution of the star formation rate from the first generations of galaxies, and establishing the evolution of the mass function of galaxies of various types.

6dF: http://www.aao.gov.au/local/www/6df/
CNOC: http://www.astro.utoronto.ca/~cnoc/
DEEP: http://deep.ucolick.org/
VIRMOS-VLT:
    http://www.oamp.fr/virmos/vvds.htm

Rees–Sciama effect

The Rees–Sciama effect, named after British cosmologists Martin Rees and Dennis Sciama who first described it in 1968, is an effect on the *cosmic microwave background (CMB) radiation by structures which are undergoing gravitational collapse as the radiation crosses them. It is one mechanism by which *cosmic microwave background anisotropies are generated. The effect applies only on rather small angular scales, and has yet to be measured.

The Rees–Sciama effect does not need a direct interaction between the CMB *photons and any material—it is a purely gravitational effect. Photons of light travelling across regions where the *gravitational potential varies experience redshifting and blueshifting, according to the gravitational redshift predicted by *general relativity. If a photon crosses a potential well which is not varying in time, this has no lasting effect on the photon, since any blueshift gained on entering the potential is cancelled by an equivalent redshift on exiting. However, as an object such as a *galaxy or galaxy *cluster forms, the gravitational potential changes. If the photons are crossing the region as the structure forms, the potential they climb out of will be different from the one they enter, and they will finish up with a net change in their energy. This is the Rees–Sciama effect.

In a *Universe which has *critical density and contains only small perturbations (known as linear *density perturbations as they obey linear differential equations), the gravitational potential does not evolve in time. There are two different regimes where variation of the gravitational potential can induce CMB anisotropies. One concerns evolution of linear perturbations, but where the critical-density condition breaks down; this is known as integrated *Sachs–Wolfe effect and is an important way of constraining models of *dark energy. The Rees–Sciama effect refers to the breakdown of the linear perturbation assumption, i.e. to the time when structures are first strongly developing. It is therefore important only on angular scales where there are substantial numbers of structures in the process of forming, which is too small an angular scale for current detectors.

The Rees–Sciama effect is distinct from *gravitational lensing, which leads to a bending of the photon trajectories in addition to the energy change. A complete analysis should include both effects simultaneously.

reheating

Reheating refers to the process by which *cosmological inflation which is trapped in the form of the inflaton is released to form a thermalized background of conventional particles such as *baryons and *photons. Its early stages may occur in a catastrophically rapid fashion known as *preheating.

reionization

Reionization is the process by which the *Universe becomes ionized once *cosmic structures begin to form. The nuclear reactions taking place within the first massive stars, both during their lifetimes and in the *supernova explosions that end their lives, release considerable amounts of energy into the *intergalactic medium (IGM), causing the *electrons to become dissociated from their atomic nuclei and recreating a plasma state. Energy released by material infalling into supermassive black holes, such as *quasars and *active galactic nuclei, will also contribute to this process. Observations indicate that this process took place at an early stage of the *structure formation process, at a *redshift of about ten.

After *recombination, shortly before the release of the *cosmic microwave background (CMB), the Universe was in a nearly neutral state, meaning that all the electrons had combined with nuclei to form atoms. Only a small *residual ionization remained. However two separate types of observation indicate that this state of affairs could not have remained right up to the present. In the *spectrum of distant quasars, the strong absorption of high-energy radiation that would be caused by neutral gas in the IGM—the *Gunn–Peterson effect—is not seen except in the most distant quasars, implying that the gas must be ionized at any subsequent epoch. And in the pattern of *CMB anisotropies, the *WMAP satellite has seen a characteristic suppression of the small-scale irregularities that would be caused by about 10% of the CMB *photons scattering from free electrons on their way towards us.

The precise details of the reionization process remain rather uncertain, and unveiling them is one of the targets of new and planned radio telescope arrays including *LOFAR (Low Frequency Array) and the *Square Kilometre Array. Standard reionization scenarios begin with the formation of

                                                    to
                                              , the
                                            g and
                                          , reion-
                                       searchers
                                    ght instead
                                  ion, where
                                y regions and
                              -density regions
                            eding in ionizing
th   . The ionizing radiation   ompete with the natural tende.   ons to bind with nuclei,
so once estab.   onization also has to be
maintained.

reionization

The fraction of the Universe which is ionized, as a function of time, is known as the
*ionization history, and in general there are
separate ionization histories for the different types of atom present (though in practice only hydrogen and helium-4 are abundant enough to need consideration). The simplest models of reionization assume that it
takes place instantaneously, specified simply
by the epoch (i.e. the redshift) at which it takes
place. Under that assumption, the three-year
results from the WMAP satellite indicate that
it took place at a redshift $z_{reion} = 12 \pm 3$. More
realistically, however, the reionization process
would happen over a somewhat extended
period of time. More complex scenarios have
also been considered where, for instance, the
Universe might be first partly ionized by an
initial energy injection, and then at a later
stage be fully ionized.

While the standard assumption is that
reionization is induced by the first structures to form, an idea in good agreement
with observations, more outlandish possibilities have been suggested. These include
that the energy of reionization was provided by the evaporation of a population of
*primordial black holes, or by the decay of
some unknown type of massive particle. However both these scenarios struggle against the
evidence that reionization does appear coincident with early structure formation.

relativistic perturbation theory
See COSMOLOGICAL PERTURBATION THEORY.

relativity
Relativity is one of the key physics ideas of
the 20th century, and synonymous with the
great physicist Albert *Einstein. In 1905 he
radically rewrote the foundations of physics
with his theory of special relativity, and

went on in 1915 to produce the theory of
*general relativity which incorporated a the-
ory of gravity. This item will cover only special relativity, as general relativity has its own
entry.

In order to define any physical laws, one
needs a reference frame, meaning a set of
coordinates in both space and time indicating where and when events happen. We will
interchange the terms 'frame' and 'observer';
by 'observer' we mean someone who uses a
particular set of coordinates to describe the
physics taking place around them. Different
observers need not choose to use the same
coordinates. Special relativity refers to a particular case of two observers moving with a
constant velocity relative to one another, and
not subject to external forces. Such observers
are known as **inertial observers** (because they
remain in motion as according to their iner-
tia), and each carries with them a coordinate system in which they are fixed. This
coordinate system is known as an **inertial
frame**.

The most fundamental expression of the
principle of relativity is that an observer working within their reference frame cannot conduct any physical experiment which would
allow them to deduce their velocity. More
concisely, each inertial observer perceives the
same laws of physics. Our most direct experience of this comes from travel; almost everyone has witnessed a train passing their own
in the opposite direction, and been momentarily unable to judge whether it is their own
train or the other which is moving. Sitting in
a car moving with constant velocity feels no
different to sitting in one which is stationary. By contrast, if there is an acceleration we
are physically aware of it, for instance being
pressed back in our seats as an aeroplane
accelerates down a runway.

[Although *physical laws* respect the principle of relativity, local physical conditions may
select out particular frames as having special significance. In *cosmology, the *cosmic
dipole of the *cosmic microwave background
(CMB) radiation indicates our velocity with
respect to the CMB (caused by the gravitational effect of local structures), and there is
a reference frame where the dipole vanishes
corresponding to an observer who is not moving relative to the CMB.]

This principle seems fairly innocuous, and
was recognized as applying to dynamical
systems already by Galileo Galilei (famous

astronomer, scientist, and inventor of the modern scientific method) in the early 17th century. But it proved highly troublesome in the context of James Clerk Maxwell's theory of electromagnetism formulated in 1873. One of the successes of this theory is that it predicts the *speed of light in vacuum, giving it the numerical value $c = 2.998 \times 10^8 \, \text{m s}^{-1}$, but the theory does not specify in which reference frame this speed is to be measured. If different reference frames saw a different speed of light, that could be used to indicate the observer's absolute velocity. Maxwell's equations were not consistent with relativity as envisaged by Galileo.

A way out of this problem was to suggest that there was a special reference frame where the speed of light matched Maxwell's prediction, and this became known as the **ether**. A famous experiment was carried out in 1887 by Albert Michelson (who had previously made what at that time were by far the most accurate measurements of the speed of light) and Edward Morley. They used what is now called a Michelson interferometer, and attempted to measure the velocity of the Earth relative to the ether by comparing the speeds of two light rays sent in perpendicular directions—see Fig. 1. The effect should have been readily detectable, but they saw nothing. Apparently there was no such thing as the ether.

There is some argument about whether Einstein was aware of the Michelson–Morley null result, but he was certainly aware of the problem of electromagnetism. In one of his famous thought experiments, he imagined himself moving at the speed of light, looking on at a now stationary light wave beside him. But Maxwell's equations did not permit such stationary solutions.

His resolution was dramatic, and was reported in one of the three landmark papers he wrote in 1905. The only way out of this paradox would be if *all* inertial observers saw the same speed of light. You see the speed of light to be c; an observer rushing past you at velocity v in the same direction still sees the speed of light as c, and not as $c - v$ (the velocity addition law we take for granted in everyday experience). Equivalently, light emitted from a source moving towards us travels no faster than if the source were stationary. This assumption allows us to retain the assumption that all inertial observers experience the same physical laws.

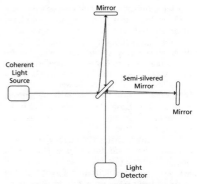

Fig. 1. A schematic of the Michelson–Morley experiment. Light emitted from a source is split into two beams going at right angles to each other by a half-silvered mirror, and then the beams are brought together and the interference pattern analysed. Such instruments are sensitive to changes in the arm lengths comparable to the wavelength of the light used. The motion relative to the ether caused by the Earth's orbit should have meant that the speed of light was different in the two arms, and rotation of the experiment (which was floated on a pool of mercury in a way that safety inspectors would be unlikely to agree to today!) would have revealed this. No effect was seen.

His theory of special relativity is therefore based on two postulates:

- All inertial observers see the same laws of physics.
- The speed of light is the same in all inertial frames and independent of the motion of the source.

From these two assumptions, all else follows. In particular, these rules ensure that no information-carrying signal can ever travel faster than the speed of light, and the speed c cannot be reached by any combination of speeds slower than c. The speed of light acts as a universal speed limit.

Typically, we want to relate the views of two different observers, so that you can predict what another observer will see using their coordinates. This requires us to transform between the coordinates used by one observer and those used by another. According to Newton's laws of motion, time is absolute and we can translate to a reference frame moving with relative velocity just by adding the accumulated distance to the spatial coordinates. But this is no longer good enough, as it violates the second postulate.

The relation between observers must be more complicated if it is to respect the postulates of special relativity. In fact, such a transformation had already been found by Dutch physicist Hendrik Lorentz, and today it bears his name as the **Lorentz transformation**. He found it as a way of keeping Maxwell's equations in the same form when viewed by both observers. Indeed Maxwell's equations were the first properly relativistic theory, in the modern sense, though it was decades before this was realized.

The Lorentz transformation mixes up space and time; what are perceived as changes in distance by one observer may be partly seen as changes in time by another. Space and time cease to have separate identities; they must be merged into a single concept, *space-time. Different observers may not agree on the separation of two events in space alone or in time alone, but they always agree on the distance between two events in space-time.

Einstein's theory has several consequences for the appearance of objects moving at significant fractions of the speed of light. Suppose someone were to fly past us at say half the speed of light, carrying a metre stick and a clock. To us, the metre stick would appear to be shorter, and the clock to tick more slowly. These phenomena are known as Lorentz–Fitzgerald contraction and time dilation respectively. [Before Einstein's paper, Lorentz had actually suggested the Lorentz–Fitzgerald contraction formula as an explanation for the null result of Michelson and Morley, by conjecturing that the motion of the apparatus against the ether shortened it. Einstein's paper completely changed the physical interpretation of that formula, but the formula itself remains the same.] However if we ourselves were holding the same items, then as far as our friend were concerned we would appear to be flying past them at half the speed of light, and they would see our ruler shorter and clock running slow. The situation would be perfectly symmetric.

Soon after his landmark paper introducing relativity, Einstein produced a second paper on the topic which led to the most famous physics equation of all, $E = mc^2$. In words, energy equals mass, times the speed of light squared. Energy and matter are really two different faces of the same coin, and one can be converted into another. Because the speed of light is such a big number, a small amount of mass can become a large amount of energy;

a nuclear weapon converts less than 1% of its total mass into energy, yet wreaks untold devastation.

[Rather disappointingly, Einstein's original paper actually called the speed of light v, rather than c as is now universally used. So the equation $E = mc^2$ never appears in that form in his paper, rather like the non-existent 'Play it again, Sam' in the film *Casablanca*. Still, that's a useful reminder that the symbols on their own don't mean anything, unless one has defined what they stand for.]

The principle of relativity is now deeply embedded throughout modern physics. All fundamental theories, such as the *standard model of particle physics, are designed to respect it, and the success of those theories already supports relativity. It is further routinely tested in many physics experiments; for instance *particle accelerators accelerate particles to speeds extremely close to that of light, and relativity is absolutely necessary to describe their dynamics.

While every test of relativity has supported it, nevertheless physicists have considered what might happen if it is valid only in the restricted circumstances where those tests have been made. As an example, physicists have considered what might happen if the speed of light is not a true constant of nature, but changes value as the Universe evolves (*see* VARYING FUNDAMENTAL CONSTANTS).

Taylor, E. F. and Wheeler J. A., *Space-time Physics*, W. H. Freeman, 1992.

Barton, G. *Introduction to the Relativity Principle*, John Wiley & Sons, 1999 [Technical].

relic particle abundances

Modern particle physics models, such as *supersymmetry and *superstring theory, predict the existence of many new types of particles. Even though they have proved difficult to create in terrestrial *particle accelerators, depending on their physical properties these particles may be created during the early stages of the *Universe's evolution. If they then survive to the present, they are sometimes known as relic particles, and the number or *density of them known as the relic particle abundance.

Some types of relic are quite desirable, for instance the *dark matter is usually thought to be made of relic fundamental particles such as *WIMPs. But in many cases particle theories predict relic particle abundances which are incompatible with present observations, because their present density would be too high and they would be the dominant constituent of the Universe. For instance, since *baryons (i.e. *protons and *neutrons) are known to comprise around 4% of the *critical density of the Universe, and the total density of material is known not to significantly exceed the critical density, any theory which predicts a relic particle whose abundance is more than twenty-five times that of baryons is in violation of observations.

Relic particle problems are common in theories which predict very heavy particles, which then are expected to become non-relativistic early in the history of the Universe. Our Universe underwent a long *radiation-dominated era. However the density of non-relativistic particles reduces more slowly with expansion than radiation, so if even only a small amount of them were formed during the radiation era, they would rapidly come to dominate over the radiation and bring the radiation era to a premature end.

One possibility of considerable historical interest is *magnetic monopoles. These were influential in leading Alan Guth to devise the *inflationary cosmology. Magnetic monopoles were predicted by *grand unification theories, which were particularly popular in the 1970s, and it was clear they would be massively overproduced. Inflation provided a mechanism by which the density of monopoles could be reduced to acceptable (indeed tiny) levels, while simultaneously solving two other concerns about the hot big bang cosmology (*see* 'Overview'), namely the *horizon problem and *flatness problem.

Nowadays the monopole problem is not taken as seriously as it was when inflation was invented, and other potential relics from *supersymmetry and *M-theory, such as gravitinos and moduli fields, are seen as potentially more damaging. For instance, the production of monopoles could be avoided simply by there never being a grand unification phase transition.

Fundamental particles are not the only relics which may be left over from the early Universe. *Primordial black holes and *topological defects are two other examples of possible relics.

residual ionization

The process of *recombination is when atoms first form in the *Universe, with atomic nuclei and *electrons combining. Until that epoch,

atoms are prevented from forming due to frequent interactions with high-energy radiation. As the strength of the radiation dies away due to *redshifting, atoms are able to form and survive, and the Universe makes a transition from a hot opaque plasma state to a transparent sea of conventional material.

However not quite all electrons do undergo recombination, even though charge neutrality of the Universe ensures a perfect balance in the numbers of *protons and electrons. The reason is that as more and more atoms form, the populations of remaining electrons and nuclei become sparser and sparser and eventually are simply unable to find each other in order to recombine, leaving a residual amount of ionization. Precise calculations of the level of residual ionization are rather hard to make, but the level could be as high as one in a thousand electrons.

The residual ionization has a small effect on the *cosmic microwave background anisotropies, because those remaining free electrons are able to scatter a small fraction of the microwave background photons. This has the same effect on the anisotropies as scattering from electrons freed by *reionization in the recent Universe.

Robertson–Walker metric

The Robertson–Walker *metric, described by Howard Robertson and Arthur Walker in separate papers in 1936, describes the *space-time of a *homogeneous and *isotropic *Universe. As such it embodies the *cosmological principle, and it plays a central role in modern *cosmology.

The standard mathematical form of the Robertson–Walker metric is as follows:

$$ds^2 = -c^2 dt^2 + a^2(t)\left[\frac{dr^2}{1-kr^2} + r^2\left(d\theta^2 + \sin^2\theta\, d\phi^2\right)\right].$$

Here spherical polar coordinates r–θ–ϕ are being used to describe the spatial dimensions, and ds is the space-time distance corresponding to shifts dt in the time coordinate and dr, $d\theta$, and $d\phi$ in the space coordinates. Robertson and Walker showed that this was the most general form of metric satisfying both homogeneity and isotropy. It can be written in slightly different but equivalent forms by changing coordinates.

The metric contains two unknown quantities, the *scale factor $a(t)$ and the spatial *curvature k. The latter is a constant, both in space and time, whose value in our Universe must be obtained from observation. The former is a function of time which measures the size of the Universe, and which is to be determined by solving the *Friedmann equation once the material content of the Universe is specified along with the constant k. *Cosmological models thus obtained are often referred to as Friedmann–Robertson–Walker or FRW cosmologies, sometimes with Georges Lemaître also named after Friedmann.

The metrics are qualitatively different depending on whether k is positive, zero, or negative, which correspond to the different possible *geometries of the Universe of spherical, flat, and hyperbolic respectively. Current observations of cosmic microwave background anisotropies indicate that k is close to or equal to zero.

It is not believed that the Robertson–Walker metric perfectly describes our Universe, which possesses *density perturbations which violate homogeneity. However at least while those perturbations are small, they can be considered to give small deviations from the Robertson–Walker metric, described by what is known as *cosmological perturbation theory. This description appears to hold good on all length scales until the epoch of *galaxy formation, and continues to apply on sufficiently large length scales right up to the present. For example, calculations of cosmic microwave background anisotropies are always made using the perturbed Robertson–Walker metric.

ROSAT

ROSAT, the Röntgen Satellite, was an *X-ray observatory developed through a cooperative programme between Germany, the United States, and the United Kingdom. Launched in June 1990, the mission ended after almost nine years, in February 1999. The satellite contained two X-ray detectors: the Position Sensitive Proportional Counter (PSPC) with a two degree field of view (FOV) and the High Resolution Imager (HRI, 40 arcmin FOV). It also contained an extreme *ultra-violet camera: the Wide Field Camera (WFC) with a 5-degree FOV.

The first six months of ROSAT operations were used to map the entire sky with the PSPC

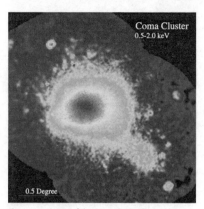

Fig. 1. X-ray image of the Coma cluster observed by ROSAT. The fainter emission to the lower-right of the centre is from a group of galaxies that are falling into the main cluster core.

in three X-ray bands centred on 0.25, 0.75, and 1.5 keV (*see* PARTICLE PHYSICS UNITS). After this the satellite made a series of deeper, pointed observations that continued for the duration of the project.

ROSAT has been used to study the X-ray variability of *active galactic nuclei, a powerful tool for helping to understand the origin of their power sources. The satellite has also been used to detect and study *clusters of galaxies. See Fig. 1 for an X-ray image of the *Coma cluster.

X-ray selection allows one to locate clusters hidden in the optical by the plane of the *Milky Way Galaxy, and to study the X-ray morphology of clusters at high redshift. Clusters at $z \sim 0.8$ show large amounts of substructure, consistent with *hierarchical structure formation models, in which massive clusters of galaxies form from mergers of less massive sub-units.

http://wave.xray.mpe.mpg.de/rosat

rotation of the Universe

Bizarre though it may sound, *cosmological models exist in which the entire *Universe is rotating. The first model of this type in Albert *Einstein's *general relativity theory was discovered in 1949 by Austrian mathematician Kurt Gödel, and also featured the alarming possibility of travel backwards in time. Other rotating Universe cosmologies have been investigated over the years, and are members of the so-called *Bianchi classification of cosmological models.

Such models are necessarily *anisotropic, as the rotation selects out special directions, and as such are strongly constrained by the observed near *isotropy of the *cosmic microwave background. Indeed, the current rotation rate is so tightly constrained by observations from the *COBE satellite that the Universe cannot have rotated by more than a ten-millionth of a full rotation since the time the microwave background was released. It has however been argued that this limit may not apply to all possible kinds of rotating cosmological model. Nevertheless, there is presently no evidence that the Universe does rotate.

running

The term 'running' refers to a possible variation of the *spectral index of *density perturbations with scale. A standard assumption is that primordial density perturbations have a *power spectrum which behaves as a power-law with scale, in which case there is no running. Otherwise, the spectrum of density perturbations is said to run with scale, and this can be quantified by $dn/d\ln k$ where n is the spectral index and k the *wavenumber.

The terminology was invented in a 1995 paper by Arthur Kosowsky and Michael Turner, who were motivated by an analogy with particle physics models in which the change of interaction strength with energy scale is called a running of the coupling constant. Almost all *cosmological inflation models predict a level of spectral index running which is much too small to be detected in the near future. The first-year data release from the *WMAP satellite actually indicated a modest preference for running at a much higher level than expected, but this was subsequently overturned by more detailed analyses incorporating extra data.

Sachs–Wolfe effect

The Sachs–Wolfe effect is one of the mechanisms responsible for creation of *cosmic microwave background (CMB) anisotropies. It is the dominant effect on the largest angular scales, in particular explaining the anisotropies detected by the *COBE satellite. It is caused by variations in the *gravitational potential at different locations on the *last-scattering surface, and was first calculated in 1967 by Rainer Sachs and Art Wolfe.

In a perfectly smooth *Universe, the microwave radiation received from different directions would be at the same temperature. However the Universe possesses *density perturbations, which are small at the time of *decoupling, subsequently growing to form *galaxies and galaxy *clusters. These density perturbations mean that the gravitational potential (strictly, its *general relativistic equivalent) at different points on the last-scattering surface is varying, signifying the gravitational forces which attract material towards the highest-density regions. When the microwave *photons are released, some are at locations where the gravitational potential is higher than average and some lower. Those at the low regions have to climb out of the gravitational potential and lose energy (are *redshifted) as they do so, while those in regions of high potential are blueshifted. The energy of the photons we receive therefore maps out the structure of the gravitational potential on the last-scattering surface. In fact, counter-intuitively the hotter (blueshifted) regions correspond to underdensities, and the colder regions are more dense than average.

Rather than considering the last-scattering surface to be a uniform sphere with a varying density across it, one can alternatively consider decoupling to be delayed in the regions of higher density and to happen earlier in regions of low density. It turns out that the two views of the Sachs–Wolfe effect are entirely equivalent.

A related effect is caused if the gravitational potential in a region of the Universe changes while the microwave photons cross it. It so happens that the gravitational potential does not change provided the Universe is dominated by matter, and provided density perturbations are small, and those two conditions are satisfied for most of the history of the Universe between the creation of the microwave background at decoupling and the present. (For the effect of large density perturbations, *see* REES–SCIAMA EFFECT.)

More interesting is the effect of the Universe not being matter dominated, known as the **integrated Sachs–Wolfe (ISW) effect** as it has to be added up along the trajectory of the light. It comes in two forms, known as early and late. The early ISW effect takes place close to the decoupling epoch; at that time the Universe is dominated by matter, but the radiation density is not yet completely negligible and drives some change in the gravitational potential. It does not have any particularly distinctive observational signature.

The late ISW effect occurs in the recent past, when matter domination gives way to domination by *dark energy. It has two observational signatures. The first is that it modifies the CMB anisotropies on very large angular scales, though this effect is likely to be concealed within the *cosmic variance. The second is that the cosmic microwave anisotropies can be correlated with the structures lying within the density perturbations which cause the ISW effect. Since dark energy only dominated relatively recently the galaxies are not particularly distant, but very large samples are needed to measure what is quite a small effect, as only a small part of the CMB anisotropies are due to the ISW effect. Using CMB maps from the *WMAP satellite, several research groups have reported low-significance detections of this correlation, and it gives an important independent line of evidence in favour of the existence of dark energy.

satellite galaxies

A satellite galaxy is simply a small, low-mass *galaxy that is in orbit about a much larger and more massive galaxy.

The best-studied satellite galaxies are those orbiting our own *Milky Way Galaxy and those orbiting our giant neighbour Andromeda. The Milky Way has 14 known satellites, some of which are shown in Fig. 1, including the relatively luminous Magellanic Clouds ($M_V = -18.1$ and -16.2 for the large and small clouds respectively), both of which are easily visible to the unaided eye from the southern hemisphere. The least luminous satellite, Ursa Major, has an absolute visual magnitude $M_V \approx -6.8$, and was discovered only in 2005 as an overdensity of red, resolved stars in *Sloan Digital Sky Survey data. All of these satellites are morphologically classified (*see* GALAXY CLASSIFICATION) as dwarf irregular (dIrr) or dwarf spheroidal (dSph).

The Andromeda galaxy (M31) also has 14 currently known satellites, the brightest and largest of which is M32 (the 32nd object in the *Messier catalogue). It was discovered in January 2006 that nine of Andromeda's satellites lie in a plane that passes through the centre of Andromeda aligned with its rotation axis (Fig. 2). It has been known since the late 1980s that the Milky Way contains two similar planes of satellites, but with the limited data then available, it was difficult to say whether or not this was significant. The Andromeda results suggest that alignment of satellite galaxies in planes may be a common phenomenon. There are several possible explanations for this alignment:

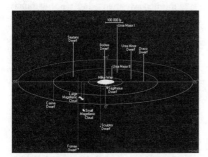

Fig. 1. The distribution of the Galactic satellites in a coordinate system centred on the Milky Way.

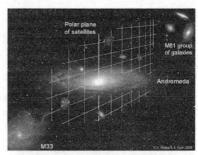

Fig. 2. Artist's impression of the Andromeda galaxy, illustrating that the satellites tend to lie in a plane (represented by the grid) that connects the giant neighbour M33 with the M81 group of galaxies. Of the brighter satellites, M32 lies below the bulge of Andromeda, M110 lies above and to the left.

- The satellites are left-overs from the break-up of a more massive galaxy, which has since been swallowed by Andromeda, in a process known as 'galactic cannibalism'. There is no strong evidence for this.
- If the *dark matter halo of a giant spiral galaxy were slightly flattened, as predicted by some *cosmological models, its asymmetric gravitational pull would tend to align any satellite galaxies in such a plane.
- The preferred orientation along a plane is a consequence of the infall of satellites along the dark matter filaments of the *cosmic web. Circumstantial evidence for this interpretation arises from the observation that Andromeda's satellite plane appears to point to the nearby giant galaxy M33 in one direction, and the M81 *galaxy group in the other direction.

Detailed studies of the orbits of satellite galaxies are required to distinguish between these possible explanations.

Satellite galaxies have been observed around numerous other giant galaxies. Dynamical studies of satellites provide a powerful way of estimating the total masses of their host galaxies.

scalar fields

Scalar fields represent particular kinds of particles, those with zero spin. Such particles have quite special properties in a cosmological context, and are widely used by *particle cosmologists; amongst other things they are postulated as being responsible for driving *cosmological inflation in the early Universe, and are candidates to be the *dark energy in the present *Universe. Regrettably, however, there is as yet no observational evidence for any fundamental zero-spin particles in Nature, though they proliferate in fundamental particle theories. The most famous example is the *Higgs boson believed responsible for giving particles their mass.

Scalar fields are one of the more enigmatic weapons in the armoury of particle cosmologists, and tend to be treated with some suspicion even by astronomers working in more traditional parts of the subject. Nevertheless, the use of scalar fields, and of fields in general, to describe particles is a well-established part of fundamental physics developed by physicists such as the legendary Richard Feynman soon after the Second World War.

Modern particle theory isn't really particle theory at all, but a theory of fields. A field is something which has a value at all points in *space-time. You might like for instance to think of the temperature; its value varies from point to point in space, and also at a given point in space can change with time. A more complicated example is the electric field, which has not just a strength at each point in space-time but also a direction—it is known as a vector field. Neither of these examples are properly consistent with *relativity theory, but electricity and magnetism combined together can be, with the electromagnetic potential being a four-dimensional vector (i.e. a vector that points partly in the time direction as well as in the space directions) that describes everything about the electromagnetic field at each space-time point.

Every type of particle is associated with a field, the value(s) of the fields at each space-time point describing the properties of the corresponding particles, rather as the Schrödinger wave function (another non-relativistic example) in *quantum mechanics does. Provided the number of particles is large, which is essentially always true in *cosmology, the quantum aspects can mostly be ignored and the system described as a classical evolution of the fields. While this sounds complicated, the mathematics is typically simpler than were one dealing with the individual particles.

By tradition, scalar fields tend to be indicated by Greek letters, the most commonly used being ϕ (Greek phi). This is then a function of both space and time coordinates.

A further important simplification is that the Universe is to a first approximation *homogeneous and *isotropic, which means that all points in space have to be equivalent. To respect this, the value of a scalar field should be independent of location in space, depending only on time. This corresponds to an even distribution of spin-zero particles throughout the Universe, whose *density varies with time. The scalar field evolution is then given by a simple equation analogous to the *fluid equation (known as the scalar wave equation or the Klein–Gordon equation), which is solved in combination with the *Friedmann equation and perhaps other fluid equations describing different types of matter coexisting with the scalar field.

The energy density of a homogeneous scalar field can be split up into two parts, known as the kinetic and *potential energies. The kinetic energy is the energy associated with the rate of variation of the scalar field, while the potential, usually indicated $V(\phi)$, measures the energy associated with the configuration of particles (including both their rest mass and any 'binding energy' that might arise from their interactions). The relative importance of these two contributions determines how the scalar field behaves in a cosmological setting.

The popularity of scalar fields with cosmologists arises because, provided the potential energy dominates over the kinetic, they can behave as if they have a negative *pressure. Such a material gains, rather than loses, energy through work done as the Universe expands, meaning that its energy density falls more slowly than other types of material. This negative pressure is just what is required to drive the Universe into acceleration, as required for *inflation or dark energy.

The behaviour of scalar fields is sometimes described in terms of an effective *equation of state, the equation of state w being the relationship between the pressure p and density ρ of any material and defined as $w = p/\rho$. A scalar field can give an effective equation of state anywhere between -1 (potential energy dominated, equivalent to a *cosmological constant) and $+1$ (kinetic energy dominated). The condition $w < -1/3$ is necessary if there is to be acceleration.

Under some circumstances it is necessary to include departures from homogeneity, for instance in the formation of *topological defects or in considering *density perturbations produced by inflation. Indeed, according to the inflationary cosmology, it is quantum fluctuations in the inflationary scalar field that are ultimately responsible for all the observed structures we see in the Universe today.

The main unknown quantity in specifying a scalar field model is the appropriate form of its potential energy density $V(\phi)$. In an ideal world, guidance as to its form would come from fundamental particle physics considerations such as *supersymmetry or *superstrings, but so far this has not been successful. From a cosmological point of view, then, the usual approach is to assume that the potential is of unknown form, and to seek to constrain it using observations. Such constraints may then help particle physicists learn how physical laws operate in our Universe, and indeed may be the only way to probe physics at very high energies.

scalar perturbations

'Scalar perturbations' is the technical name for the dominant type of *density perturbations, which are responsible for *structure formation in the *Universe. Indeed, the term 'density perturbations' is often used to indicate just the scalar perturbations. The terminology comes from Albert *Einstein's *general relativity, in which scalar refers to quantities whose value is unchanged under a change of coordinate system (put another way, all observers agree on the value of a scalar at a given point in space-time, whereas they do not agree on the components describing a vector).

A general perturbation can be decomposed into scalar, vector, and *tensor perturbations, each of which evolves independently of the others. The scalar perturbations experience *gravitational instability leading to structure formation. Vector perturbations are not often discussed, as they are not generated by inflation and because they are predicted to decay quickly in any case even if they were formed in the young Universe. Tensor perturbations are *gravitational waves.

Scalar perturbations may be *adiabatic perturbations or *isocurvature perturbations, with observations demanding that the adiabatic ones dominate.

scale factor

The scale factor $a(t)$ of the *Universe expresses the factor by which it has expanded at time t. In an expanding Universe, such as ours, $a(t)$ is increasing as a function of time. Since $a(t)$ is zero at the time of the *big bang, the scale factor is normally expressed relative to its present-day value $a(t_0)$ which is abbreviated to a_0.

The physical separation of objects fixed in *comoving coordinates scales with $a(t)$. Mathematically, the relation between *proper distance r and comoving distance x is given by $r = a(t)x$. Since the wavelength of light received by us from a distant object at *redshift z is stretched exactly by the change in scale factor between redshift z and redshift zero, the ratio of scale factors is equal to the reciprocal of one plus redshift: $a/a_0 = 1/(1 + z)$. One can thus say that the Universe was a factor $a/a_0 = 1/(1 + z)$ smaller at redshift z than it is today.

In conventional *cosmological models, the evolution of the scale factor is determined by the *Friedmann equation.

See also HUBBLE EXPANSION.

scale-invariant spectrum

Scale-invariance is a property of a particular type of *density perturbation, in which the typical sizes of the perturbations, as measured by their *power spectrum, are independent of length scale. The phrase is most commonly used in connection with *adiabatic perturbations; such a scale-invariant spectrum is known as the *Harrison–Zel'dovich spectrum and discussed in detail at that entry. This type of spectrum is popular because it gives a good fit to observational data, and because the simplest *cosmological inflation models predict nearly scale-invariant adiabatic perturbations.

The term 'scale-invariant spectrum' is also used more widely. It might be used, for instance, to refer to *isocurvature perturbations, to *scalar field perturbations generated during inflation, or to the spectrum of *gravitational wave perturbations.

SCUBA

SCUBA stands for Sub-millimetre Common-User Bolometer Array. It is an instrument installed on the UK's James Clerk Maxwell Telescope in Hawaii, which, with an aperture of 15 metres, is the largest astronomical telescope in the world designed specifically to operate in the sub-millimetre (450 and 850 micron) region of the *electromagnetic spectrum. The two wavelengths have their own arrays, consisting of 91 and 37 hexagonal close-packed elements respectively, and may be used simultaneously by means of a beam-splitter. The detectors operate at a temperature of just one tenth of a degree above absolute zero. This cooling is necessary as otherwise the detectors would be swamped by their own thermal energy.

The combination of field of view (2.5 arcmin) and sensitivity was sufficient to enable the first searches for sub-millimetre-wave emission from previously unknown distant *galaxies. Almost two hundred, very luminous, high-redshift galaxies have been detected with SCUBA at 850 microns. It is thought that the majority of these galaxies are likely to lie at *redshift $z > 1$, and the median redshift is probably in the range 2–3. However, it is still extremely difficult to obtain spectroscopic redshifts for galaxies this faint, and so only a handful have confirmed redshifts. Comparison of the integrated light from these galaxies with the *cosmic infrared background suggests that these sub-millimetre galaxies are responsible for the release of a significant fraction of the energy generated by all galaxies over the history of the *Universe.

SCUBA is better at detecting galaxies at high redshift than low, because the energy output of long-lived stars peaks in the near-infrared at about 2 microns, and they emit only very weakly at 850 microns. High-redshift galaxies tend to be younger and dustier than their low-redshift counterparts. The *dust in these young objects absorbs starlight, is heated to about 40 degrees above absolute zero, and then re-radiates the energy at longer wavelengths, around 200 microns. This infrared radiation is then redshifted into the sub-millimetre regime.

The observed number density of sub-millimetre galaxies implies very strong evolution in their luminosities. Without evolution, we would expect only 1 SCUBA source every 4 square degrees rather than the observed density of several hundred per square degree.

The original SCUBA instrument unfortunately stopped working during 2005. However, a new-generation replacement, SCUBA-2, should have been commissioned by the time this book goes to press. SCUBA-2 will utilize the first large-format 'CCD-like' camera for sub-millimetre observations. It will be able to map large areas of sky 1 000 times faster than the original SCUBA instrument, thus kindling a new era of sub-millimetre astronomy.

Blain, A. et al., *Physics Reports, 369, 111*, 2002.

selection function

Most astronomical surveys select objects by their observed flux (or equivalently, apparent *magnitude). Since observed flux decreases as the square of distance, only the most intrinsically luminous objects will appear at large distance in such a survey, and, since such luminous objects are rare, the overall space density of objects decreases with distance. The selection function of a survey gives the probability that an object within a given luminosity range will be included in the survey, as a function of distance or redshift. The selection function generally decreases smoothly from one at low redshift, where all objects in the luminosity range will be included, to zero at large redshift, where there are no objects of sufficiently high luminosity to be included.

Knowledge of the selection function, which may be obtained by integrating the *luminosity function, is necessary in order to estimate the *correlation function of a flux-limited sample.

Once distances and luminosities of all of the objects in a sample are known, a **volume-limited** sub-sample may be selected. This sub-sample is defined by a maximum redshift z_{max} and minimum luminosity L_{min}, such that an object at redshift z_{max} and with luminosity L_{min} would just be included in the sub-sample. A volume-limited sub-sample has the desirable property that the selection function is unity throughout, and so the space density of objects is independent of distance.

semi-analytic galaxy formation

Semi-analytic galaxy formation models, or simply semi-analytic models (SAMs), are 'recipes' for forming *galaxies within *numerical simulations. They follow the evolution of galaxies within the context of *hierarchical structure formation, aiming to include all relevant physical processes in a sometimes approximate fashion. SAMs are then able to predict the properties of galaxies at any chosen epoch.

The history of SAMs dates back to a classic paper by Simon White and Martin Rees from

1978. They considered the effects on galaxy formation of energy dissipation along with purely gravitational processes. Other physical processes, such as star formation and heating by young stars, were later added. A major breakthrough, in 1993, was the development of models of the merger of individual galaxies within *dark matter halos. These allowed the properties of *individual* galaxies to be followed as a function of time.

There are several steps in the development of a semi-analytic model:

Choice of *cosmological model—All recent SAMs use the *cosmological parameters of the 'concordance' model: $\Omega_m = 0.3$, $\Omega_\Lambda = 0.7$, and $H_0 = 70$ km/s/Mpc. The other important parameter affecting galaxy formation is the *baryon fraction $\Omega_b / \Omega_m \approx 0.18$.

Evolution of the dark matter halos—Two assumptions made here are that the dark matter halos are spherical and have density profiles which may be described in technical language as 'singular isothermal spheres', that is that the density varies inversely with the square of the distance from the centre. Neither of these assumptions is strictly valid. Various extensions of the *Press–Schechter theory have been used to allow for merging of the halos. These can only assign probabilities to a galaxy belonging to a particular halo, and so a *Monte Carlo algorithm is required to actually assign galaxies to halos. Recently it has become possible to model halo evolution directly using *N-body simulations, avoiding some of these problems.

Gas cooling—Gas must be able to cool for stars to be able to condense out. Again it is usually assumed that the gas has an initially uniform (isothermal) temperature distribution. The gas is then allowed to cool if the timescale for cooling is smaller than some characteristic value of order the age of the object.

Star formation—It is assumed that the star formation rate is proportional to the mass of cooled gas available, amongst other factors.

***Feedback and chemical evolution**—Cooled gas will be reheated by the radiation from young stars and *supernovae will pollute the gas with heavy elements, affecting subsequent cooling timescales.

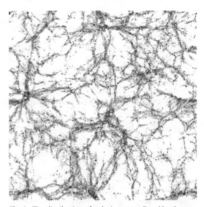

Fig. 1. The distribution of galaxies as predicted by the semi-analytic model of Andrew Benson and collaborators.

Merging, starbursts and morphological evolution— Several assumptions are made here:

- halo gas will be shock heated when halos merge;
- galaxies will merge on a timescale determined by dynamical friction;
- cold gas in merging subclumps forms a disk around the central galaxy;
- mergers of galaxies of similar mass will form elliptical galaxies.

Stellar populations—The *initial mass function (IMF) of the stellar population, that is the distribution of masses of young stars, is assumed to be the same everywhere, i.e. universal. Different forms of the IMF are assumed.

Normalization—The above steps involve setting the values of many parameters. In order to help constrain the models, they are normalized to either agree with (1) the *luminosity density of the Universe, or (2) to agree with the observed properties of the *Milky Way.

It is also becoming apparent that heating by *active galactic nuclei is an important feedback mechanism, which is required to suppress an otherwise predicted over-abundance of very luminous galaxies.

Successful SAMs are able to reproduce the following observations: the *luminosity functions of galaxies measured in various bands, the observed *Tully–Fisher relation, the observed *clustering of galaxies and the *star-formation history of galaxies. An example plot of the predicted galaxy distribution is shown in Fig. 1.

As observations improve, and as we learn more about the evolution of the above galaxy properties, we will be able to place more constraints on the prescriptions within the SAMs, and so gain an insight into the important physical processes in galaxy formation.

The above description was summarized from the excellent lecture notes by Scott Trager, available from his website:

http://www.astro.rug.nl/~sctrager/

Shapley supercluster

The Shapley *supercluster (Fig. 1), a cluster of *clusters of galaxies, also known as the Shapley concentration, is named after Harlow Shapley who discovered it in 1930. It was then largely forgotten until its 'rediscovery' in the late 1970s when many *X-ray sources were observed in this part of the sky. It lies at a *redshift of around 0.05, and comprises 25 known *Abell clusters along with five major *galaxy groups. It is by far the richest supercluster within our neighbourhood.

http://www.anzwers.org/free/universe/superc/shapley.html

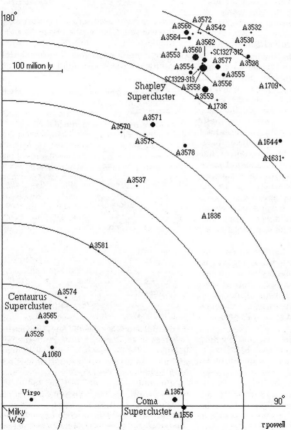

Fig. 1. The distribution of major galaxy clusters in the direction of the Shapley supercluster. The Centaurus supercluster, part of the Coma supercluster and individual Abell clusters are also shown.

significance

The significance of an observation or result tells us how much weight it should be given when attempting to decide whether a given *cosmological model is correct or not.

Because it is impossible for us to observe the entire *Universe, and because any measurement we make will have an associated uncertainty, we can never be 100% certain that a particular set of measurements definitively rules out any given theory. Instead, given a quantitative theoretical prediction and a set of observations, we can estimate the probability that the theory is correct. This is generally done by assuming that if we repeat an experiment (or observation) a number of times, the measured value of some parameter will follow a *Gaussian distribution, also known as a normal distribution. This distribution is characterized by two parameters: the mean μ and the standard deviation σ. The smaller the standard deviation, the more accurately a single measurement will reflect the true mean value. One can show that, for a Gaussian distribution, a series of independent measurements will lie within plus or minus one standard deviation σ of the mean value 68.27% of the time, within 2σ 95.45% of the time, and within 3σ 99.73% of the time.

Say a particular theory predicts that parameter x should have value x_0, and that a set of measurements finds a mean value μ_x with a standard deviation σ_x. One can then say that the measurement is $n_\sigma = (\mu_x - x_0)/\sigma_x$ standard deviations away from the theoretical value. If the absolute value of n_σ is less than two, then one normally regards the observations as being consistent with the theory (there is a 4.55% probability of making the observed measurements if the theory is correct). Otherwise, one can say that the theory is rejected by the observations at a significance of $n_\sigma\sigma$. As a rule of thumb, a theory is rejected if it disagrees with observations at a significance of 3σ or higher (less than 0.3% probability).

In practice, the situation is often much more complicated than this. For instance, measurements may not result in a truly Gaussian distribution of a measured parameter, or there may be a systematic error associated with a measurement, which will not be reflected in the estimated standard deviation. In addition, one is often trying to constrain more than one parameter simultaneously. In this case, any covariance between the parameters must be taken into consideration (see DEGENERACY).

Silk damping

Silk damping refers to the dissipation of small-scale perturbations caused by *photons' random walking out of overdense regions. It was first described by astrophysicist Joseph Silk in 1968.

Before *decoupling, the photons that will ultimately become the *cosmic microwave background (CMB) are interacting strongly with *electrons, scattering from them frequently and executing a complicated path known as a random walk. Any irregularities in the *density distribution on scales smaller than the length they can travel will dissipate; the net flux is away from high-density regions simply because there are more photons there.

Silk damping sets the scale for the thickness of the *last-scattering surface, and is responsible for the *CMB anisotropies being suppressed on small angular scales.

simulations

See NUMERICAL SIMULATIONS.

singularities

A singularity refers to a location in *spacetime where physical laws break down, typically because physical quantities such as the *density become infinite. This might happen at a specific location in space, as at the centre of a *black hole, or may occur everywhere in space at a specific time, as at a *big bang or *big crunch. Most commonly, singularities are discussed in the context of Albert *Einstein's theory of *general relativity, where they correspond to the *curvature of spacetime becoming infinite.

Many of the simplest solutions to the equations of general relativity feature singularities. The standard models of the expanding *Universe feature an initial singularity, where the density and space-time curvature are both infinite. However such solutions are highly idealized, as they assume that material in the Universe is perfectly evenly distributed whereas in practice it has inhomogeneities. Until the 1960s, it was thought that the singularities might be an artefact of that idealization, and wouldn't occur in practice because of this unevenness of the Universe, which might prevent everything coming together at a well-defined instant.

This idea was overturned by powerful theorems proved in the 1960s by Roger Penrose and Stephen Hawking, known as the singularity theorems, which showed that singularities were inevitable even in the presence of inhomogeneities, provided some simple and plausible conditions held. These results made both cosmological and black hole singularities seem inevitable. As physical laws break down at the singularity, that appears to make predictions of physical quantities in the vicinity of the singularities impossible.

Concerning black hole singularities, these are indeed predicted by general relativity, but in fact may not be as dangerous as they sound. In all known models of black holes, the singularity is surrounded by an event horizon, which marks out a region of space from which no signal can penetrate, even travelling at the *speed of light, due to the strength of gravitational interaction. Events occurring at the singularity therefore cannot be seen by observers outside the black hole event horizon, and so physical laws can be successfully applied. It is believed that singularities are always hidden behind event horizons—the so-called **cosmic censorship hypothesis**—which protects any external observers from their effects.

The situation regarding cosmological singularities is less clear, because a popular model of the early Universe is *cosmological inflation, and this can only occur if one of the conditions needed to prove the singularity theorems (known as the strong energy condition) breaks down. It is indeed possible to construct *cosmological models where the Universe starts from a very large (or infinite) size and collapses, but whose collapse stops and alters to expansion before the point of infinite density is reached. However it is unclear whether such solutions can describe our Universe, with some researchers claiming that an initial singularity is still inevitable.

All of the above presumes that general relativity is a genuinely fundamental theory of Nature, applicable to arbitrarily high densities and space-time curvature. However few if any physicists believe that this will be true, because general relativity is incompatible with *quantum mechanics and that incompatibility should manifest itself at high densities. The implication is that general relativity and quantum mechanics are theories which work well within the realm of their validity (in the same way that Newton's laws of motion do not need to be replaced by Einstein's *relativity unless objects are moving close to the speed of light), but which need to

be replaced by a more complete theory at high densities. Such theories are generically known as *quantum gravity, and one promising candidate for such a theory is *superstring theory.

Whether singularities will still exist in a quantum gravity theory is unknown, though it is certainly tempting to think that if particles are really made of strings, that this stringiness may prevent them collapsing to infinite density. What might happen instead is however unclear.

There is one quite well-established (theoretically anyway) consequence of quantum gravity, which is black hole evaporation, discovered by Stephen Hawking in 1975. This considers a situation where quantum gravitational effects are small but not completely negligible; in that case different quantum gravity theories make the same predictions. Such a situation can apply not at the singularity, but rather at the event horizon where gravitational forces are weaker, and causes a black hole to appear to radiate away energy, hence losing mass and shrinking. Eventually the black hole will disappear entirely, perhaps leaving behind a singularity without an event horizon. However only a full theory of quantum gravity will be able to address what happens in the final instants of evaporation.

SKA

See SQUARE KILOMETRE ARRAY.

Sloan Digital Sky Survey (SDSS)

The Sloan Digital Sky Survey collaboration was formed in 1988 with the aim of constructing a definitive map of the local *Universe, which would then be used to constrain *cosmological models and models for the formation and evolution of *galaxies. The survey would incorporate CCD imaging in five passbands denoted *ugriz* from the near *ultraviolet to the near *infrared and *spectroscopy of around one million galaxies and 100 000 *quasars over one quarter of the sky. In order to complete such an ambitious project over a reasonable timescale, it was decided to build a dedicated 2.5-metre telescope equipped with a large CCD array imaging camera and multi-fibre spectrographs. The survey itself began in April 2000, and the first instalment of the survey (SDSS-I) officially ended in June 2005. The survey covers a contiguous area of just under 8 000 square degrees in the North Galactic hemisphere and three disjoint stripes, each 2.5 degrees wide,

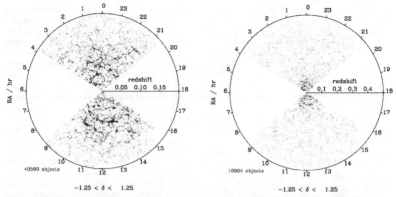

Fig. 1. Distribution of SDSS galaxies in right ascension (RA) and redshift. The left plot shows 40 590 galaxies from the main flux-limited galaxy sample within a redshift z = 0.2. The right plot shows 10 904 galaxies from the luminous red galaxy sample to z = 0.5. Only a small fraction of all galaxies observed, those within a 2.5 degree wide stripe centred on the celestial equator, are plotted, as otherwise the clustering pattern is 'washed out' by projection.

and covering a combined area of 750 square degrees, in the South Galactic hemisphere.

The survey telescope is situated at Apache Point Observatory, New Mexico, US. Survey hardware comprises the main 2.5-metre survey telescope, a 0.5-metre photometric telescope, a 120 mega-pixel imaging camera, and a pair of dual beam spectrographs, each capable of observing 320 fibre-fed spectra over the wavelength range 3 800–9 100Å. The data are transported on magnetic tape to Fermilab, near Chicago, where they are reduced by a series of automated pipelines and the resulting data products are made available via a web interface.

The effective integration time for the imaging observations is 55 seconds, allowing objects down to an r-band magnitude $r \approx 22$ to be detected. Spectroscopic integrations last for 45 minutes, and are carried out whenever observing conditions are not adequate for imaging, that is when atmospheric turbulence leads to stellar image sizes larger than 1.5 arcsec or when skies are not completely clear. Several distinct spectroscopic samples are observed, all of which are selected from SDSS imaging data. The largest is the **main galaxy sample** which is flux-limited to an r band apparent magnitude, $r < 17.77$, yielding about 90 galaxy targets per square degree with a median *redshift $\langle z \rangle \approx 0.1$. A sample

of **luminous red galaxies** (LRGs), selected by their red colours and luminosities estimated from photometric redshifts, form a sample of approximately constant number density out to a redshift z = 0.38. LRGs are extremely powerful for studying *clustering on the largest scales and for investigating *galaxy evolution. The final extra-Galactic targets are a sample of quasars, which are distinguished from stars by their distinctive colours in the SDSS passbands.

The distribution of SDSS galaxies is shown in right ascension (RA) versus redshift wedge plots in Fig. 1. Galaxy clustering is clearly visible in this plot: the galaxies appear to lie within filamentary structures enclosing regions of substantially lower density. The drop in main galaxy density with redshift (distance from the centre of the plot) is entirely due to the fact that this sample is limited by apparent flux: only the most luminous galaxies, which are rare, can be seen beyond a redshift $z \gtrsim 0.15$.

By contrast, the luminous red galaxy sample is designed to be volume-limited, i.e. to be of uniform density, out to redshift z = 0.38. This sample also includes additional galaxies to $z \sim 0.5$, although these high redshift galaxies do not form a complete subsample. At $z <$ 0.15, the simple linear colour cut used allows less luminous galaxies to enter the sample,

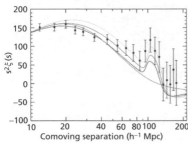

Fig. 2. SDSS LRG correlation function $\xi(s)$ scaled by the square of the galaxy separation s. The peak near $100h^{-1}$ Mpc is due to baryon oscillations in the pre-recombination Universe. The lower curve shows the model prediction for a Universe without baryons, for which this feature is absent. The remaining three curves show the predictions for a Universe containing baryons, with a combination of matter density Ω_m and Hubble parameter $h = H_0/100$ km/s/Mpc of $\Omega_m h^2 = 0.12$, 0.13 and 0.14 respectively from top to bottom.

hence the increase in LRG density at these low redshifts.

Although the primary science driver behind the Sloan Digital Sky Survey was characterization of the *cosmic structure of the Universe, the survey has had a significant impact on several branches of astrophysics, from the investigation of asteroids in our own Solar System to the discovery of the most distant known objects in the Universe.

Fig. 2 shows the SDSS LRG *correlation function $\xi(s)$ scaled by the square of the galaxy separation s. The expected peak near $100h^{-1}$ Mpc due to *baryon oscillations in the pre-recombination Universe is clearly detected, providing important confirmation of our paradigm of structure formation via *gravitational instability seeded by small fluctuations in the matter density in the early Universe. The spatial scale of this feature also provides a *standard ruler, allowing one to accurately measure the ratio of distances between the median redshift of the SDSS LRG galaxies ($\bar{z} = 0.35$) and the redshift of the *last-scattering surface ($z = 1\,089$). Such distance measurements can provide important constraints on the *equation of state of *dark energy.

The SDSS has broken the $z = 6$ redshift barrier, with the discovery of 12 quasars with redshifts $z \geq 5.7$. The highest redshift quasar is at $z = 6.42$; light from this quasar was emitted when the Universe was only about 5%

of its present age. Although intrinsically very blue objects, the expansion of the Universe stretches the light emitted by them so far into the red part of the spectrum that they are only detected in the reddest z band of the survey. More numerous L and T dwarf stars are intrinsically very red and may also only be detected in the z band. These nearby objects were eliminated using follow-up near-infrared photometry and confirming spectra were obtained with larger telescopes. The SDSS has now observed a well-defined sample of 19 luminous quasars at redshift $z > 5.7$. The luminosities of these quasars are consistent with a central *black hole of mass several times 10^9 M$_\odot$, and with host *dark matter halos of mass $\sim 10^{13}$ M$_\odot$. The existence of such mass concentrations at redshifts $z \approx 6$, when the Universe was less than 1Gyr old, provides important constraints on models of the formation of massive black holes.

Another interesting feature of quasars with $z \gtrsim 6$ is the almost total absorption of flux blueward of the *Lyman alpha line at 1 216Å due to the presence of neutral hydrogen in the intergalactic medium (IGM), the *Gunn–Peterson effect. This contrasts strongly with quasars at lower redshift where flux *is seen blueward of the Lyman alpha, since, at redshifts $z \lesssim 6$, hydrogen in the IGM is almost completely ionized and thus transparent to light at 1 216Å. These observations suggest that the epoch of *reionization of the Universe, due to radiation from the first massive stars, took place relatively recently, at $z \sim 6$.

The second instalment of SDSS, SDSS-II, began in July 2005. This comprises three parts. The first is the **Legacy Survey**, which will obtain imaging and spectroscopy for areas of sky left incomplete at the end of SDSS-I. The second part is the **SDSS Supernova Survey**, which, by repeatedly imaging the SDSS Southern equatorial stripe, aims to detect several hundred Type Ia *supernovae in the redshift range $z \sim 0.1$–0.4. The third and final part of SDSS-II is the **Sloan Extension for Galactic Understanding and Exploration** (SEGUE), a survey of stars in the *Milky Way Galaxy.

http://www.sdss.org

smoothed particle hydrodynamics (SPH)

To follow the formation and evolution of structures in the *Universe, it is necessary to carry out large-scale *numerical simulations.

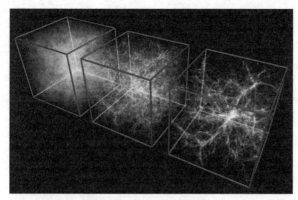

Fig. 1. The gas and galaxy distribution in a cosmological simulation by the GADGET2 code.

Simulations which follow only the evolution of the *dark matter are known as *N-body simulations. However full modelling of *galaxy formation requires that the *baryons be tracked as well, a more difficult task as they have *pressure forces as well as gravitational ones, and may be subject to other physical processes such as heating by radiation. A cosmological simulation must be able to follow the evolution of material across a very wide range of densities, and the most common technique for achieving this is smoothed particle hydrodynamics.

The method was introduced in 1977 independently by Leon Lucy and by Bob Gingold and Joe Monaghan. The distribution of baryons is modelled as a gaseous fluid, and approximated by breaking it up into a large number of particles (usually millions). The particles move under gravitational and other forces, and also carry information such as the temperature of the gas that they are representing. Because the particles themselves are moving, they naturally introduce higher numerical resolution in regions of high density because there are more particles there. Other physical processes can readily be incorporated, and the technique can be applied to a wide range of problems including star formation and *cosmology. Physical properties such as the gas temperature in a given region can be recovered by smoothing over a number of neighbouring SPH particles.

A cosmological SPH simulation requires an N-body component to follow the dark matter (which only experiences gravitational forces),

as well as the SPH component following the gas. Naturally, gravitational evolution causes the two populations to closely follow one another, but the extra forces experienced by the SPH particles means the correspondence won't be exact. Close groupings of SPH particles can be identified as galaxies.

A popular SPH code for cosmology is GADGET2, written by Volker Springel, which can be downloaded from the WWW site below, though a serious application requires access to considerable amounts of supercomputer time. Fig 1 shows an example of its use.

GADGET2 Home Page at:
http://www.mpa-garching.mpg.de/gadget/

space-time
In everyday experience, space and time are two quite distinct topics. However, the main lesson from Albert *Einstein's theories of *relativity is that in fact there are no separate concepts of space and time, and different observers may have different perceptions of each. Instead, space and time must be united into a single entity, space-time.

In the special theory of relativity, space-time is a passive environment within which physical events happen. In *general relativity, Einstein's theory of *gravity, space-time is an active participant, with the presence of massive objects distorting the shape of space-time and hence influencing the motion of objects. The geometry of space-time is described by the *metric, which measures the distance between neighbouring space-time points, and the typical goal of a general relativity calculation is to obtain the form of the metric corresponding to a particular distribution of matter.

Except perhaps at the earliest stages of the *Universe, where *extra dimensions of space may make an appearance, the standard assumption in *cosmology is that space-time is four-dimensional, with three space dimensions and one time direction. This means that each point in space-time can be specified by four coordinates. According to relativity theory, there is no unique choice of coordinate system and different people may make different choices as they see fit, the rules of relativity however guaranteeing that calculations will give the same ultimate physical answer regardless of the coordinate system used.

Space-time diagrams
Some key aspects of space-time can be summarized in a space-time diagram, most easily represented by suppressing one or two of the spatial dimensions. Each point in space-time corresponds to an **event**. Different observers (i.e. people using different coordinate systems) do not agree on whether two different events might have happened at the same location, or at the same time; all they agree on is the distance between two events in space-time, which is measured by the metric. Ignoring space-time curvature for the time being, this distance Δs is given by

$$\Delta s^2 = \Delta x^2 + \Delta y^2 + \Delta z^2 - c^2 \Delta t^2$$

where x, y, and z are the three spatial coordinates and c the *speed of light. Here Δ indicates the difference between the coordinate values at the two points ($\Delta x = x_2 - x_1$, etc.). This is simply Pythagorus' theorem, except that the difference in time appears with a negative sign; this negative sign is precisely the reason why we perceive time and space as such different things even though they are unified into space-time.

A light ray moving in the x-direction has velocity $\Delta x/\Delta t = c$, which corresponds to a zero distance in space-time. In fact, rays of light form a *light cone, which divides space-time into separate regions, those with Δs^2 positive and those with Δs^2 negative. The former region is known as the **space-like region**, and it is not possible to send any signal between space-like separated regions as the signal would have to travel faster than the speed of light. The latter region is known

as **time-like**, and two events separated by a time-like interval can communicate (*see* LIGHT-CONE, Fig. 1).

Space-time diagrams are also used for general relativistic situations in which space-time is curved, such as black holes, where they are sometimes also known as Penrose diagrams after British mathematician Roger Penrose. These are generally much harder to construct and interpret, particularly as often distances which are truly infinite have to be represented as finite distances in the diagram.

Slicing up space-time
Despite the unification of space and time into space-time, it is still often useful to break them apart again for actual calculations. This can be done by choosing a set of space-like slices, and stacking them together into a time sequence. This process is called a **foliation** of space-time into space-like slices. There is no unique way to do this, and relativity tells us that we must get the same physical outcome regardless of which slicing we use. However it may well be that calculations are much simpler with a cleverly chosen slicing that respects some symmetry of the problem in mind.

Such a slicing allows one to reintroduce the idea of dynamical evolution into relativistic calculations. While standard practice is to obtain a solution describing the geometry of the complete space-time, one can instead specify the geometry on some initial space-like slice, and then use evolutionary equations to build up the whole space-time. This is much more akin to how one usually thinks of dynamical systems developing with time. In doing so, one must also establish a threading of space-time that indicates how points on the adjacent spatial slices are to be joined up. The combination of slicing and threading is shown in Fig. 1. This approach is known as the ADM formalism, after its creators Richard Arnowitt, Stanley Deser, and Charles Misner.

Use of space-like slicing is particularly powerful in studying the evolution of cosmological *density perturbations, where different choices are usually referred to as **gauges**. Different choices of slicing may prove convenient, two popular ones being the uniform-density slicing, defined so that the total density is constant on each slice, and the spatially-flat slicing, defined so that the *spatial curvature vanishes.

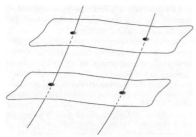

Fig. 1. An illustration of how a space-time can be split up into a spatial slicing and a time-like threading.

Space-time singularities

A space-time *singularity is a location where the structure of space-time breaks down. The formal definition of a singularity in general relativity comes from an idea known as geodesic incompleteness; put simply, the space-time is singular if there exist trajectories which come to an end after only a finite time, as experienced by an observer moving along that trajectory. They correspond to a breakdown in predictability of the theory. Typically, singularities are associated with the curvature of space-time becoming infinite.

It is common for calculations in general relativity to lead to space-times with *singularities, for example those at the centre of *black holes and the initial singularity of the *big bang cosmology (*see* 'Overview'). These are usually taken as indications that general relativity is an incomplete theory, which needs to be modified in the vicinity of space-time singularities.

Ellis, G. F. R. and Williams, R. M. *Flat and Curved Space-times*, Oxford University Press, 2000 [Moderately technical].

spatial curvature

The curvature of space, not to be confused with the curvature of *space-time. *See also* CURVATURE.

special relativity

See RELATIVITY.

spectral energy distribution (SED)

The spectral energy distribution of an object describes how much energy it emits as a function of wavelength or frequency. It is synonymous with the word *spectrum except that spectra are not always calibrated to give physical units of energy. An SED may be obtained by calibrating a spectrum using observations of a spectrophotometric standard star taken with the same observational set-up. SEDs are required in order to calculate the *K correction of an object and in order to estimate its *photometric redshift.

spectral index

The spectral index measures the scale dependence of the size of *perturbations in the *Universe. Perturbations are described by a *power spectrum, which measures their typical amplitude as a function of wavelength or of *wavenumber k. In particular, if the perturbations are assumed to have a power-law form in wavenumber, $P(k) \propto k^n$, then the spectral index is the power n which appears. If the power-law assumption is not made, it can be defined more generally as $n = d\ln P/d\ln k$, and then itself depends on k.

The simplest type of *density perturbations are known as *adiabatic perturbations, a particularly important version being the *Harrison–Zel'dovich spectrum which corresponds to the perturbation in the curvature of space being independent of scale. By unfortunate historical convention the spectrum is defined so that the spectral index n equals one (rather than the more logical zero) for this case. The simplest models of the *inflationary cosmology predict adiabatic perturbations with spectral index close to one.

If the spectral index is greater than one, this is sometimes called a **blue spectrum**, because it implies that the shorter wavelength perturbations will be dominant in amplitude (by analogy to the *electromagnetic spectrum in which blue light has the shortest wavelength amongst visible light). If the spectral index changes significantly with scale (i.e. the assumption of a power-law spectrum breaks down), then this is known as spectral index *running.

Spectral indices can also be defined for other types of perturbation, such as *gravitational wave perturbations. In such cases the value zero is normally used to indicate perturbations whose magnitude is independent of scale.

The spectral index of density perturbations is one of the key *cosmological parameters that cosmologists seek to extract from observations. As such density perturbations are *scalar perturbations, this is also known as the scalar spectral index, sometimes indicated with a subscript to read n_s.

Thus far, the Harrison–Zel'dovich spectrum has remained compatible with observations, and if the usual assumption of purely adiabatic perturbations is made then current limits are approximately $0.92 < n_s < 1.00$. If gravitational wave perturbations are also included then larger values of n_s are allowed. In order to meaningfully probe the dynamics of the inflationary cosmology, deviations of n_s from one would need to be discovered in the future.

The term 'spectral index' is also occasionally used in a different context, referring to the frequency dependence of radiation from sources, particularly radio sources.

spectroscopy

Spectroscopy is the process of obtaining or analysing the wavelength dependence of the *electromagnetic radiation emitted by a source. A plot of measured flux as a function of wavelength is referred to as a *spectrum. The most important use of spectroscopy in *cosmology is in determining *redshifts.

Originally, spectroscopy was performed one object at a time by dispersing the light falling through a single slit—a 'long slit' spectrograph. The development of fibre-fed spectrographs and slit-masks containing many small slits has led to multiplexing spectroscopy and has enabled large *redshift surveys such as the *two degree field and *Sloan Digital Sky Surveys. More recently, integral field units have become available. These obtain spatially resolved spectroscopy over a contiguous field, and are valuable for studying the dynamics of *galaxies.

Other uses of spectroscopy include identification of *active galactic nuclei and *starburst galaxies, measurement of chemical abundances in *quasar absorption systems and measuring the star formation rate in galaxies.

spectrum

The spectrum of an object (not to be confused with the term *'power spectrum') is a plot of the measured flux of *electromagnetic radiation as a function of wavelength. The most important use of spectra in cosmology is in determining *redshifts.

An optical spectrum is obtained by dispersing light using a diffraction grating, a set of fine, narrowly spaced lines etched on a smooth, reflective surface, or a grism, a combined diffraction grating and prism. In either case, light is separated into its component wavelengths. The spectrum thus obtained consists of a component varying smoothly with wavelength, the continuum, upon which are superposed a number of absorption and/or emission features. The absorption features are due to relatively cool material absorbing radiation at certain wavelengths, and the emission features are due to hot material emitting radiation at other characteristic wavelengths. The wavelengths of both absorption and emission features are

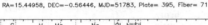

RA=15.44958, DEC=−0.56446, MJD=51783, Plate= 395, Fiber= 71

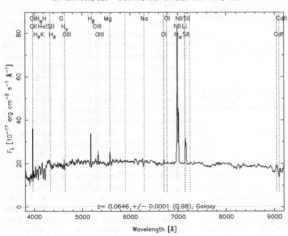

z= 0.0646 +/− 0.0001 (0.98), Galaxy

Wavelength [Å]

Fig. 1. A galaxy spectrum as observed by the Sloan Digital Sky Survey. The vertical dotted lines indicate the expected wavelengths of spectral lines at the redshift of this galaxy, $z = 0.0646$.

predicted by the *quantum mechanical theory of atoms, and form recognizable patterns in spectra, see Fig. 1 (previous page).

Two techniques are used to extract redshifts from *galaxy and *quasar spectra. The first technique identifies particular emission lines in the spectrum. By comparing the observed wavelength of each line with its known rest wavelength, a redshift may be determined for each line present. If the lines have been correctly identified, then these redshifts will be the same, and yield the redshift of the object. The second technique may be applied to objects without emission lines. In this case, one cross-correlates the observed spectrum with a number of template spectra. If one has a template spectrum which is similar in shape to the observed spectrum, then the cross-correlation will produce a large peak in the redshift difference between the two spectra.

Both techniques are in common use, and are capable of producing galaxy radial velocities accurate to ∼ 10 km/s. Optical spectra are the most common means of obtaining galaxy and quasar redshifts, although radio telescopes are also used to measure the redshift of the 21-cm hyperfine transition of hydrogen.

Spectra are frequently obtained without being properly flux calibrated (that is the actual energy emitted at any given wavelength is not known). Such spectra are perfectly adequate for estimating redshifts and even for estimating the relative strengths of individual features. However, in order to calculate a *K correction, a flux-calibrated spectrum is needed. This is known as the *spectral energy distribution.

speed of light
According to the theory of *relativity, the speed of light (and of all *electromagnetic radiation) in vacuum has a constant value. All inertial observers (that is observers who do not experience any external forces), will measure the same value for the speed of light, irrespective of their relative velocities. By the modern definition of the metre unit of length, the speed of light is defined to be exactly $c = 299\,792\,458$ m/s, a value which is frequently approximated to 3×10^8 m/s. No information-carrying signal can ever travel faster than the speed of light; c acts as a universal speed limit.

Philosophers since ancient times have disagreed on whether or not light travels at a finite speed. Proponents of a finite speed included the Greek philosopher Empedocles, and the medieval Islamic philosophers Avicenna and Alhazen, along with the Indo-Aryan school of philosophy. Prominent opponents of a finite speed included Aristotle, Hero of Alexandria, Johannes Kepler, and René Descartes. Attempts in the early 17th century to measure the speed of light over distances of order one mile were unsuccessful, due to the extremely short travel time. The first successful estimate of the speed of light was made in 1676 by Ole Rømer. By studying the motion of Jupiter's satellite Io with a telescope, he estimated a value for c of about 75% of the modern value. James Bradley, in 1728, made an improved estimate of c of 298 000 kilometres per second by measuring the aberration of light due to the Earth's orbit about the Sun. In California in 1926, Albert Michelson used a rotating prism experiment, devised in the previous century by Leon Foucault, to measure the time it took light to travel from Mount Wilson to Mount San Antonio and back. Michelson measured a speed of 299 796 kilometres per second.

It should be noted that the speed of light slows down when travelling through media other than a vacuum, due to the effects of light's oscillating electric field on the surrounding atoms. This slowing-down, which is given by the refractive index of the medium, depends on the frequency or wavelength of the light. For optical light (of wavelength around 600 nm), the decrease in speed in air is less than 0.03%. When passing through water or glass, light slows down to speeds of roughly three-quarters and two-thirds of its speed in vacuum. It is this large change in refractive index that is responsible for the bending of light when it enters water or a block of glass at an angle.

Some cosmologists believe that c may not be a true fundamental constant, but may vary as the Universe evolves (*see* VARYING FUNDAMENTAL CONSTANTS).

spiral galaxies
See GALAXY CLASSIFICATION.

Spitzer Space Telescope
The Spitzer Space Telescope (Fig. 1, over page), originally named SIRTF, the Space InfraRed Telescope Facility, and now named after astronomer Lyman Spitzer, was launched in August 2003 with a planned

Fig. 1. A composite image showing the Spitzer Space Telescope seen against a background of the infrared sky at 100 microns.

mission lifetime of 2.5 years. It is currently anticipated that the telescope will in fact continue to operate at least until the end of 2008. Its instruments are sensitive to *infrared radiation with wavelengths between 3 and 180 microns. Most of this infrared radiation is blocked by the Earth's atmosphere and thus cannot be observed from the ground.

The telescope has a 0.85-metre diameter primary mirror and three instruments.

InfraRed Array Camera (IRAC)—A four-channel camera that provides simultaneous 5.12×5.12 arcminute images at 3.6, 4.5, 5.8, and 8 microns. Each of the four detector arrays in the camera are 256×256 pixels in size.

InfraRed Spectrograph (IRS)—This enables the detailed astrophysics of infrared sources to be studied by obtaining their *spectra. The IRS consists of four modules which provide various combinations of wavelength coverage and resolution: 5.3–14, 10–19.5, 14–40, and 19–37 microns.

Multiband Imaging Photometer for Spitzer (MIPS)—It provides far-infrared imaging and *spectroscopic capability. MIPS comprises three detectors: a 128×128 array for imaging at 24 microns; a 32×32 array at 70 microns, which can also take spectra from 50–100 microns; and a 2×20 array at 160 microns.

The near-infrared wavelengths as detected by IRAC are important for detecting radiation emitted by long-lived stars at redshifts $z \sim$

1–2. The far infrared bands detected by MIPS are important for observing objects whose optical radiation is obscured by *dust. Dust absorbs *ultra-violet and optical radiation from young, massive stars and re-radiates at far infrared wavelengths.

One of the major surveys conducted with the Spitzer Space Telescope is the Spitzer Wide-area InfraRed Extragalactic survey (SWIRE). SWIRE imaged a total area of nearly 50 square degrees in six different directions on the sky, detecting over two million galaxies. It surveyed a volume of the high-redshift Universe at $z \sim 1$ comparable to that surveyed by the *Sloan Digital Sky Survey in the local Universe at redshift $z \sim 0.1$.

http://www.spitzer.caltech.edu

Square Kilometre Array (SKA)

The Square Kilometre Array is a planned radio telescope with a collecting area of one square kilometre, equal to one million square metres. It will be sensitive to the frequency range 0.1–25 GHz, and it will be 100 times more sensitive than current radio telescopes. As with *LOFAR, which may be considered a prototype, SKA will consist of a large array of small individual antennae. They may be combined to view different areas of sky simultaneously, or to make more sensitive observations in a single area. Although the collecting area will be one square kilometre, the antennae will be spread out over an area several thousand kilometres across, allowing extremely high resolution radio-images to be taken. (The

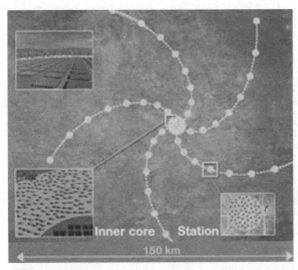

Fig. 1. Artist's impression of the central portion of the SKA array.

angular resolution of a telescope is roughly equal to the wavelength of radiation divided by the diameter of the collecting area. In the case of SKA this will be roughly 0.2 arcseconds.)

The final design and location are to be decided in 2007; the sites being considered are Argentina, Australia, China, and South Africa. There will be a compact core of elements containing about half of the collecting area within an area 5 km across, another quarter of the collecting area within 150 km (see Fig. 1), and the remaining elements spread up to a few thousand kilometres away. Once the site is chosen, construction of SKA is scheduled to begin in 2010, with initial observations in 2015. It is intended to be fully operational by 2020. The estimated cost is US $1 billion. It will be the most sensitive radio instrument ever conceived, being able to detect every *active galactic nucleus (AGN) out to a *redshift of 6, when the *Universe was less than one billion years old. It would have the sensitivity to detect television broadcasts on nearby extrasolar planets.

SKA has five key science drivers:

1. The cradle of life: a search for Earth-like planets.
2. Strong field tests of *gravity using pulsars and *black holes: testing *general relativity.

3. The origin and evolution of cosmic magnetism: origin of magnetic fields in the Universe.
4. *Galaxy evolution, *cosmology and *dark energy: measurement of galaxy redshifts to $z \sim 5$ via neutral hydrogen.
5. Probing the Dark Ages: observations of the first objects to form in the Universe.

http://www.skatelescope.org/

standard candles

A standard candle is an astrophysical object whose luminosity is assumed to be known. By comparing the measured brightness of a standard candle with its assumed luminosity, and allowing for any *extinction by intervening material, its *luminosity distance may be determined. Standard candles thus provide steps on the *distance ladder, and some examples are as follows.

Cepheid variables

Cepheid variables are a class of pulsating variable star, whose light output varies periodically in a characteristic way: a rapid brightening phase is followed by a more gradual fading. The peak luminosity of Cepheids correlates tightly with their period of variability, and they thus are a useful standard candle. For more detail, *see* CEPHEID VARIABLES.

RR Lyrae variables
Another class of pulsating variable star, these are named after the prototype, the variable star RR in the Lyra constellation. They have periods of 0.2–1.2 days and an amplitude of variability of 0.2–2.0 mag. They are older and more common, but less luminous, than Cepheid variables. There are two sub-types of RR Lyraes: RRAB pulsating in the fundamental mode and RRC pulsating in the first overtone. RRAB variables all have approximately the same luminosity (within 0.5 mag), making them particularly useful standard candles within the *Milky Way.

Supernovae
Type Ia supernovae are the most luminous known standard candles, and provide our only distance estimates beyond *redshift ∼ 0.1. In brief, one compares the observed peak brightness of the supernova with that predicted from the decay time of its light curve, and so estimates the distance to the supernova. Supernovae provided the first evidence that the expansion of the *Universe is actually accelerating today, contrary to previous expectations. *See also* SUPERNOVAE.

standard cosmological model
By standard cosmological model, we refer to the best current description of the *Universe. This means not just the general choice of the hot big bang cosmology (*see* 'Overview'), but also the favoured values of the various *cosmological parameters. The different values of these parameters describe all the possible *big bang models, but only one can match our actual Universe, with observations seeking to determine which.

Underlying assumptions
The standard cosmological model, also sometimes known as the **concordance cosmology**, features four main epochs. The first of these is the *inflationary epoch, which is responsible for ensuring the correct global properties of the Universe and for laying down the seed *density perturbations that will later form *cosmic structure. This is followed by a long *radiation-dominated era, during which for instance atomic nuclei form through the process of *nucleosynthesis. This gives way to a *matter-dominated era during which *structure formation takes place, and which is eventually supplanted by *dark energy domination. It is not known what, if anything,

might have happened before inflation, and nor is it known whether domination by dark energy will last forever.

The general assumptions of the standard cosmological model are that the Universe is described by a hot big bang cosmology, and that *Einstein's theory of *general relativity describes how *gravity works. The material content and the way particles interact is specified by the *standard model of particle physics, enhanced by the inclusion of *dark matter and dark energy.

For most purposes it suffices to consider that the Universe contains five ingredients, *baryons, dark matter, radiation in the form of *photons, *neutrinos, and dark energy which is presumed to be in the form of a *cosmological constant. The neutrino masses are currently assumed small enough to have no impact on observations even if, as particle physics experiments indicate, they are nonzero. This assumption may however need to be reconsidered as observational precision grows.

The *spatial curvature of the Universe is taken to be zero, so that the sum of these five densities adds up to the *critical density. Finally, the seed density perturbations which initiated structure formation need to be specified, the standard cosmological model adopting the choice of *adiabatic perturbations of nearly *scale-invariant form.

The very simplest form of the standard cosmological model takes the density perturbations to be precisely of the scale-invariant *Harrison–Zel'dovich form. Further, in assuming that the standard particle physics model applies, it is possible to predict the density of neutrinos from that of the photons, as their relative abundances were fixed during the early thermal equilibrium phase before neutrino decoupling. In combination, all the above assumptions mean that a complete specification of the model requires only five fundamental parameters, which can be taken to be the *Hubble parameter giving the present expansion rate, plus the current baryon, photon, and dark matter densities (the dark energy density then being determined by the condition that the densities add to the critical density), plus the amplitude of the Harrison–Zel'dovich perturbations.

Although these are the only fundamental parameters required to fit observational data, for some data types a complete calculation

Table 1. The parameter values of the standard cosmological model (2007 edition). The densities are indicated by their density parameters, but however are best determined in the combination of the density parameter multiplied by the square of the Hubble constant. The constraints come from observations of the cosmic microwave background anisotropies by the WMAP satellite; combination of these observations with others such as galaxy clustering gives similar and consistent results.

Parameter	Symbol	Value and uncertainty
Hubble constant	h	0.73 ± 0.03
Dark matter density	$\Omega_{dm} h^2$	0.104 ± 0.006
Baryon density	$\Omega_b h^2$	0.022 ± 0.001
Radiation density	$\Omega_r h^2$	$(2.47 \pm 0.01) \times 10^{-5}$
Density perturbation amplitude	δ_H	$(1.9 \pm 0.1) \times 10^{-5}$
Spectral index	n_s	0.95 ± 0.02

of model predictions has proved too complex, and further parameters have to be included. These are phenomenological parameters, meaning that they could in principle be calculated from the fundamental ones, if only we were smart enough. However at present these calculations are too difficult, and the physical processes must be modelled with extra parameters.

The main two phenomenological parameters are the *optical depth to *reionization and the galaxy *bias parameter. The former is required to model *cosmic microwave background (CMB) anisotropies, such as those measured by the *WMAP satellite. After the Universe is reionized, microwave photons can scatter from the reionized *electrons which modifies the anisotropy pattern. Present understanding is not sophisticated enough to accurately model the reionization process for different *cosmological models.

The galaxy bias parameter is needed to compare observations of galaxy *clustering with theoretical predictions that usually refer to the clustering of dark matter. While galaxies and dark matter roughly follow one another, the correspondence is not perfect and depends on the details of how *galaxy formation takes place. Rather than attempting to model this process in detail, the usual approach is to introduce an extra parameter known as the bias parameter to account for this relation.

Parameter values

Of the above parameters, only the photon density is determined with such accuracy, based on direct measurement of the *cosmic microwave background temperature, that it can be considered a known fixed quantity in

comparing to other types of observations. The others are determined by requiring that the model predictions then match a wide range of cosmic observations. Usually such a compilation of observations will include, as a minimum, measures of cosmic microwave background anisotropies, the galaxy *power spectrum, and the luminosity–redshift relation of type Ia *supernovae.

Table 1 shows the preferred values of the cosmological parameters. Conventionally, the densities of the different components are shown using the *density parameter, measuring the fraction of the critical density; however, the combination of this times the Hubble constant squared is better determined and so that is shown in the table. To obtain the best value of the density parameter itself you can use the best-fit value of the Hubble constant, but you then need to incorporate its uncertainty too. As the model is assumed to be spatially flat, the density parameters, including the dark energy, must add up to one. This gives a best fit value for the dark energy density parameter of 0.76.

There are many different terminologies in use for the amplitude of density perturbations. The one we have chosen to quote, known as δ_H, is an approximate measure of the magnitude of the large-angle CMB anisotropies, showing that they are of order one part in a hundred thousand.

The *spectral index n_s measures how the size of perturbations varies with length scale, with $n_s = 1$ being the Harrison–Zel'dovich spectrum corresponding to the perturbations being independent of scale. Current observations suggest that if n_s is allowed to vary, then its best-fit values are all smaller than one. However according to *model selection analyses, a model where n_s is fixed at one (a justifiable assumption as it is a special parameter value) also gives an acceptable description of the data, with its poorer fit quality being compensated by the improved model simplicity.

To summarize, we see that the majority of the density in the present Universe is in the form of dark energy. Of the remainder, it is mostly dark matter, with the baryons being roughly 20% of the dark matter density. In the present Universe the radiation density is much lower, but nevertheless it would have dominated in the young Universe.

Future cosmological models
The standard cosmological model as described above represents the best understanding of observations as of 2007. However, the standard cosmological model may continue to evolve, in response both to observational and theoretical developments; indeed, most cosmologists like to believe that there are exciting future discoveries ahead of us as to how the physics of the Universe works. Such discoveries would introduce new cosmological parameters, whose values would then need to be determined from observations. We can only guess where future observations might take us, but some example discoveries that might lie in the future include the following. A time variation of the dark energy density may be discovered, requiring the cosmological constant to be replaced with a more general dark energy model. Cosmic *non-Gaussianity may be discovered, requiring a more detailed understanding of the mechanism by which the initial perturbations were generated. The assumption that the dark matter is in the form of *cold dark matter may prove oversimplistic.

Lahav, O. and Liddle, A. R. *The Cosmological Parameters*. Read online at http://pdg.lbl.gov.

standard model of particle physics
The standard model of particle physics, known to particle physicists simply as the Standard Model, is a classification of fundamental particles and the interactions between them. It ignores *gravity but includes the three other *fundamental forces of Nature. Until the mid 1990s, it gave a complete description of all available particle physics data, but it is no longer able to explain all data (specifically, it cannot explain that *neutrinos

have a non-zero mass). Even while it did, particle physicists have always believed it to be an approximation which would require revision as higher and higher energies were probed.

According to fundamental particle physics, a key discriminator of particles is their spin. Particles are classified into either *bosons, whose spin is an integer value 0, 1, 2, etc., or *fermions, whose spin is a half-integer $1/2$, $3/2$, etc. Whether particles are fermions or bosons has a major impact on how they behave; for instance only fermions obey the Pauli exclusion principle which forbids more than one particle occupying a given quantum state. The spin value s indicates the maximum value of the spin: the actual value for a given particle runs in integer steps from $-s$ to s.

Quarks and leptons
The fundamental constituents of matter are *quarks and *leptons. These particles are all fermions, with spin equal to one-half, meaning that they have two possible spin states, spin-up $(s = 1/2)$ or spin-down $(s = -1/2)$. The standard model describes how these particles interact via three of the forces of Nature, the electromagnetic interaction and the weak and strong nuclear forces.

Each class of quark or lepton comes in three different types, known as flavours, and further they can be grouped in pairs, known as families or generations. In total therefore there are twelve fundamental fermions, as shown in Table 2, each also coming with its corresponding *anti-particle. The structure of the standard model dictates the pairing of the quarks and leptons, but does not say how many generations there should be. A range of both *particle accelerator experiments and cosmological observations point to there being just three, in which case all of the fermions have been directly detected by experimentalists.

Interactions: gauge bosons
The standard model describes three kinds of interactions between the particles: electromagnetism, the weak nuclear force, and the strong nuclear force. At a fundamental level these forces are associated with the exchange of a particle, which is always a boson of spin-1. Whether or not particles are affected by a given force depends on whether they interact with the force-carrying bosons, which are known as **gauge bosons**. Particle physicists

Table 2. The fundamental particles of Nature according to the standard model, with their electric charges.

Fermions				Bosons	
Leptons		*Quarks*			
Flavour	*Charge*	*Flavour*	*Charge*	*Particle*	*Charge*
ν_e: electron neutrino	0	u: up	2/3	γ: photon	0
e: electron	-1	d: down	$-1/3$	W^-	-1
ν_μ: muon neutrino	0	c: charm	2/3	W^+	$+1$
μ: muon	-1	s: strange	$-1/3$	Z^0	0
ν_τ: tau neutrino	0	t: top	2/3	g: gluon (8 types)	0
τ: tau	-1	b: bottom	$-1/3$	H: Higgs boson	0

attribute the forces to particular types of symmetry, known as gauge symmetries, obeyed by fundamental particles.

Electromagnetism is the simplest force, and is mediated by a single particle, the familiar *photon of light. Particle physics jargon for this force is that it is a $U(1)$ force, the terminology describing the type of symmetry it corresponds to. The particles which interact electromagnetically are those which carry an electric charge; almost all the quarks and leptons do, the exception being the neutrinos. If the charges are opposite the force is attractive, otherwise it is repulsive. Because the photon has zero mass, it can be exchanged between particles of arbitrary separation without violating energy conservation and the Heisenberg uncertainty principle (which permits short-term violations of energy conservation). The electromagnetic force therefore has an infinite range, though its strength falls off as an inverse square law with the distance between the particles.

The weak nuclear force is more complicated, as it is mediated not by one particle but by three, known as W^-, W^+ and Z^0, where the superscript indicates the electric charge of the particles. The corresponding symmetry is known as $SU(2)$. Further, these particles have a mass, which means that the weak interaction has a finite range (less than 10^{-16}cm). The particles were first proposed in the late 1960s, and discovered at the CERN particle accelerator in the early 1980s. In *particle physics units, the mass–energies of the W particles are about 80 GeV and of the Z^0 is 91 GeV, both roughly one hundred times the mass of a *proton. It is known as the weak nuclear force because the strength of the interaction is much less than the other two forces. One consequence

of this is that neutrinos, the unique standard model particles which only interact via the weak force, very rarely interact with other particles, for instance passing through entire planets and stars without noticing they are there.

The strong nuclear force is yet more complicated. It is mediated by particles known as gluons (the original intent being that they are responsible for 'gluing' quarks together), and there are eight different types, the symmetry this time known as $SU(3)$. The gluons have no electric charge, but do carry a different charge (sometimes called a quantum number) conventionally known as colour, of which there are three types red, green, and blue. This of course doesn't mean that the quarks actually have colours, though it tends to be quite convenient to think of them as if they do. The colour charge of a quark dictates which types of gluon it can interact with. Yet more confusingly, anti-quarks carry anti-colour. Gluons carry both a colour and an anti-colour and change the colour of quarks on interaction; an example would be

red quark + green–anti-red gluon $\longrightarrow$ green quark

Overall, the theory of quarks and gluons is known as quantum chromodynamics, or QCD for short.

The structure of QCD dictates that one can never see an isolated colour charge; colour charge obeys a property called **confinement** whereby the strength of the interaction becomes smaller the closer the particles are, rather than the further they are. Accordingly, one can never see an isolated quark or gluon, only combinations of them where the colours cancel out. Hence quarks must come either in groups of three, one of each colour (in this scheme, red + blue + green = colourless), or as a quark–anti-quark pair where the colour and anti-colour cancel. The former are known as *baryons, amongst which protons and *neutrons are the most prominent examples, and the latter as **mesons**. It has also been speculated that there should be particles made entirely of gluons of cancelling colours, known as glueballs, though evidence is tentative at best so far. Incidentally, although quarks carry a fractional electric charge of either 2/3 or −1/3, this rule guarantees that the physical combinations always have an integer electric charge.

The principal interactions responsible for the nature of matter in the *Universe are the strong and electromagnetic forces. The strong force is responsible for holding together the groupings of three quarks that form protons and neutrons, and after doing that also provides the residual force that holds those particles together in atomic nuclei. The electromagnetic force is responsible for confining electrons to those nuclei in order to form atoms. By contrast, the weak nuclear force plays a minor role, its effects primarily restricted to being responsible for radioactive decays by converting a neutron to a proton, electron, and electron anti-neutrino. One shouldn't belittle it entirely, however, as these are exactly the sorts of interactions which provide the fuel for the Sun to burn, a rather crucial requirement for our form of life, and less positively also provide the energy released in nuclear explosions.

Each of the three interactions has its own symmetry group, and the standard model brings these together simply by multiplying them. The standard model is also sometimes known by the messy nomenclature $SU(3) \times SU(2) \times U(1)$.

The Higgs particle

There is one remaining particle in the standard model, which plays a very special role. The Higgs particle, indicated usually by the symbol H, is a spin-zero boson (also known as a *scalar field). Its role is to give mass to the other particles, via a mechanism known as spontaneous symmetry breaking. The particle's existence was postulated by Scottish physicist Peter Higgs, following on from work by Tom Kibble, in the mid 1960s. It is the only standard model particle whose existence has not been verified experimentally, and its

discovery is a key goal of the Large Hadron Collider experiment beginning operations at CERN around 2008.

The Higgs particle is essential because, in its absence, the gauge bosons are all predicted to be massless. This is fine for the photon and the gluons, but most definitely is not true of the W and Z particles mediating the weak force. The weak interaction $SU(2)$ corresponds to a symmetry which is not actually present in our Universe, the boson masses being the symptom of the loss of symmetry. The basic idea is that in the present Universe the Higgs particle has a nonzero value. Depending on the way the gauge bosons interact with the Higgs, this nonzero value may generate an effective mass for them.

The Higgs particle can also give a mass to the fermions, via a mechanism too subtle to describe here. Due to a curiosity about the way neutrinos appear in the particle classification, it turns out that they do not get a mass via the Higgs mechanism, and thus are predicted to be massless. This, unfortunately, is now in contradiction with observations, indicating the need to extend the standard model.

Electro-weak unification

While presently in the Universe the Higgs particle has a non-zero mass, thereby giving the weak interaction its nature, it may not always have been so. Indeed, theory predicts that during the early stages of the Universe, when the typical particle energies were greater than the current masses of the weak gauge bosons, the Higgs particle would have a zero value and the weak bosons would be massless, so that the weak interaction would also be long-range. In fact, things are even slightly more subtle than that; at these energies the electromagnetic and weak interactions would lose their separate identities and become unified into a single force, known as the **electro-weak interaction**. This would be mediated by four particles, the photon, two Ws, and the Z, and the electromagnetic and weak interactions would be of the same strength.

The epoch of the Universe's evolution where the electromagnetic and weak interactions first gain their separate identities, and the Higgs particle attains its present non-zero value, is known as the *electro-weak phase transition. At that time the W and Z bosons gain their mass, but the photon avoids that fate and stays massless to cause the electromagnetic interaction.

Beyond the standard model
The standard model is not regarded as the final word on particle interactions, and various extensions have been proposed both with experimental and theoretical motivation. Amongst common objections are the following:

- How are you going to explain the observation that neutrinos are not massless?
- $SU(3) \times SU(2) \times U(1)$ is not really a unified theory of fundamental forces if all you do is stick the separate pieces together.
- The theory is rather complicated and arbitrary, in fact featuring 21 separate constants of Nature that are not predicted but which must rather be fitted from observations.
- Since quarks and leptons are in separate sectors, there is no explanation of why the electric charge of a proton is precisely the opposite of the charge of an electron.
- What's the point of having three generations, when one would seem quite enough for a satisfactory Universe?
- What about gravity?

An extension to include neutrino masses is now mandatory. More elaborate ideas, not yet experimentally supported, include the following:

Supersymmetry—Introduced to explain why particle masses are much smaller than the characteristic energy scale of gravity, the *Planck scale.

Grand unification theories—Introduced to unify the electro-weak and strong interactions into a single symmetric framework.

Supergravity—Introduced to incorporate gravity into the description of fundamental particles.

Superstrings and M-theory—Candidate theories unifying particle physics and gravity in a full quantum theory.

These are all discussed within their separate entries.

Weinberg, S. *Dreams of a Final Theory*, Vintage Books, 1994.

standard rulers
A standard ruler is an object, or a phenomenon, of known physical size. Measuring the apparent angular size of such a ruler then allows one to determine its *angular-diameter distance. Although standard rulers that may be resolved at cosmological distances are rare, they have an advantage over *standard candles in that their apparent size is unaffected by any *extinction by intervening material.

In practice, cosmological standard rulers are based on correlations between objects, rather than individual objects. Examples include the correlation length (the separation at which the *correlation function is equal to one) and the scale of *baryon acoustic oscillations in the galaxy distribution. Use of the correlation length as a standard ruler is complicated by uncertainties in the evolution of *clustering with *redshift.

star-formation history
One of the goals of observational *cosmology is to determine the star-formation history of the *Universe, that is, the mean rate at which stars are formed as a function of *look-back time or *redshift. There are many ways of measuring star-formation rate.

The light from recently formed *galaxies is dominated by *ultra-violet (UV) radiation from the hottest and most massive stars. Thus in principle, UV luminosity provides a good tracer of star formation. However, the Earth's atmosphere strongly absorbs UV radiation, and so these measurements must be made from space. In addition, UV light is strongly attenuated by *dust that is frequently found around galaxies at high redshift. Observed UV flux must therefore carefully be corrected for dust extinction.

In the optical part of the spectrum, the first Balmer line of hydrogen, denoted Hα, provides a very good indicator of

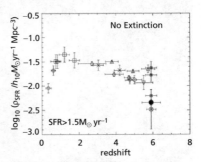

Fig. 1. Star formation rate plotted as a function of redshift.

star-formation rate. However, the rest-wavelength of Hα is 6563Å, at the red end of the optical region, and so *infrared spectroscopy is required to measure the Hα line in galaxies at redshift $z \gtrsim 0.2$.

In dusty environments, such as very young galaxies, the UV radiation from stars is absorbed by dust grains. The grains thus heat up, and re-radiate in the mid–far infrared part of the spectrum. Thus for these galaxies, infrared luminosity traces star-formation.

These and other techniques may be combined to make a plot of star-formation rate as a function of redshift or look-back time, a plot also known as a Madau–Lilly diagram, after the first two astronomers, Piero Madau and Simon Lilly, and their collaborators, to independently make such plots in 1996. An example is shown in Fig. 1 (previous page). Star-formation rate is seen to rise steeply between redshifts zero and one, where it flattens off to an approximately constant or slowly declining rate. Note the large uncertainty beyond redshift $z \sim 4$, due to uncertain extinction corrections in these dusty, high-redshift galaxies. It is apparent, however, that significant star formation is already taking place by redshifts $z \sim 5$.

starburst galaxy

'Starburst' is the name given to a *galaxy observed to be undergoing a rapid burst of star formation. Whereas normal spiral galaxies such as the *Milky Way are forming stars at the rate of only about one Solar mass per year, starburst galaxies form stars at a much higher rate, a rate that, if maintained, would exhaust the available supply of gas in much less than the *Hubble time.

Starbursts are revealed by a very large *infrared luminosity (since they are often very *dusty environments) or by very strong emission lines in the galaxy's *spectrum. These emission lines are due to transitions in the energy level of *electrons in atoms of hydrogen, nitrogen, and oxygen. The electrons are excited to higher energy levels by the intense radiation emitted by massive young stars. The electrons rapidly drop down to a lower energy state, emitting radiation of a well-defined wavelength, hence giving a narrow emission line in the spectrum. The observed wavelength of any such emission line will have been increased by a factor of $(1 + z)$ for a galaxy of *redshift z. Since these emission lines are few in number, their observation provides a simple way to determine the redshift of a starburst galaxy. Another point worth noting is the fact that, due to the finite *speed of light, we always observe a galaxy as it was when the light was emitted, not as it is today. For this reason, a greater fraction of high-redshift than low-redshift galaxies are classified as starbursts. Most starbursts are

Fig. 1. Composite image of the starburst galaxy M82. The nearly vertical disk shows starlight. The diffuse, patchy, nearly horizontal light shows the distribution of hot hydrogen gas ejected by a 'superwind' travelling at more than a million miles an hour due to supernova explosions within the galaxy. The starburst is thought to have been triggered by an interaction with its neighbour M81 around 300 million years ago.

thought to be triggered by tidal interactions or mergers with other galaxies.

Another type of object, the *active galactic nucleus (AGN), also exhibits strong emission lines. These, however, are due to accretion of material on to a central *black hole, rather than bursts of star formation. Starburst galaxies and AGN may be distinguished by comparing the strengths of the emission line ratios [NII]/Hα and [OIII]/Hβ.

In nearby galaxies, the phenomena can also be distinguished by the fact that the AGN radiation comes from the nucleus of the galaxy, whereas starburst activity tends to be concentrated in 'knots' away from the centre.

The starburst phenomenon occurs almost exclusively in 'late-type' spiral galaxies (Sc and later), dwarf, and irregular galaxies. The *Messier galaxy M82 is a prototypical starburst galaxy, see Fig. 1.

steady-state cosmology

Steady-state cosmology was proposed in 1948 by Hermann Bondi, Thomas Gold, and *Fred Hoyle. It was introduced in order to satisfy the perfect *cosmological principle, which requires that the *Universe looks the same on large scales at any point in space and time. Clearly, with the discovery of the expanding Universe by Edwin *Hubble a couple of decades earlier, one would expect the *density of matter in the Universe to decrease as the Universe expanded, thus contradicting the perfect cosmological principle. In order to get around this problem, proponents of the steady-state model postulated a Universe which had always been expanding (no *big bang event) and in which a constant mean density was maintained by the continuous creation of matter. The amount of matter required is undetectably small, just 0.3 atoms per cubic kilometre per year.

Steady-state cosmology is a good theory in that it makes definite, testable predictions. The first problems with the theory emerged in the late 1960s when observations began to yield evidence for the evolution of extra-Galactic radio sources: *quasars and *radio galaxies were found only at large distances, and thus, given their large *look-back times, in the past. It also had problems predicting the abundance of light elements (particularly deuterium and helium), which the theory maintained were produced within stars.

Steady-state theory was ruled out for most cosmologists in 1965 when Arno Penzias and Robert Wilson discovered the *cosmic microwave background (CMB), the leftover radiation from the big bang. Steady-state theory attempted to explain the CMB as due to light from ancient stars that had been scattered by intergalactic *dust. However, it is hard in this interpretation to explain:

- the extreme smoothness of the CMB;
- its lack of strong *polarization expected in scattering processes;
- the closeness of the CMB spectrum to that of a *black-body spectrum.

In order to get around these problems facing steady-state cosmology, Hoyle, Geoffrey Burbidge, and Jayant Narlikar developed quasi-steady-state cosmology (QSSC) in the 1990s. In this model, a single big bang event is replaced with a series of mini big bangs: the sources of energy in quasars and *active galactic nuclei result from mini creation events. One can think of a QSSC Universe as undergoing long-term, steady-state expansion interspersed with short-term oscillations, with a maximum observable *redshift $z \sim$ 5–6. Explanation of the CMB requires the existence of whiskers of heavy elements such as carbon which are about 1 mm in length. A density of 10^{-35} g cm^{-3} would be sufficient for these whiskers to absorb optical starlight and re-radiate at millimetre wavelengths. These whiskers would also absorb light from distant sources, and thus be a possible explanation for the dimming of type Ia *supernovae, avoiding the need for an accelerating expansion of the Universe.

QSSC has many problems, and is thus not taken too seriously by the majority of cosmologists. Amongst its many problems are:

- matter is created on an ad hoc basis;
- generation of CMB radiation is rather artificial, and QSSC does not predict the secondary peaks in the CMB power spectrum observed by the *WMAP satellite;
- there is evidence of a galaxy at redshift $z \approx 10$ (*see* VERY LARGE TELESCOPE), while no objects at redshift $z \gtrsim 6$ are allowed in QSSC.

See also ALTERNATIVE COSMOLOGIES.

Narlikar, J. V. and Padmanabhan, T. *Annual Review of Astronomy and Astrophysics*, 39, 211, 2001.

string cosmology
String cosmology refers to attempts to model *cosmology within the context of *superstring

theory or its descendent, *M-theory. The distinctive effects of those theories, namely that fundamental particles are not point-like but rather have at least one dimension, are expected to be important only during the very *early Universe. String cosmology models are usually concerned with phenomena such as *cosmological inflation, explaining the dimensionality of *space-time, or the generation of primordial *gravitational waves.

The most important property of superstring theory in this context is known as **string duality**: it turns out that in string theory there is a mapping between the physics describing very short and very large length scales, so that the same physical laws apply in each regime. In a sense, this means that there is a minimum distance in string theory; trying to go to a shorter length scale from there is just the same as going to a longer length scale.

A consequence of this is that for every expanding Universe model there is a corresponding contracting one, and this is exploited in models known as **pre big bang models**, introduced by Maurizio Gasperini and Gabriele Veneziano. Here the Universe goes through a contracting phase which can be thought of as preceding the *big bang, and then passing through a 'bounce' phase to re-emerge as a standard expanding Universe. In this way the initial *singularity is avoided. It turns out that the contracting phase acts just like a period of inflation. Unfortunately, however, known physics breaks down at the bounce which therefore cannot be modelled, and this difficulty has prevented the scenario entering the mainstream. A similar but more elaborate proposal known as the *ekpyrotic Universe has been made in the context of *braneworld cosmology.

string landscape

The string landscape is the set of possible physical laws that might arise within *superstring theory or *M-theory. While those theories take on a simple form in their full higher-dimensional version, before they can be compared to observations they must be reduced to a four-dimensional theory, via *compactification or a *braneworld construction. Physical laws are predicted to vary depending precisely how this happens, and recent calculations have indicated that the number of possible configurations is vast,

probably more than 10^{500}. The configuration in our region of the *Universe may arise just through random dynamical chance, meaning that it is not predictable even in principle.

The mental picture which goes with the concept is of a landscape of complex topography, with the valleys corresponding to different states that the Universe might find itself in, each potentially having different physical laws. Widely separated regions of the Universe may correspond to different such regions. This might manifest itself, for instance, in different strengths of forces such as electromagnetism, different masses and lifetimes for fundamental particles like the *neutron, or different values for the *density of *dark energy. In reality, the string landscape will have many dimensions, rather than the two that we can readily visualize.

The string landscape has recently been used as a way of motivating the *anthropic principle, which relies on the idea that while all these different possibilities might arise throughout the Universe, only special configurations, corresponding to particular physical laws, provide suitable conditions for life to arise.

Susskind, L. *The Cosmic Landscape: String Theory and the Illusion of Intelligent Design*, Little Brown, 2005.

string theory

See SUPERSTRING THEORY.

structure formation

Structure formation is a generic term referring to the formation of objects in the Universe due to amplification of initial *density perturbations by *gravitational instability. Examples include *galaxies and *galaxy clusters. The term is also often used to include *cosmic microwave background anisotropies, although they are not structures in the traditional sense. In the *standard cosmological model, structure formation proceeds hierarchically with the smallest objects forming first and later being assembled into ever larger ones.

Sunyaev–Zel'dovich (SZ) effect

The Sunyaev–Zel'dovich effect is the heating of *photons of the *cosmic microwave background (CMB) when they scatter from hot *electrons in the *intra-cluster medium (ICM) of *clusters of galaxies. It was first described by Rashid Sunyaev and Yakov Zel'dovich in

1970, and comes in two forms, **thermal** and **kinetic**. In individual galaxy clusters it can be used as a probe of the types of physical processes occurring in the ICM. SZ surveys are a powerful new method of detecting distant clusters with the potential to strongly constrain *cosmological models.

The thermal Sunyaev–Zel'dovich effect

When material falls into the deep gravitational well of a galaxy cluster, it is strongly heated as its initial *gravitational potential energy is converted to kinetic energy. It may additionally be heated due to radiation from *supernovae, massive stars and *active galactic nuclei. The temperature of the cluster gas, composed mainly of hydrogen and helium, may be tens or even hundreds of millions of degrees. Such hot gas can be seen directly due to strong emission in the *X-ray part of the *electromagnetic spectrum.

Such hot gas is in a fully ionized state, meaning that the electrons are all dissociated from their corresponding atomic nuclei. These electrons can interact with (scatter) any incoming radiation. The SZ effect concerns such scattering of the photons of the CMB. Since the CMB has a temperature of only a few degrees above absolute zero, the effect of the scattering is typically to transfer energy from the hot gas to the CMB photons, shifting them to higher frequency (or equivalently energy, as for photons frequency is proportional to energy). However the interaction probability is rather low; of those photons traversing a cluster perhaps only one in ten thousand scatter from an electron, the remainder passing through unaffected.

Since the scattering process does not create or destroy photons, the overall result of the SZ effect is to increase the number of photons at high frequencies, and therefore to *decrease* the number at low frequency (in each case as compared to a *black-body spectrum). This gives a characteristic spectral signature of the SZ effect, whose form turns out to be independent of the gas temperature. This is known as a y-distortion, quantified by the *y-parameter.

The crossover point, where the number of photons is unaffected by the scattering (due to a balance between those lost by scattering to higher energies and those gained by scattering from lower energies), is known as the SZ null point, and lies roughly in the middle of the range of frequencies traditionally used by

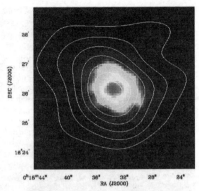

Fig. 1. The galaxy cluster CL0016, imaged via the SZ effect using the Berkeley–Indiana–Maryland array instrument.

CMB experiments. Those operating at lower frequencies will see the SZ effect as a reduction in the intensity of the CMB, while those operating above will see an increase. Experiments tuned to operate at the SZ null should see no effect. Combining the three types of measurement is a powerful way of separating the SZ effect from *anisotropies in the CMB itself.

In the most massive galaxy clusters, the electrons are mildly *relativistic (i.e. have speeds a significant fraction of the *speed of light). Allowing for this modifies the spectral distortion, and this may eventually be detectable.

The kinetic Sunyaev–Zel'dovich effect

The thermal SZ effect arises from the thermal motions of the individual electrons within the cluster. The kinetic SZ effect comes from the collective motion of the cluster. Like any cosmological object, galaxy clusters do not perfectly follow the *Hubble expansion law, but exhibit some velocity relative to it due to the local inhomogeneities in the density field. These motions are known as *peculiar velocities or *bulk motions. The scattering of CMB photons picks up a *Doppler shift due to the overall cluster movement in the radial direction towards or away from us.

Unlike the thermal SZ effect, the kinetic SZ effect does not possess a characteristic spectral signature. Instead, it simply gives a fractional shift in temperature related to the cluster velocity, while preserving the black-body spectrum of the radiation.

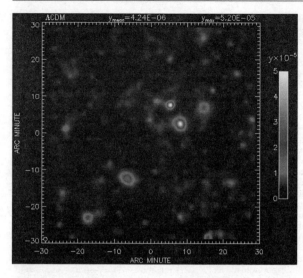

Fig. 2. A simulation of an SZ survey. A hydrodynamic simulation of structure formation has been used to create an image of area one square degree. Many clusters can be seen, indicating that a survey of many square degrees will pick up large numbers of clusters.

Accordingly, the kinetic SZ effect cannot be distinguished from genuine *CMB anisotropies by studying the spectrum alone. It is also a much smaller effect, by an order of magnitude or more, than the thermal SZ effect. So far it has not been convincingly detected for any cluster. In future it may be detected by using spatial information; for instance the typical angular size of distant galaxy clusters is much smaller than the predicted CMB anisotropy pattern. One could also use locations of measured thermal SZ effect to select regions where a kinetic SZ effect is expected.

Sunyaev–Zel'dovich observations

The SZ effect was first detected in 1983, but only relatively recently has it been possible to make maps of individual clusters. An example is given in Fig. 1 where a nearby cluster is shown. Several tens of galaxy clusters have now been mapped in this way.

The future of SZ observations will be large-area surveys aiming to find new clusters, rather than to study known clusters. The SZ effect is a powerful tool for finding distant clusters; since it relies on scattering rather than emission it has a much weaker dependence on distance, allowing clusters to be seen at considerable distances across the Universe. Fig. 2 shows a computer simulation

of what such a survey should see, indicating that many clusters per square degree should be found. [As we write, the first surveys are underway, by experiments known as SZA and AMI, but no results have yet been published.]

Rather than measuring the SZ effect in specific directions, one can ask what is the magnitude of the effect averaged over *all* directions. This is known as the mean y-distortion, and gives a global measure of the amount of heating that has been experienced by gas in the Universe. The mean y-distortion was strongly constrained by the FIRAS experiment on the *COBE satellite, but this limit remains a factor of a few higher than the level expected in the *standard cosmological model.

Historically, one motivation of SZ studies was as a way of measuring the *Hubble constant, which gives the present expansion rate of the Universe. The SZ properties of a cluster can be compared with the X-ray emission properties coming from the same hot cluster gas, and the two signals have different dependence on the Hubble constant so that simultaneous measurements can constrain it. In practice, however, calculations were plagued with uncertainties relating to short-scale clumpiness in the gas distribution and more generally to physical processes in the ICM, as well as to asphericities in the shape of clusters. Because of these problems, the technique has never been competitive with other ways of measuring the Hubble constant.

Finally, there are predicted to be *polarization signals induced by the kinetic SZ effect, caused by the motion of clusters transverse to the line of sight rather than radially as with the kinetic effect. These will be extremely challenging to detect.

superclusters

A supercluster is a grouping of *clusters of galaxies, smaller *galaxy groups and individual *galaxies. Unlike clusters, which are at least partially *virialized, superclusters tend to be elongated structures up to ~ 100 Mpc in length. The superclusters are separated by

vast voids of diameter ~ 100 Mpc containing very few galaxies (see Fig. 1).

Our own Galaxy and the *Local Group belong to the Local Supercluster (also known as the Virgo Supercluster), a flattened structure around 40–50 Mpc in diameter with the *Virgo cluster near its centre and the Local Group near one end. The plane of the Local Supercluster is readily visible in the distribution of nearby galaxies in Fig. 2. This figure shows galaxies plotted in supergalactic coordinates, which are defined so that the plane of the Local Supercluster lies along the centre of the projection and with the centre of the supercluster (approximately the location of the Virgo cluster) near the centre of the plot.

In the *hierarchical structure formation paradigm, superclusters are the last

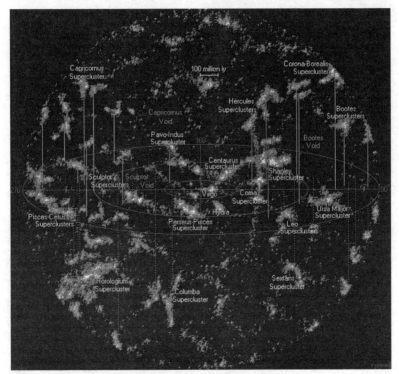

Fig. 1. A three-dimensional map of superclusters within a sphere of radius one billion light years centred on the Milky Way. Each point represents a group of galaxies.

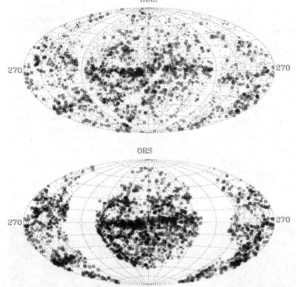

Fig. 2. The distribution of nearby galaxies seen in the infrared (top) and in the optical (bottom), each plotted in a Hammer–Aitoff projection in supergalactic coordinates. Objects within 2 000 km/s are shown as circled crosses; objects between 2 000 and 4 000 km/s are indicated as crosses, and dots mark the positions of more distant objects. The blank regions in these plots are due to obscuration by the Galactic plane.

structures to form, after galaxies, groups and clusters have formed. Superclusters are thus recently formed and are presently collapsing under *gravity to form increasingly dense superstructures. In 1999, the ROSAT *X-ray satellite detected intergalactic winds blowing through superclusters along the supercluster axis. It appears that these winds are transporting stars and gas towards the centre of the supercluster. *Numerical simulations also show a flow of matter along filaments towards high-density regions.

supergravity
See SUPERSYMMETRY AND SUPERGRAVITY.

supermassive black holes (SMBHs)
A supermassive *black hole has a mass significantly larger than any star, and is usually used to refer to a black hole with a hundred thousand or more times the mass of the Sun. There is evidence for the existence of black holes with masses up to about ten billion times that of the Sun.

Since nothing, not even light, can escape from a black hole, how can we tell that they are there? Supermassive black holes contain an enormous amount of mass within a volume of the size of the Solar System, and so

they exert an extremely strong gravitational pull on any nearby material, such as stars and clouds of gas. The material will be accelerated towards the SMBH, and in the case of gas clouds, the gas will be compressed and heated as it is pulled into a smaller volume. When the gas is hot enough to be ionized (for *electrons to be stripped from the atoms) the accelerating charged particles radiate energy. Once material crosses the *event horizon then no more energy can escape, but huge amounts of energy can be radiated before this happens, particularly when gas is orbiting the SMBH rather than falling directly into it, much as the Earth orbits the Sun.

It is thought that SMBHs provide in this way the power source for *active galactic nuclei (AGN), including *quasars. Indeed, the first evidence for the existence of SMBHs came from observations of AGN. The widths of the emission lines in the *spectra of some AGN indicate that gas clouds are travelling at speeds of $\sim 10\,000$ km/s—much faster than typical stellar velocities in *galaxies (~ 200 km/s). The compact size of the active region, with an upper limit of order the size of the Solar System as derived from the timescale of order days on which AGN vary in strength,

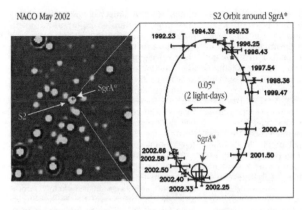

Fig. 1. The image on the left shows a 2 × 2 arcsecond region at the centre of our Galaxy taken in the near-infrared with the Very Large Telescope. When this image was taken (May 2002) the star labelled S2 lay just below and to the left of the compact radio source SgrA*. The position of this star at the different times indicated is shown on the diagram on the right. The star is on a ~ 15 year elliptical orbit, with SgrA* at one focus.

suggested that the only feasible power source was a SMBH.

More recently, studies of the dynamics of stars and gas near the centres of quiescent (non-active) galaxies suggests that SMBHs lie at the centres of most galaxies. Fig. 1 shows the orbit of a star orbiting within 17 light-hours of the compact radio source SgrA* at the centre of our Galaxy. By studying the motion of the star over time (it takes just over 15 years to make one complete orbit), it has been estimated that the radio source is about 2.6 million times more massive than the Sun. Only a black hole can contain that much mass in such a small volume of space.

Dynamical evidence for the existence of SMBHs with masses ~ 1 billion times that

of the Sun has been found in other galaxies, including *Messier objects M84, M87, and M104. As of June 2003, SMBHs with masses ranging from about one million to three billion Solar masses have been identified in 40 galaxies of varying types. Rather than identifying the orbit of a single star, one measures the *Doppler shift of the integrated light from many stars or gas clouds very close to the galaxy's centre. In the case of M84, the velocity within 26 light years of the centre reaches 400 km/s towards us. At a similar distance the other side of the centre, the sign of the velocity reverses (i.e. is moving away from us), resulting in a characteristic 'S' shape in a plot of recession velocity versus distance from the centre (see Fig. 2).

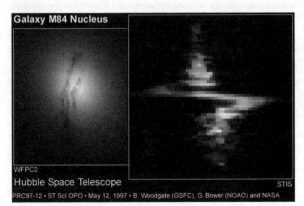

Fig. 2. The image on the left shows a Hubble Space Telescope (HST) image of the nucleus of M84. The right hand image plots measured velocity along the horizontal axis against position along the spectrograph slit (which is indicated in the left hand image) along the vertical axis. The velocity changes from −400 km/s (toward us) to +400 km/s (away from us) within about 50 light years, resulting in this characteristic 'S' shaped curve.

It seems likely that SMBHs do not accrete matter steadily but in bursts. Soon after formation, a galaxy and its central black hole accumulate matter. Eventually, radiation pressure suppresses the infall of further matter until the hot gas cools and resumes falling into the black hole, resulting in further growth of the black hole. Thus whether one classifies a galaxy as active or quiescent may just depend on the current state of accretion of matter onto the central black hole.

The origin of SMBHs remains obscure. One idea is that an individual starlike black hole forms and swallows up enormous amounts of matter over the course of millions of years to produce a SMBH. Another possibility is that a cluster of starlike black holes forms and eventually merges into a single SMBH. Finally, an SMBH might form from the collapse of a single large gas cloud.

Estimates of the mass of the spheroid component (or 'bulge') of the host galaxy and of the central SMBH suggest that the two are closely correlated: a galaxy bulge twice as massive as another would typically have a central black hole that is also twice as massive. This discovery suggests that the growth of the black hole is linked to the formation of the galaxy in which it is located.

Kormendy, J. in *Carnegie Observatories Astrophysics Series, Vol. 1*, Cambridge University Press, 2003.

supernova surveys

Type Ia *supernovae have become vital cosmological tools for probing the *acceleration of the Universe. Because supernovae reach their peak luminosity within about a month of the initial explosion, dedicated surveys are needed to identify and observe them at or before their peak. These surveys repeatedly observe the same areas of sky on timescales of the order of a few nights, and specially written software is used to identify objects that change in brightness. The task of identifying supernovae is made harder by the large number of variable stars that change in brightness. Follow-up spectroscopy is needed in order to confirm that a candidate object is indeed a supernova, and in order to identify its type.

The first systematic search for low-*redshift supernovae was the Calan/Tololo survey carried out in Chile in the early 1990s by Mario Hamuy and collaborators. They obtained light curves and spectra for about 30 Type Ia supernovae at redshifts $z < 0.1$. Although not at high enough redshifts to detect the accel-

eration of the Universe, these supernovae have proved extremely important in providing a low-redshift baseline for observations of supernovae at higher redshift, and in calibrating the decay-time of a supernova light curve as a function of peak luminosity (*see* SUPERNOVAE, Fig. 2).

The first supernovae above redshift $z = 0.1$ were observed in the late 1980s by Norgaard-Nielsen and collaborators. They found just two supernovae, at redshifts of $z = 0.31$ and $z = 0.28$, the latter of which was probably a type II supernova.

Two major searches for high-redshift supernovae were initiated in the early 1990s: the **Supernova Cosmology Project**, led by Saul Perlmutter from Lawrence Berkeley Laboratories in California, and the **High-z Supernova Search Team** led by Brian Schmidt of Mount Stromlo and Siding Spring Observatories in Australia. These teams have found and measured supernovae to redshifts $z \approx 1$, and have made the surprising discovery that distant supernovae are fainter than expected, unless the expansion of the Universe is accelerating (see Fig. 1). These observations have led astronomers to seriously consider the possibility of a non-zero *cosmological constant or the existence of *dark energy.

The Supernova Cosmology Project is now using the *Hubble Space Telescope to detect supernovae in *clusters of galaxies above a redshift of one.

The High-z Supernova Search Team has evolved into the ESSENCE project, a rather tortured acronym which stands for Equation of State: SupErNovae trace Cosmic Expansion. They have been using the Blanco 4-m telescope at Cerro Tololo Inter-American Observatory (CTIO) to find about 200 Type Ia supernovae in the redshift range 0.15–0.75, with the primary goal of constraining the *equation of state of dark energy.

Part of the second instalment of the *Sloan Digital Sky Survey is a search for intermediate-redshift supernovae in the range $z \sim 0.1$–0.4. After only one season of operation, the Sloan Supernova Survey detected 139 confirmed Type Ia supernovae, one of the largest supernova samples ever compiled in a single observing program.

Most ambitiously of all, the Supernova Legacy Survey is using the Canada-France-Hawaii Telescope to measure several hundred high-redshift supernovae.

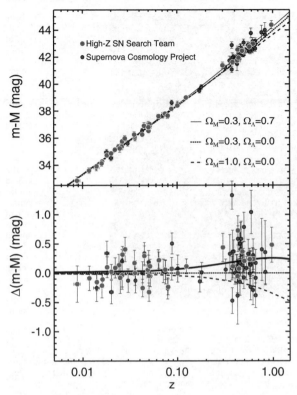

Fig. 1. Hubble diagram (distance versus redshift) from supernovae. These results appear to be inconsistent with a critical density, $\Omega_m = 1$, Universe (dashed line) and instead favour an accelerating Universe (solid line).

The SkyMapper telescope, due to start observing at Siding Spring Observatory in New South Wales, Australia, in early 2008, is expected to find up to 50 000 $z \lesssim 0.1$ supernovae over its first five years of operation.

Future planned supernova surveys include the Supernova / Acceleration Probe (SNAP) satellite.

Leibundgut, B. *Annual Reviews of Astronomy and Astrophysics, 39, 67,* 2001.

High-z Supernova Search Team:
http://cfa-www.harvard.edu/cfa/oir/Research/supernova/ HighZ.html
Supernova Cosmology Project:
http://www-supernova.lbl.gov/
ESSENCE project:
http://www.ctio.noao.edu/wproject/
Supernova Legacy Survey:
http://www.cfht.hawaii.edu/SNLS/

SkyMapper Telescope:
http://www.mso.anu.edu.au/skymapper/
SNAP satellite:
http://snap.lbl.gov/

supernovae

A supernova (plural *supernovae*) is the phenomenon that occurs when a massive star ends its life in a violent and spectacular explosion that shines for a period of a few weeks as brightly as an entire *galaxy; see for example Fig. 1. Supernovae play a vital role in creating elements heavier than iron and in distributing these heavy elements widely through space; our bodies contain elements synthesized in past supernova explosions. There are two main classes or types of supernovae, defined observationally by the absence (Type I) or presence (Type II) of hydrogen Balmer lines in their *spectra.

Fig. 1. Hubble Space Telescope image of supernova 1994D. This type Ia supernova is clearly visible to the lower-left of its host galaxy.

Type I supernovae

Type I supernovae are subdivided into Types Ia, Ib, and Ic. Type Ia exhibit a strong absorption line due to ionized silicon, which Types Ib and Ic lack. Type Ib have a strong helium absorption line which is lacking in Types Ia and Ic. It is thought that Types Ib and Ic supernovae are the death-throes of massive stars, like Type II, but which have lost their outer layers of hydrogen before they explode, either in a stellar wind or by transfer to a companion star. Type Ic supernovae appear to have lost most of their helium, as well as hydrogen.

In contrast, Type Ia supernovae are thought to result from the accretion of material from a companion red giant star by a white dwarf star. White dwarfs are evolved, moderate-mass stars which have burned most of their fuel into carbon and oxygen and which are held up by **electron degeneracy pressure**, a *quantum-mechanical effect which prevents *electrons from being squeezed too closely together. As mass is accreted onto the white dwarf from its companion, it shrinks in size. If the mass of the white dwarf reaches a critical mass known as the Chandrasekhar limit (about 1.4 times the mass of the Sun),

the increased pressure ignites carbon fusion in the interior of the white dwarf. Because the white dwarf is degenerate, it cannot expand, and instead the increasing temperature causes a runaway thermonuclear reaction, resulting in a Type Ia supernova which blows the white dwarf apart. White dwarfs contain essentially no hydrogen or helium, and the characteristic silicon feature in the Ia spectrum is a product of carbon fusion.

Type Ia supernovae are important to *cosmology as *standard candles. Since they form from the collapse of a stellar core of a particular mass (the Chandrasekhar mass), rather than the range of core masses possible for other supernova types, Type Ia supernovae are expected to have the same luminosity. In fact, Type Ia supernovae show a range of luminosities, but fortunately the peak luminosity can be predicted from the rate at which the supernova brightens and then fades—the more luminous ones take longer to brighten and then fade, see Fig. 2. Thus by measuring the light curve for a supernova, and taking its spectrum to confirm that it is indeed of Type Ia, one can compare the observed peak brightness with that predicted

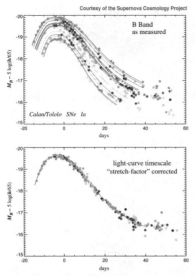

Courtesy of the Supernova Cosmology Project

Fig. 2. Light curves (plots of absolute magnitude as a function of time) for fifteen Type Ia supernovae from the Calan/Tololo supernova survey. The top panel shows the light curves as measured; note that the brighter supernovae have wider light curves. The bottom panel shows the light curves after applying a 'stretch correction' that corrects the magnitudes depending on the width of each light curve.

from the light curve, and so estimate the distance to the supernova. Because they are such spectacular events, supernovae can be seen to huge distances, and so provide our only distance indicators to galaxies beyond *redshift ~0.1.

Type II supernovae

Stars that begin their lives with masses more than about eight times that of the Sun end their lives as a Type II supernova. These stars spend most of their lives fusing hydrogen into helium and then into heavier elements. Energy released by these thermonuclear reactions balances the force of *gravity and supports the outer shells of the star from collapse. Once the stellar core is converted into iron, after a few million years, no further energy-producing nuclear reactions are possible. The core then contracts and heats up to a temperature of around five billion degrees within a tenth of a second. Within another tenth of a second, the core becomes so dense that

electrons are forced to combine with *protons in the nuclei of the atoms, a process which releases large amounts of energy in the form of *neutrinos, and the core attains **nuclear density** of around 10^{17} kg/m^3. (If the Earth had this density, it would be only 300 m in diameter.) Matter this dense cannot be compressed further, and so core contraction suddenly stops. The cooling of the core from neutrino loss decreases the energy available to hold up the surrounding shell of the star against gravity, and material falls in under the intense gravitational field of the dense core at speeds of order one tenth the speed of light. The material hits the rigid core and bounces off, creating a shock-wave, a supernova explosion. This process releases around 10^{46}J—one hundred times the energy the Sun has emitted over the past five billion years. Type II supernovae are distinguished from Type I supernovae by the presence of hydrogen lines in their spectra. Even though all the hydrogen in the core of the progenitor star has been fused into heavier elements, there is still a high abundance of hydrogen atoms in the stellar atmosphere. As they are heated in the explosion, the hydrogen atoms are excited and emit light at characteristic wavelengths.

Supernova surveys

Several *supernova surveys have been undertaken or are underway in order to measure supernovae at cosmological distances. These teams have found and measured supernovae to redshifts $z \approx 1$, and have made the surprising discovery that distant supernovae are fainter than expected, unless the expansion of the Universe is accelerating (*see* SUPERNOVA SURVEYS, Fig. 1). These observations have led astronomers to seriously consider the possibility of a non-zero *cosmological constant or the existence of *dark energy.

See also DISTANCE LADDER.

http://www.supernovae.net/isn.htm

superstring theory

One of the most vexing questions in modern physics is to resolve a deap-seated incompatibility between the two big physics ideas of the 20th century, *Einstein's theory of *gravity, known as *general relativity, and the theory of *quantum mechanics. While each is resoundingly successful within its own realm of validity, the theories prove incompatible in regimes where both types of effect should play a role. Superstring theory, often shortened to string theory, is the leading candidate to unify gravity with the other fundamental forces in a well-defined quantum framework. It is sometimes called a Theory of Everything.

Basics of string theory

The assumption of superstring theory is that the fundamental constituents of matter are not point-like particles, but one-dimensional strings. An *electron, for instance, appears point-like to us in experiments only because we have not been able to peer closely enough at it. In some versions of the theory, the strings are all in the form of loops, known as closed strings, while in others the strings can be short segments with ends, known as open strings. The idea is that as we move into the realm of *quantum gravity, the inherent stringiness of the particles becomes apparent. This keeps the theory well-defined.

The initial ideas of string theory go way back into the early 1970s through the work of Gabriele Veneziano and Leonard Susskind, amongst others. The key breakthrough, however, came in 1984 when Michael Green and John Schwarz showed that string theory was only consistent if based on one of two very special symmetries. Almost immediately, a substantial fraction of particle physics theorists around the world dropped whatever they were doing and began to work on string theory, recognizing its potential to give a unique theory of quantum gravity. The Green–Schwarz discovery is often referred to as the first superstring revolution.

Superstrings and extra dimensions

One minor drawback. The Green–Schwarz mechanism predicts that the *Universe should be ten-dimensional, with nine space dimensions and one time dimension. A glance around you will convince you that this is not the case, and that there are only three dimensions (up–down, left–right, forwards–backwards). Superstring theory appears immediately to be in conflict with observations.

However superstring theory is not the first to consider the possibility that *space-time has *extra dimensions. The idea goes all the way back to the 1920s, when Theodor Kaluza and Oskar Klein postulated a single extra dimension in an attempt to unify gravity with electromagnetism. In principle, this extra dimension can be hidden from our view

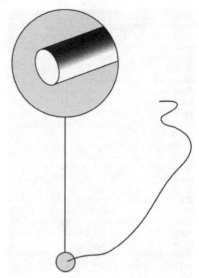

Fig. 1. Seen from a distance a hosepipe looks like a one-dimensional line, but close up its second dimension is revealed.

if it is very small, a mechanism known as *compactification.

A useful analogy (see Fig. 1) is to consider a thin hose-pipe. Seen from close up, the hose-pipe is a two-dimensional structure—a long cylinder with a hole running through it to carry the water. However if we step far enough back, we are no longer able to see the second dimension and the hose looks like a one-dimensional line. The same might be true for our Universe; in addition to the usual three large dimensions, there may be other dimensions (in the case of superstring theory, six of them) which are too small for us to see. Only particles with very high energies would be able to probe the extra dimensions, which might have been possible during the very *early Universe.

Very recent work in superstring theory has brought to light a possible alternative to compactification, known as the *braneworld scenario, which is discussed fully at that entry.

Superstring phenomenology

The ultimate goal of superstring theory is to match experiments, and in particular to reproduce the *standard model of particle physics at sufficiently low energies, perhaps via a *grand unification scenario. This goal has been quite frustrated, and it remains unclear how the connection to normal physics will be achieved. One problem is that even if the string theory itself were unique (we will see shortly that it is not), there is an untold number of ways in which compactification could take place, with each potentially corresponding to different physical laws for the Universe. The set of possible compactifications is known as the *string landscape, and there are no convincing proposals for how this landscape should be navigated in order to yield acceptable physical models.

Cosmology of superstrings

Much research has been devoted to incorporating ideas from string theory into *cosmology, but so far the subject remains in its infancy. Primarily the aim has been to discover whether string theory can predict a period of *cosmological inflation. This work has several strands.

One strand is to find string-theory motivated models of inflation. This is mainly done by looking at the low-energy limit of string theory, and does not try to incorporate any genuinely 'stringy' effects. As just discussed, superstring phenomenology remains an uncertain business and appears to offer plentiful opportunities for inflation model building.

A second strand attempts to use fundamental properties of string theory to motivate models. (For a full discussion, *see* STRING COSMOLOGY).

A third strand is to consider regimes where the 'stringiness' of matter makes itself apparent, which happens only at the highest attainable energies and hence right at the beginning of the Universe. Most recent discussion of this has focused on the braneworld scenario and is explored at that entry.

As a final quirk, it has been argued that the present *acceleration of the Universe may be incompatible with superstring theory if the acceleration continues indefinitely. This is a subtle technical argument; string theory is only well-defined provided it can model interactions where the strings ultimately become arbitrarily well separated, but such separations cannot arise in an accelerating Universe as it has a future *event horizon which gives a maximum attainable distance. Superstring

theorists, convinced that their models are on the right track, have cited this as evidence against present acceleration, but given that the acceleration is a fairly well-established observational fact it is tempting to instead view it as an argument against string theory. In any case, the conflict can be evaded if the present burst of acceleration proves to be a transient one.

Beyond string theory

Given that string theory was supposed to provide a unique theory unifying gravity and quantum mechanics, it was rather embarrassing when it transpired that there were five different string theories, all internally consistent. The next breakthrough came in the 1990s, when Edward Witten discovered that those five string theories could all be viewed as limiting cases of a single more powerful theory, known as *M-theory. This discovery is sometimes called the second superstring revolution.

Greene, B. *The Elegant Universe*, Vintage, 1999 [accessible for a general readership].

Green, M. B., Schwarz, J. H. and Witten, E. *Superstring Theory* (two volumes), Cambridge University Press, 1988 [highly technical].

supersymmetry and supergravity

Supersymmetry, sometimes shortened to 'susy', is a proposal for new physics beyond that given by the *standard model of particle physics. Its most striking prediction is that every known particle has a partner, immediately doubling the number of fundamental particles. It has not yet (at time of writing) seen any experimental verification, but provides the basic framework for current ideas in fundamental physics including *superstrings and *M-theory. One of the key goals of the Large Hadron Collider (LHC), due to begin operations at CERN around 2008, is to seek evidence for supersymmetry.

The motivation for supersymmetry is to try to explain why the characteristic energy of the electro-weak interactions (given by the masses of the W and Z particles, around 100 GeV in *particle physics units) is so much less then the characteristic scale of *grand unification and that of gravity, the *Planck scale. This is unexpected because quantum effects are thought to give additional contributions to the mass of any particle at the level of the highest energy scale, making it hard to understand the vast difference in these energy scales. This discrepancy is known as the **hierarchy problem**, and supersymmetry is by far the most widely accepted proposal for resolving this.

Supersymmetry achieves this by associating a new particle with each existing particle, with the rule that the partner of each *boson is a *fermion and vice versa, their properties otherwise being identical. When quantum effects of those particles are computed, the bosons and fermions make equal and opposite contributions and hence neatly cancel, protecting low energy scales against quantum corrections from much higher ones. This applies also to the *cosmological constant, though in that case it cannot explain the observed small value.

Supersymmetry as it stands is more of an idea than a model; it is something that can be applied to models. One option is to create a supersymmetric version of the standard model, and the simplest such version is called the Minimal Supersymmetric Standard Model (MSSM). There are however more complex variants. Supersymmetric versions of grand unification theories can also be constructed.

The major problem with this proposal is, of course, that there is no sign of all these extra particles. Supersymmetry therefore does not exist in the present *Universe. In particle-physics parlance, if it is to be true it must be a **broken symmetry**, which holds only at high energies. Although this might sound artificial, the set-up is essentially the same as the breaking of electro-weak symmetry in the standard model, in order to account for the difference between the weak and electromagnetic interactions. The breaking of supersymmetry will give each supersymmetric particle a mass of order of the energy scale of supersymmetry breaking, and if this is high enough it will explain why such particles have yet to be seen even in particle accelerators. This works provided the energy scale is at least the electro-weak scale, and at the same time it cannot be much higher or the hierarchy problem will no longer be solved. The prediction therefore is that the supersymmetric particles should have masses within an order of magnitude or so of the electro-weak scale, bringing them within reach of the LHC.

In the simplest supersymmetry theories, the supersymmetric particles carry a quantum number called supersymmetric charge, which is conserved in all interactions. This restricts the types of interactions that the supersymmetric particles can display, and in particular guarantees that the lightest supersymmetric particle (LSP) must be stable (as there is no lighter particle into which it can decay that also possesses the supersymmetric charge). That is rather interesting, because such a particle is an ideal candidate to be the *dark matter in the Universe. Even better, back of the envelope calculations of the likely abundance of LSP particles left over from the hot stages of the *early Universe, although highly model dependent, indicate that it is in the general vicinity of the *critical density, as required. Accordingly, the LSP is regarded as the leading candidate to be the dark matter, and the results of direct dark matter search experiments are often phrased in terms of how they constrain supersymmetric model-building.

Supergravity, sometimes shortened to 'sugra', takes the supersymmetry idea one step further and incorporates *gravity, which is a major advance on the standard model which ignores gravity completely. Supergravity theories are not normally regarded as fundamental theories in their own right, but rather as a stepping stone to the more sophisticated scenarios of superstrings and M-theory.

surface brightness
The surface brightness of an extended source, such as a *galaxy, is the observed flux per unit solid angle. In general, the surface brightness may vary over the the image of a galaxy, with surface brightness generally being higher near the centre of the galaxy and fading away from the centre. When quoting a single, representative surface brightness of a galaxy, it is common to quote the **effective surface brightness**, that is the average surface brightness with a circular aperture containing half of the total light of the galaxy.

If the observed flux F is measured in physical units, such as Janskys, then the surface brightness I of an object of area A arcsec^2 is simply given by $I = F/A$ with units of Janskys per square arcsecond. A large value of I denotes an object of high surface brightness. If, however, we use the *magnitude m of an object, then its surface brightness is given by the formula $\mu = m + 2.5 \log_{10} A$, where μ has units of magnitudes per square arcsecond. In this case, a large value of μ denotes an object of *low* surface brightness.

Surface brightness has the useful property that it is independent of *distance, at least when expansion of the *Universe is neglected, a reasonable approximation for galaxies at *redshift $z \lesssim 0.1$. The is because both flux and solid angle decrease with distance squared, and so the ratio of these quantities, the surface brightness, is constant.

This rule breaks down for galaxies at high redshift because the Universe is expanding, and in fact the observed surface brightness I scales as the fourth power of the *scale factor $a \propto 1/(1 + z)$, so that $I \propto (1 + z)^{-4}$. The first two powers of $(1 + z)$ come from the fact that *angular-diameter distance and transverse *comoving distance differ by a factor of $(1 + z)$, and so the apparent solid angle scales as $(1 + z)^2$. The third power of $(1 + z)$ comes from *time dilation in receiving the *photons and the fourth power comes from the loss of energy per photon.

This $I \propto (1 + z)^{-4}$ redshift-dependence of surface brightness was proposed by Richard Tolman in 1930 as a test of the expanding Universe. It is a generic prediction of the *Robertson–Walker metric in an expanding Universe, and is independent of the correctness of *general relativity and of the *cosmological parameters. In the 'tired-light' *alternative cosmology one would expect $I \propto (1 + z)^{-1}$.

When *galaxy evolution has been accounted for, *Hubble Space Telescope observations of the surface brightness of distant galaxies are consistent with the Tolman test, and rule out the tired light model at a *significance of more than 10σ.

Lubin, L. M. and Sandage, A. *Astronomical Journal*, *122, 1084*, 2001.

symmetry breaking
See PHASE TRANSITIONS.

SZ effect
See SUNYAEV–ZEL'DOVICH EFFECT.

temperature–time relation

The temperature–time relation for a *cosmological model describes how the *Universe's temperature changes as it ages. Since the temperature of a gas of particles is a measure of the energy of the individual particles, this relation also indicates the types of particle interactions that are prevalent at a given epoch.

If cosmological models are extrapolated all the way to the instant of the *big bang the temperature would be infinite then, though few if any cosmologists believe that such an extrapolation is possible since known physical laws are predicted to break down at high temperatures. The temperature–time relation therefore first becomes valid sometime after the big bang. Moreover, the 'time' which appears as the age of the Universe as if its behaviour really could be extrapolated to the instant of the big bang, defined as time zero.

The temperature–time relation is most useful during the *radiation-dominated era of the Universe's evolution, lasting for the first few thousand years. During that epoch the *density of material in the Universe, responsible for driving the expansion via the *Friedmann equation, can be directly related to the temperature of that material, giving an accurate temperature–time relation. Where T is the temperature and t the time, it can be written approximately as

$$\frac{T}{10^{10}\,\mathrm{K}} \approx \left(\frac{1\,\sec}{t}\right)^{1/2}.$$

The constant of proportionality has been written in a suggestive way to make it easy to generate particular examples. We learn that at an age of one second the Universe had a temperature of about 10^{10} kelvin. By 100 seconds it had fallen to 10^9 kelvin, whereas at 10^{-10} seconds old the temperature would have been 10^{15} kelvin.

The above law embodies the redshifting relation $T \propto 1/a$ (a being the *scale factor), since during the radiation era $a \propto t^{1/2}$. This

relation holds most of the time, but can be briefly violated if the Universe departs from *thermal equilibrium. This can happen if, for instance, the cooling of the Universe means that massive particles can no longer be readily produced in interactions, while those already existing decay or annihilate. The most recent example was the annihilation of *electrons and positrons when the Universe was around one second old.

The characteristic energy of an individual particle in thermal equilibrium at temperature T is given by $k_B T$, where k_B is a fundamental constant of Nature known as the Boltzmann constant. This is really just telling us that the macroscopic property of temperature is derived from the individual energies of the constituent particles. Using *particle physics units, we can rewrite the temperature–time relation in terms of the characteristic particle energy as

$$\frac{k_B T}{1\,\mathrm{MeV}} \approx \left(\frac{1\,\sec}{t}\right)^{1/2},$$

where MeV stands for mega-electron-volt.

This version is the key to determining the types of physical process happening at different stages of the Universe's evolution. For example, the binding energy of typical light nuclei, such as helium-4, is around 1 MeV per particle. Hence when the Universe was younger than one second and the ambient energy higher than 1 MeV, any nucleus that might form would rapidly be destroyed. After one second they can begin to form. The epoch starting around one second is therefore that of *nucleosynthesis, where the elemental content of the young Universe was fixed.

By contrast, the binding energy holding electrons in atoms is typically about an electron-volt. According to our equation, electrons therefore cannot settle in atoms until around 10^{12} seconds (i.e. tens of thousands of years) after the big bang. This is the epoch of *recombination, when the Universe converts from a plasma to a neutral atomic state. [Actually the above equation isn't strictly applicable, as it shows that recombination happens after the radiation era ends. However it is not too bad as an estimate.]

The highest energy particles created on Earth, using *particle accelerators, have energies around 10^6 MeV (known as a teraelectron-volt). This is the highest energy at which we can consider fundamental physical processes to be well tested and understood, and was achieved when the Universe was about 10^{-12} seconds old. Any study of earlier epochs necessarily requires some hypotheses as to which physical laws might be appropriate.

During the *matter-dominated era which follows radiation domination, one can no longer write such a simple relation, since the dominant material driving the expansion is no longer the radiation. Indeed, after *decoupling the *baryonic material and the radiation begin to develop different temperatures, which they can do as they no longer interact. Perhaps the most useful guide here is to take the accurately measured present temperature of the microwave background, 2.725K, combined with the redshifting law $T \propto 1/a$ which has held ever since electron–positron annihilation during the radiation era.

Fig. 1 illustrates the evolution of temperature with time from the radiation-dominated epoch onwards. The present epoch of *dark

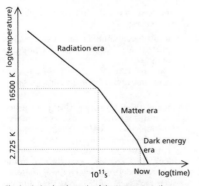

Fig. 1. A simple schematic of the temperature–time relation for our Universe, showing the three mains eras of radiation domination, matter domination, and now dark energy domination. A more detailed analysis would reveal some substructure in this simple curve.

energy domination is characterized by an accelerating expansion leading to a sharp drop in temperature.

One can also consider the temperature history during *cosmological inflation. Whatever the temperature might have been when inflation started (if indeed it had a start), it is rapidly supercooled by the expansion and indeed predicted to closely approach absolute zero. At the end of inflation, the processes of *preheating and *reheating recreate a thermal bath of conventional particles, returning the Universe to a high temperature and a radiation-dominated epoch.

Liddle, A. R. *An Introduction to Modern Cosmology*, 2nd ed., John Wiley and Sons, 2003 [undergraduate level].

tensor perturbations

Tensor perturbations in the *metric of *spacetime is another name for *gravitational waves, used particularly for the gravitational waves predicted to be produced by a period of *cosmological inflation. The term 'tensor' comes from the way the quantities describing the gravitational waves change under coordinate transformations, as according to the theory of *general relativity. These perturbations may interact with corresponding tensor perturbations in the matter distribution (known as *anisotropic stress), or may travel freely, the latter case being the one normally associated with gravitational waves.

A general perturbation can be decomposed into tensor, vector, and *scalar perturbations, each of which evolves independently of the others. The scalar perturbations are the dominant type and are responsible for *structure formation. Vector perturbations are not often discussed, as they are not generated by inflation and because they are predicted to decay quickly in any case, even if they were formed in the young Universe.

Tensor perturbations do not lead to structure formation, but can create *cosmic microwave background anisotropies, though such an effect has yet to be detected. In particular, they are responsible for generating one of the types of CMB *polarization anisotropies, known as the B-mode polarization, which cannot be generated by scalar perturbations.

thermal equilibrium

If different types of particle (for instance *baryons and *photons of radiation) are interacting rapidly with one another, they approach a state known as thermal equilibrium. A thermal state is determined by specifying the temperature, and perhaps also the value of any conserved quantity such as baryon number. The process of thermalization removes the memory of initial conditions from a system.

Particles in a thermal distribution have a range of different energies, but with mean value of order $k_B T$ where T is the temperature and k_B is a fundamental constant of Nature known as the Boltzmann constant. The particle energies determine the kinds of interactions that the thermalized particles can have.

Thermal equilibrium is actually a combination of two conditions. Kinetic equilibrium requires that the individual particle species each have a thermal energy distribution, and can be established through particle collisions alone. Chemical equilibrium requires that the number density of each particle species also reaches equilibrium with respect to the others, which requires the existence of interactions that can create and annihilate particles. Under some circumstances it might be possible to establish kinetic but not chemical equilibrium, an example being an energy injection into the *cosmic microwave background that might create what is known as a Compton μ-distortion.

In the early *radiation-dominated era, baryons, photons, and *neutrinos are expected to be in thermal equilibrium with each other. Additionally *dark matter particles might or might not participate in the thermal equilibrium, depending on their interaction properties. *Dark energy was probably never part of a thermal equilibrium state.

As the *Universe expands and cools, thermal equilibrium can be lost for some particles because interaction rates may become too slow to sustain it. The process of exiting thermal equilibrium is known as *decoupling. Photons decoupled when the Universe was around one thousandth of its present size, forming the cosmic microwave background, while neutrinos (which interact much more weakly) did so at around one second into the Universe's existence. Decoupling from thermal equilibrium is the mechanism used in the *WIMP theory of *cold dark matter.

Despite no longer having significant interactions, photons and neutrinos still have thermal energy distributions today as those are preserved by the expansion cooling of the Universe. The cosmic microwave background is known to accurately follow a thermal distribution, which is interpreted as showing that it was in thermal equilibrium with other particles during the Universe's early history. This was the decisive observational evidence in favour of the hot *big bang cosmology (see 'Overview').

In the context of *density perturbations in the Universe, the existence of a thermal state would be important as it forces the perturbations in any type of material participating in the thermal equilibrium to become *adiabatic perturbations.

time dilation

Time dilation in *cosmology is the apparent slowing down of temporal phenomena at high *redshift, and is a result of the expansion of the *Universe. Imagine a distant source emitting one *photon every second. In a static Universe, we could (in principle) detect each photon at some time t later, where t is given by the *distance to the source divided by the *speed of light. The gap between each photon would still be one second. Now in an expanding Universe, we detect light from a source at redshift z emitted when the *scale factor of the Universe was $(1 + z)$ times smaller than its present value. Thus in the time the light has taken to reach us from the source, the Universe has expanded by the same factor of $(1 + z)$. Since the number of photons is conserved, the time delay between photons reaching us also increases by a factor $(1 + z)$.

Time dilation has the following observable consequences:

- It introduces an extra factor of $(1 + z)$ in the *surface brightness redshift dimming of *galaxies.
- The time taken for a *supernova to fade from maximum brightness increases as $(1 + z)$.

The name 'time dilation' is also given to the apparent slowing of a clock as a consequence of *Einstein's theories of *relativity. In special relativity, time dilation occurs for clocks moving with respect to the observer. In *general relativity, time dilation occurs for clocks at a lower potential in a *gravitational field. In most cosmological applications, time dilation due to redshift is the dominant effect.

top-down structure formation

In the top-down scenario of *structure formation, large structures, such as *superclusters and *clusters of galaxies, form early on, and subsequently fragment to form smaller structures such as *galaxies. Such a mode of structure formation is expected in a *Universe dominated by hot *dark matter: the near-relativistic velocities of the hot dark matter particles prevent small structures from forming early on. This contrasts with *hierarchical structure formation, also known as bottom-up structure formation, in which small, subgalactic objects form first, and later merge to form galaxies and clusters.

The top-down scenario is now disfavoured for both observational and theoretical reasons. Observationally, very few clusters and superclusters appear as relaxed, spherical objects. Instead they tend to be elongated, suggesting that they are still in the process of formation. On the contrary, many galaxies are in a relaxed state, and age estimates for some of them are a significant fraction of the *age of the Universe.

From the theoretical perspective, there are many indications that matter in the Universe is predominantly in the form of *cold dark matter (CDM). In CDM-dominated *numerical simulations, structure is seen to form hierarchically.

topological defects

Topological defects, such as *cosmic strings, are regions of trapped energy thought to form when the *Universe undergoes *phase transitions during which its properties change dramatically. Phase transitions are very common throughout physics; an example is the sudden change in properties of water when heated above 100 degrees Celsius to become steam. Other examples include the way in which some metals become superconducting when cooled sufficiently, and the way that the magnetic properties of materials can change dramatically with temperature. Phase transitions are usually caused by changes in temperature, though they can also be induced by pressure or magnetic fields. Since the Universe is believed to have started out in an extremely hot dense state and to have subsequently cooled to only three degrees above absolute zero, it is expected that it has undergone a series of phase transitions during its evolution. Examples include the *electroweak phase transition and the *quark–hadron phase transition.

Phase transitions commonly leave behind defects, meaning that the phase transition hasn't occurred perfectly smoothly. For example, when water freezes to become ice, there are commonly dislocations where separately formed cubic lattices meet and are misaligned. Once such defects form they are extremely hard to remove, and a class known as topological defects are impossible to remove without heating up the material to reverse the phase transition.

The nature of defects formed depends on the type of phase transition that takes place. Most dramatically, the defects can take the form of thin walls of energy, known as domain walls, dividing regions with slightly different physical conditions. The defects can be in the form of lines of energy called cosmic strings, or they can be point-like in which case they are known as monopoles. *Grand unification theories predict monopoles carrying a magnetic charge (something otherwise forbidden in Nature) known as *magnetic monopoles. More exotic possibilities include pairs of monopoles joined by cosmic strings.

While many models of fundamental physics predict phase transitions and hence topological defects, none have been observed, which places tight constraints on the types of phase transition which can have occurred in the Universe. Domain walls are particularly strongly constrained, since walls of energy passing through our observable Universe ought to have striking effects both on the *galaxy distribution and on the *cosmic microwave background; *Zel'dovich and collaborators were the first to realize this in the mid 1970s and it represents one of the first uses of astrophysical considerations to constrain very high energy particle physics. Magnetic monopoles are also potentially very dangerous observationally, as described in their own entry. By contrast, cosmic strings have interesting properties which may help an understanding of the formation of structure in the Universe.

topology of the Universe

The topology of an object is a global property which measures its connectedness. Topology needs to be specified in addition to *geometry in order to completely specify a *space-time. It is not presently known what topology our *Universe possesses.

What is topology?

Topology is a mathematical concept which is notoriously difficult to define in a comprehensible way. The relevant *Oxford*

English Dictionary definition is 'The branch of mathematics concerned with those properties of figures and surfaces which are independent of size and shape and are unchanged by any deformation that is continuous, neither creating new points nor fusing existing ones; hence, with those of abstract spaces that are invariant under homomorphic transformations.' In essence, it refers to properties of an object which are unchanged by geometrical deformations. For instance, a soccer ball and a rugby ball have the same topology, as a smooth deformation (in this case a squeezing) can turn one into another. By contrast, a sphere and a torus have different topologies, as they cannot be turned into one another without tearing and gluing together in a different way. Perhaps the most famous example of two objects with the same topology is a coffee cup and a doughnut; each has one hole through them and a smooth mapping of one to another is possible—see Fig. 1.

A useful way to think about topology is via identification of boundaries. For instance, take a square and impose the condition that if you pass from the right edge you reappear on the left (and vice versa), and if you pass through the bottom you reappear at the top (and vice versa). If you physically join the sides you make a cylinder, and then if you join the ends you will find you have made a torus (i.e. a doughnut shape), but the topology is just as well represented by the flat square with the edges identified. In fact, this type of set-up is used in many simple video games, which therefore are effectively taking place on the surface of a torus!

The terminology is that the region used to build up the space, here a square, is called the fundamental domain. Choosing a fundamental domain is the first part of defining the topology. Having then chosen the way in which the edges are identified, we can view

the complete space, known as the **covering space**, by 'tiling' it by placing the fundamental domains next to one another. The space will make sense provided the tiles fit together properly; in two dimensions squares will and so will hexagons, but octagons for instance will not.

Topology in cosmology
The above discussion used two-dimensional examples, but for the Universe we need to consider more dimensions. By 'topology of the Universe', cosmologists usually mean that they are discussing the topology of space, not of space-time. *General relativity only describes the geometry of space and space-time, i.e. their curvature, and does not tell us anything about the topology. The topology can be independently chosen amongst all the available possibilities, and it is left as an observational question which topology actually does describe the Universe we live in. However general relativity does dictate that the topology of space does not change as the Universe evolves; it is simply carried along with the expansion of the Universe (i.e. the topological structure is fixed in *comoving coordinates).

Which topologies are permitted does depend on the geometry of space, which for *homogeneous Universes has the three possibilities of flat, spherical, or hyperbolic. The simplest topology, usually called the trivial topology, is that there are no identifications, with the flat and hyperbolic spaces continuing infinitely.

The flat case is the simplest, and can have many different topologies corresponding to choices of the fundamental domain and the way identifications are made. The simplest is to take a cube, and identify top and bottom, left and right, and front and back. The covering space is then a three-dimensional Universe broken up into the replicated cubes. This is known as a three-torus, the 'three' indicating the dimensionality. More generally, one could use a cuboid rather than a cube so that the identification distance is different in the three dimensions, and one or two of those dimensions could even be infinite (the latter case known as slab topology as the Universe is built up from a stack of slabs which are finite only in one direction). For further complication, the basic domain could, instead of a cube, be a hexagonal prism, and additionally it is possible to introduce twists before

Fig. 1. Smooth deformations transform a coffee cup to a doughnut.

identifying the edges. In all, there are 18 possible topologies for the flat geometry.

For the two curved spaces, spherical and hyperbolic, things become much more complicated, but non-trivial topologies are possible here too. These cases are quite interesting because while the flat case allows topologies on any scale, in the curved cases there is a minimum allowed topology scale, otherwise the fundamental domains will not fit together properly on the curved space they inhabit. In the spherical case, an interesting possibility is to build up the space from a fundamental domain which is a dodecahedron, the space being known as Poincaré's dodecahedral space after its discoverer Henri Poincaré.

Observational consequences of topology

Our observations are limited by the size of the *observable Universe, delimited by the distance light can have travelled since the Universe began. If the scale of any topology is much greater than this, then we have no way of measuring it. If the scale were much less than the observable Universe, then our observable Universe would contain vast numbers of copies of the fundamental domain, and this would have long ago been noticed by astronomers. The interesting case, therefore, is when the topology is on a scale comparable to the size of the observable Universe.

The most powerful probe is the *cosmic microwave background (CMB), as it comes from the greatest distance. Its point of origin is the *last-scattering surface which is almost at the limit of the observable Universe. It has so far allowed three main tests:

Reduced large-scale power—Many models with topology predict that there should be a reduced level of very large-scale *density perturbations in the Universe, leading to a suppression of the lowest few multipoles of the *power spectrum of the *CMB anisotropy. This is because the Universe is not as big as it appears, and there cannot be any perturbations on scales larger than the topology scale. Both the *COBE satellite and the *WMAP satellite did see a suppression of the lowest multipoles, though there are many possible interpretations including statistical fluke.

Breakdown of statistical isotropy—Once topology is imposed, the Universe is no longer *isotropic, since there are special directions corresponding for instance to the orientation of the planes where identifications are made. This leads to a particular type of cosmic *non-Gaussianity, characterized by the anisotropy multipoles losing their statistical independence and instead becoming correlated to each other. This can manifest itself, for instance, in alignments of the orientation of the low multipoles. Such alignments have been seen in data from WMAP, but topologies have not been found which are capable of explaining them.

Circles in the sky—When the last-scattering surface is larger than the fundamental domain, it necessarily intersects with itself, in the simplest case doing so along circles as shown in Fig. 2. The anisotropy pattern along those circles is expected to match, though not perfectly as part of the signal is velocity-dependent (the Doppler effect) and part of it is acquired after light leaves the last-scattering surface (the integrated *Sachs–Wolfe effect). Searching for matched circles is computationally challenging. Most searches thus far have been null, though a low-significance detection was made of a match of the type predicted by the dodecahedral Universe.

In conclusion, some of these tests have shown tentative indications that cosmic topology might be present, but none in such a convincing manner as to bring topology into the mainstream of cosmology studies.

While the CMB is the most promising observational probe of topology, it is also possible to look for repetitions of the fundamental domain by seeking distant regions of the Universe which appear to repeat the structures seen locally. This is however difficult as the regions tend to be seen at different epochs, and hence objects that might be studied, such as *quasars, may have evolved considerably in their appearance and be difficult to connect to one another.

An observation of cosmic topology would pose considerable difficulties for theories of *cosmological inflation, since the rapid inflationary expansion would increase the scale of the topology. The expectation would be that, if there were a topology scale, it would end up much larger than the observable Universe, otherwise the *horizon problem would not be solved.

Levin, J. *How the Universe Got its Spots*, Weidenfeld & Nicholson, 2002.

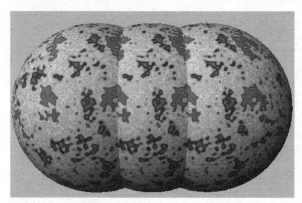

Fig. 2. A complicated but pretty image showing how cosmic topology leads to repeated structures in the CMB. We can consider ourselves to be in any of the fundamental domains, and shown is the last-scattering surface viewed from three such points. Those three surfaces are seen to intersect along circles, which by definition must show the same anisotropy pattern as they are at the same physical location. Accordingly, the observer should see 'matched circles' on the sky with the same anisotropy pattern, indicating the presence of non-trivial topology.

Tully–Fisher relation

This is an empirical relationship between the rotation speed of a spiral *galaxy and its luminosity, derived by Brent Tully and Richard Fisher in the 1970s and 1980s. Since rotation speed may be measured without knowing the *distance to a galaxy, this relationship provides an important way of estimating luminosities of spiral galaxies, and hence their distances.

The rotation speed of a nearby spiral galaxy may be measured by the technique of **long-slit spectroscopy**. One aligns a spectrograph slit along the (apparent) major axis of the galaxy and thus simultaneously measures the *Doppler shift of starlight along the major axis. If the galaxy is not face-on to the observer, then the side of the galaxy rotating toward the observer will show a blueshift relative to the centre of the galaxy, and the other side will show a *redshift. From the maximum blue and redshifts, and by estimating the inclination of the galaxy's rotation axis to the line of sight (this may be done by assuming that the galaxy's disk is intrinsically circular and measuring the apparent ellipticity), the rotation speed of the galaxy may be determined.

For more distant galaxies, rotation speed is more easily measured using a single dish radio telescope with a beam large enough to include all of the galaxy's neutral hydrogen (HI) gas. By measuring the amount of HI flux at each velocity, radio astronomers can obtain the HI **global profile**, a plot of

HI flux density as a function of velocity (see Fig. 1). For spiral galaxies, this has a characteristic 'twin-horned' profile, since the rotation speed of the galaxy is approximately independent of distance from the centre for a large range of radii. The total velocity range Δv is related to the maximum rotation speed of the galaxy V_{max} and inclination angle i by $\Delta v = 2V_{max} \sin i$. By comparing the luminosities of independently calibrated galaxies with rotation speed, Tully and Fisher noticed a tight relation between them: $L \propto \Delta v^4$. In other words, luminous galaxies rotate faster than dim ones. Once this relation was calibrated, it could be applied to other galaxies at unknown distances. By comparing the predicted luminosity with observed brightness, and allowing for interstellar *extinction,

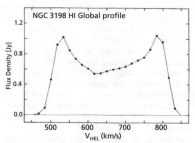

Fig. 1. The HI global profile (HI flux versus velocity) for galaxy NGC3198.

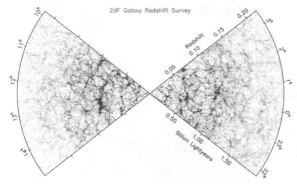

Fig. 1. Distribution of galaxies in the 2dF galaxy redshift survey. Note that galaxies tend to be distributed along filamentary structures, surrounding underdense regions known as voids. The lower galaxy density at high redshifts is due to the observational selection function of the survey.

the distances to these galaxies could thus be determined.

As well as providing distance estimates, the Tully–Fisher relation also tells us that the luminosity of a galaxy is tightly correlated with its total mass, which is dominated by *dark matter. (One expects mass to increase with rotation speed since centripetal acceleration $a = v^2/R$ must balance the force of *gravity $F = GM/R^2$, so that $v^2 = GM/R$, where G is Newton's gravitational constant, M is mass and R is radius from the galaxy centre.) The relation works best in near-*infrared light, since bluer light is dominated by hot, young stars rather than the overall stellar mass, and also near-infrared light is less strongly absorbed by interstellar *dust than shorter-wavelength optical light.

See also FABER–JACKSON RELATION.

two degree field galaxy redshift survey (2dFGRS)

The two degree field galaxy redshift survey maps the distribution of almost a quarter of a million *galaxies. It was carried out using the two degree field (2dF) multi-object spectrograph on the Anglo-Australian Telescope, New South Wales. This 2dF facility features a robotic positioner to place 400 optical fibres at the locations of galaxy and *quasar targets within the telescope's two degree field of view. The fibres carry the light from these distant objects to one of two spectrographs which break the light down into its component wavelengths. Due to the expansion of the *Universe, light from a distant object will be stretched to redder wavelengths. By measur-

ing the *redshift of spectral features of known rest wavelength, a distance to the object may be estimated. This enables observers to map out the three-dimensional distribution of galaxies in space.

The 2dFGRS, carried out in conjunction with the *two degree field quasar redshift survey (2QZ), has measured redshifts of 221 414 galaxies selected from the *APM galaxy survey. The position of these galaxies, as a function of redshift (or distance) and right ascension coordinate is shown in Fig. 1.

Important scientific results to emerge from the 2dFGRS include:

- an accurate measurement of the *power spectrum of galaxies on very large scales, placing tight constraints on the total mass density of the Universe and the fraction of that mass due to *baryons;
- an upper limit on the total *neutrino mass;
- an accurate determination of the number density of galaxies as a function of their luminosity.

http://www2.aao.gov.au/2dFGRS

two degree field quasar redshift survey (2QZ)

The two degree field quasar redshift survey maps the distribution of 23 424 *quasars in two 75 degrees by 5 degrees stripes in the northern and southern Galactic caps (Fig. 1, over page). It was carried out using the two degree field (2dF) multi-object spectrograph on the Anglo-Australian Telescope, New South Wales, concurrently with the two degree field galaxy redshift survey (2dFGRS).

The 2dF Quasar Redshift Survey

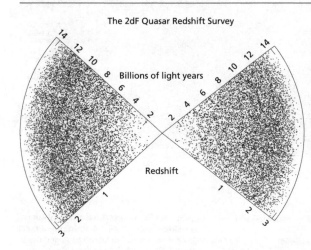

Billions of light years

Redshift

Fig. 1. Distribution of quasars in the 2dF quasar redshift survey. Note that this plot goes out to significantly higher redshifts than the corresponding plot for the 2dF galaxies (Fig. 1 on previous page). For this reason, the quasars *appear* to be more uniformly distributed than the galaxies, although in fact they are not. The lack of quasars at low redshift is partly a result of candidate selection.

Quasar candidates were identified by multicolour selection techniques using digitized, wide-field photographic plates in *ultra-violet (U), blue (J) and red (R) passbands taken with the UK Schmidt Telescope. In any single passband, quasars are indistinguishable from stars in our own Galaxy, but by comparing the brightness in different passbands, quasars stand out from the stars. Fig. 2 shows the distribution of point sources in a single Schmidt field in ($U-J$) versus ($J-R$) colour. Most stars have colours in a well-defined region of this colour–colour space; quasar candidates are those objects with colours that lie away from the stellar region. It is estimated that only 7% of quasars with redshifts in the range $0.3 < z < 2.2$ are missed because they have similar colours to stars. More than half of the quasar candidates were *spectroscopically confirmed as quasars.

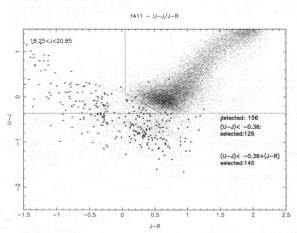

f411 — U–J/J–R

18.25<J<20.85

detected: 156
(U−J)< −0.36:
selected:126

(U−J)< −0.36+(J−R)
selected:145

Fig. 2. A plot of ($U-J$) versus ($J-R$) for point sources in a single Schmidt field (small points). The majority of stars lie in a well-defined locus running from near the centre of the plot to the top right-hand corner. The large points are known quasars in this field. 2QZ quasar candidates are those objects that lie below and to the left of the dotted line.

Fig. 1. This image shows the distribution over the entire sky (in a Hammer–Aitoff projection in equatorial coordinates) of over 1.6 million galaxies as detected by the Two Micron All Sky Survey Extended Source Catalogue (2MASS XSC). The diffuse band running almost vertically through the centre of the image and along the edges represents the light from stars near the plane of the Milky Way Galaxy. The Galactic plane obscures our extra-Galactic view in optical light.

Important scientific results to emerge from the 2QZ include:

- an accurate measurement of the *power spectrum of quasars on very large scales of ∼ 500 Mpc, placing tight constraints on the matter and *baryon content of the *Universe;
- an estimate of the masses of the central *black holes that power these quasars;
- an accurate determination of the number density of quasars as a function of their luminosity.

http://www.2dfquasar.org/

Two Micron All Sky Survey (2MASS)

The Two Micron All Sky Survey provides a uniform survey of the entire sky (see Fig. 1) in three near-*infrared passbands, observed using two automated 1.3-m telescopes, one in Arizona, and one in Chile. Each telescope was equipped with a three-channel camera, each channel consisting of a 256×256 array of mercury-cadmium-telluride (HgCdTe) detectors, capable of observing the sky simultaneously at J (1.25 microns), H (1.65 microns), and K_s (2.17 microns). Each part of the sky was observed for 7.8 seconds.

Survey observations began in June 1997 in Arizona and March 1998 in Chile. Observations were completed in February 2001, and all data were published in final form in March 2003. 2MASS yielded three data products:

Digital atlas of the sky—This comprises approximately 4 million 8×16 arcmin images at 4 arcsec spatial resolution in each of the three wavelength bands.

Point source catalogue—This contains accurate positions and fluxes for ∼ 300 million stars and other unresolved objects.

Extended source catalogue—This contains positions and total *magnitudes for more than one million *galaxies and other *nebulae.

Near-infrared photometry is of particular cosmological interest because main-sequence stars such as the Sun radiate most of their energy at around 2 microns in wavelength, within the K band. Luminosity in the K band is thus an accurate tracer of stellar mass. Light in the optical part of the *electromagnetic spectrum is more sensitive to the fraction of recently formed stars, and is thus a poorer tracer of stellar mass.

By combining 2MASS near-infrared photometry with galaxy *redshifts obtained from the *two degree field galaxy redshift survey and *Sloan Digital Sky Survey, astronomers have been able to estimate the stellar mass function in low-redshift galaxies. Due to its whole-sky coverage and accurate photometric calibration, 2MASS observations are also being used to calibrate data from deeper near-infrared surveys such as UKIDSS and VISTA.

http://www.ipac.caltech.edu/2mass/

ultra-violet (UV)

The term ultra-violet refers to radiation from that part of the *electromagnetic spectrum that is of shorter wavelength than can be perceived by the human eye (the visible part of the spectrum), about 400 nm, but with wavelength longer than about 10 nm, shorter than which radiation is classified as an *X-ray. It is emitted by hot stars (including the Sun; UV light is responsible for sunburn) and other hot objects in the *Universe. It is particularly useful as a tracer of young, recently formed *galaxies which contain a large fraction of hot, massive stars.

Universe

The Universe, a word of long history with a Latin origin *universum* meaning 'turned into one', traditionally refers to the totality of all that exists. As far as *cosmology is concerned, it therefore refers to all the planets, stars, interstellar material, particles, radiation, etc. that there is. In practice, the scientific study of the Universe means that all the components are taken as a single combined whole, and does not include separate investigation of individual objects.

A description of the Universe requires knowledge of the physical laws which govern it, such as the laws of *gravity and of particle interactions. However the physical laws themselves are not sufficient to uniquely describe the Universe; there are many possible universes consistent with known physical laws, differing for example in their material composition, rate of *Hubble expansion, and *geometry. The main goal of observational cosmology, beyond verifying that known physical laws do describe the properties of the Universe, is to determine which of the possible universes corresponds to our own, by determining the *cosmological parameters which measure those various properties. As well as determining the present state of the Universe, cosmologists seek to learn about its early stages by studying various

kinds of relic evidence, such as the abundance of light chemical elements, believed to have been left by physical processes taking place in the young Universe.

The standard view amongst cosmologists is that the Universe is described by the *big bang cosmology, where the Universe has expanded from a hot dense initial state. The distribution of material within the Universe at its early stages exhibited mild deviations from uniformity (the leading theory for the origin of these deviations being the *inflationary cosmology), which grew under the influence of gravity to form the structures we see today, such as *galaxies. *Dark matter played a crucial role in providing the gravitational attraction which would have been needed to assemble structures, such as galaxies, by the present day. The expansion of the Universe is observed to have recently begun to accelerate, and *dark energy has been postulated as the cause of this unexpected phenomenon. These ingredients combine to give the *standard cosmological model, almost universally accepted by cosmologists as the best existing explanation of the observed Universe.

A more limited sense of the word Universe is given by the *observable Universe, which refers to the portion of the entire Universe that we can in principle observe. This is a finite region, because the Universe has a finite age and the *speed of light gives a maximum rate to the spread of information. While cosmologists may speculate as to what lies outside the observable Universe, there is no way of testing their ideas in practice.

Although the standard definition of the word Universe precludes there being more than one, in some contexts cosmologists like to talk about multiple Universes, sometimes called the *multiverse. For example, some versions of the inflationary cosmology predict that the Universe becomes highly irregular on scales much larger than our observable Universe, and one can consider different regions of this multiverse to be separate 'baby' or 'pocket' Universes, in which physical conditions and even the basic laws of physics may be quite different from our own. This viewpoint may provide a physical basis for the *anthropic principle. An even more radical view of multiple Universes comes from a possible interpretation of *quantum mechanics, called the many-worlds interpretation, which suggests that all quantum possibilities simultaneously exist, and that each time a measurement is made the Universe splits into two or more separate Universes containing all possible outcomes of that measurement.

vacuum energy
See COSMOLOGICAL CONSTANT.

varying fundamental constants
Almost all physical laws feature what are known as **constants of Nature**, being quantities whose numerical values dictate the magnitude of physical effects. For instance, the strength of *gravity is determined by Newton's gravitational constant, usually indicated by the symbol G, while the strength of *relativistic effects is judged by comparing velocities to the universal *speed of light in a vacuum, denoted c. Such numbers are usually regarded as fixed quantities throughout space and time, hence the name 'constants of Nature'. The apparently self-contradictory term 'varying fundamental constants' refers to the possibility that these numbers are not in fact fixed, but may be different at separate locations or different times.

The first person to consider the possibility that the 'constants' might vary was British physicist Paul Dirac, who in the late 1930s noted a curious coincidence: the ratio of the size of the *observable Universe to the size of an *electron, and the ratio of the strength of electromagnetism to that of gravity (between an electron and a *proton), are both huge numbers but are very similar, about 10^{40}. He postulated that this was more than coincidence, being true for some unknown reason. But since the Universe expands, maintenance of this relation requires that the strength of at least one of the forces also changes. This coincidence is known as **Dirac's large numbers hypothesis**, the large numbers themselves usually being attributed to earlier work by Arthur *Eddington and sometimes called the Eddington number. Nowadays such arguments are usually considered to be meaningless numerology.

The most well-developed theories of a varying 'constant' concern the variation of the strength of gravity, the simplest example being the *Jordan–Brans–Dicke theory, in which the value of Newton's constant G becomes a dynamical variable (represented by a *scalar field). Such a theory predicts deviations from the predictions of *general relativity, with gravity within the Solar System predicted to be slightly weaker the closer to the Sun one gets. Such deviations are now strongly constrained by observation. Nevertheless, it remains possible that G has varied significantly with time throughout the Universe's evolution, and some studies of early Universe physics attempt to exploit this possibility.

Another possibility is variation of the speed of light, which for instance could be considered to be much larger in the early epochs of the Universe. Such varying-speed-of-light (VSL) theories have been suggested as a possible alternative to *cosmological inflation for solving the *horizon problem, the idea being that in the early Universe causal interactions over large distances are made possible by the speed of light being much greater. Such ideas have yet to enter the cosmological mainstream, in part because it is difficult to construct explicit models permitting a suitable variation.

A further fundamental constant is the fine-structure constant. This is interesting because it is a dimensionless constant, whose value determines the detailed pattern of spectral lines from atoms, and because there are claims that a variation has been detected, by looking at the *spectra associated with distant *quasars. There are indications in some analyses that the lines are systematically displaced from those measured in the laboratory, implying that the fine-structure constant was smaller in the past. This however is presently quite controversial, to say the least.

Although the constants discussed above are the most familiar, they are not the only ones. For instance, the *standard model of particle physics features around twenty such constants, specifying the masses and interaction rates of the various particles.

One motivation for variation of fundamental constants comes from theories featuring *extra dimensions, such as *superstring theory. In such theories, the true fundamental constants are presumably those associated with the full number of dimensions, rather than the reduced number visible to us. The values we see, for instance of G, are then related to this fundamental value via the size of the extra dimensions. If that size is changing, then the constants will appear to us to be varying, even though from a higher-dimensional perspective they are fixed.

Another view of varying constants arises from the eternal inflation picture, which predicts that on extremely large scales, much larger than the observable Universe, our Universe becomes quite irregular. Such irregularities might manifest themselves by these widely separated regions possessing different values of the fundamental constants, leading to different physical laws. Indeed, this may even be predicted by superstring theory, in which the physical laws operative at low energies are determined in part by the configuration that the extra dimensions find themselves in. It is known that there are a vast number of possible configurations (referred to as the *string landscape), only a tiny fraction of which correspond to the known physics of the standard model of particle physics.

In such a scenario, one can actually hope to explain why the physical constants take the values they do. For instance, some particular values might be especially amenable to inflation taking place, and those regions undergo such vast expansion as to dominate the present volume of the Universe. Alternatively, one can ask whether particular combinations of the constants are required in order for life to develop (for instance, if the *density perturbations in the Universe had been of somewhat lower magnitude, then stars and galaxies would never have formed, making our notion of life rather difficult). Attempting to predict physical laws on the basis of the existence of life is known as the *anthropic principle, and is explored more fully at that entry. Like most other things discussed within this topic, it is quite controversial!

Magueijo, J. *Faster than the Speed of Light*, William Heinemann, London, 2003.

velocity dispersion

The velocity dispersion provides a way of quantifying the spread of velocities within a system such as a *galaxy or a *cluster of galax-ies. For a system that has reached dynamical equilibrium, the kinetic energy due to the internal velocities is equal to minus one half of the *gravitational potential energy (*see* VIRIALIZATION). Since the gravitational potential depends on mass, one may use the measured velocity dispersion to infer the mass of the system.

For a system of N particles each with a velocity v_i, the velocity dispersion σ^2 is mathematically defined by

$$\sigma^2 = \frac{1}{N} \sum_{i=1}^{N} (v_i - \bar{v})^2,$$

where $\bar{v}$ is the mean velocity of the system given by

$$\bar{v} = \frac{1}{N} \sum_{i=1}^{N} v_i,$$

and the symbol $\sum$ denotes a sum over all particles.

Of course, velocity is a vector quantity (that is, it has a direction as well as an amplitude). In practice, it is only possible to measure radial (line-of-sight) velocities on extra-galactic scales. For galaxies, this is done by obtaining a *spectrum and measuring the broadening of spectral features due to Doppler broadening. For clusters of galaxies, radial velocities of member galaxies are measured and the radial velocity dispersion calculated from the above equation. Once the radial velocity dispersion has been measured, the total velocity dispersion must be inferred. For a spherically symmetric system, with no preferred axes, the total velocity dispersion is equal to three times the radial velocity dispersion, since there are three perpendicular directions.

Velocity dispersions provide useful estimates of the total mass, including *dark matter, in elliptical galaxies and galaxy clusters.

Very Large Telescope (VLT)

The Very Large Telescope is actually the collective name given to four 8.2-m reflecting Unit Telescopes and four moving 1.8-m Auxiliary Telescopes, the light beams of which can be combined in the VLT Interferometer (VLTI). The VLT is situated at the Paranal Observatory (Atacama, Chile), and run by the European Southern Observatory (ESO), see Fig. 1. With a total light-collecting area equivalent to a single 16-m telescope (the

Fig. 1. An aerial view of the four main VLT telescopes at Paranal Observatory, Chile.

light-collecting area scales as the square of the primary mirror diameter), the VLT is the world's largest and most advanced optical telescope.

The first Unit Telescope began astronomical observations in April 1999, with the remaining three units completed by August 2001. The VLT is equipped with a state-of-the-art **adaptive optics** system to obtain images of comparable resolution to space telescopes. Adaptive optics systems work by means of a computer-controlled deformable mirror that counteracts the image distortion induced by atmospheric turbulence. It is based on real-time optical corrections computed from image data obtained by a special camera at very high speed, many hundreds of times each second. This system has enabled the VLT to obtain images at the diffraction limit of a single unit telescope, an image resolution of 0.032 arcsec, compared with an uncorrected resolution of 0.9 arcsec.

Amongst the scientific discoveries with this system has been the detection in the near-*infrared of a star orbiting within 17 light-hours of a *supermassive black hole at the centre of our Galaxy. By studying the motion of the star over time (it takes just over 15 years to make one complete orbit), it has been estimated that the central black hole is about 2.6

million times as massive as the Sun. Only a black hole can contain that much mass in such a small volume of space.

To obtain even higher resolution, the light from two or more Unit Telescopes may be combined using the VLT Interferometer. With a separation of about 100 metres, two VLT Unit Telescopes combined in this way have been used to measure the angular diameter of our nearest stellar companion. Proxima Centauri was found to have a diameter of 1.02 ± 0.08 milli-arcseconds, that is about one thousandth of an arcsecond.

Of more direct relevance to *cosmology, however, is the sheer light-collecting area of the VLT, enabling light to be detected from the most distant objects in the *Universe. In March 2004, Roser Pelló and collaborators used the VLT to discover a galaxy named Abell 1835 IR1916, which is claimed to be at a *redshift of 10. If this redshift is correct, the galaxy is seen at a time when the Universe was merely 470 million years old, that is, barely 3% of its current age. Even with the power of the VLT, this galaxy is only visible because it is *gravitationally lensed by a factor of 25–100 in brightness by the *Abell cluster number 1835. This claim is controversial, as it relies on the weak detection of a single *Lyman alpha emission line. A re-reduction of the VLT data by Stephen Weatherley and colleagues does not find this emission line.

http://www.eso.org/outreach/ut1fl/

Virgo cluster

The Virgo cluster is the closest *cluster of galaxies to us. It is situated in the constellation of Coma Berenices, south of the *Coma cluster, at a distance of roughly 15 Mpc (50 million light years). It contains more than 2 000 *galaxies. It is classified as an irregular cluster, being elongated in shape and having no obvious central concentration. It covers a region of sky roughly ten by twelve degrees (and thus too spread out to be included in Abell's catalogue), with a physical diameter of about 3 Mpc (9 million light years). Charles Messier described the Virgo cluster as a 'cluster of nebulae' and identified sixteen *nebulae within it.

The *Local Group of galaxies feels a gravitational pull towards the Virgo cluster (an effect known as 'Virgo infall'), and it is likely that it will eventually become part of the cluster.

http://www.seds.org/messier/more/virgo.html

virial theorem

See VIRIALIZATION.

virialization

A gravitationally bound system such as a *galaxy or a *cluster of galaxies is held together by its *gravitational potential energy, a negative energy that binds the stars within a galaxy. Opposing this gravitational pull is the **kinetic energy** of the stars. This is a positive energy due to the stars' motion which increases with the square of velocity, and is what prevents the stars in an elliptical galaxy all falling towards the centre. For such a system that has reached steady-state equilibrium (e.g. it is no longer collapsing) and that has no external forces acting on it, then the **virial theorem** states that the kinetic energy of the system is equal to minus one half of its potential energy. Virialization is the name given to the process by which a collapsing proto-galaxy or proto-cluster reaches equilibrium. Once an object has reached equilibrium, it is said to be **virialized**.

The power of the virial theorem is that it allows cosmologists to estimate the total mass, including that of *dark matter, of systems such as elliptical galaxies and virialized clusters. The kinetic energy of the system may be estimated by measuring its *velocity dispersion, the relative velocities within the

system. For an elliptical galaxy this may be done by measuring the Doppler broadening of spectral features, and for clusters by measuring the recession velocities of the member galaxies. The gravitational potential and hence the total mass of the system may then be obtained via the virial theorem.

Such estimates show that typical elliptical galaxies are dominated by dark matter, the visible stars accounting for less than 10% of the total mass. Galaxy clusters contain an even larger fraction of dark matter. However, it is debatable how many clusters have yet virialized, with many showing evidence of ongoing merger activity, and so mass estimates for clusters are less reliable than those for galaxies.

visibility function

As light traverses the *Universe, it encounters material from which it may scatter or be absorbed, and hence we may not see it as directly emanating from its true source. The visibility function indicates the range of distances at which the radiation received from a particular direction was last scattered (more precisely, it is the probability distribution of the position of last scattering). It therefore tells us where the radiation we see actually came from.

An example is radiation from the Sun. The *photons of light originate deep within the Sun where nuclear reactions are taking place. However these do not emerge directly, but rather undergo a long random walk of collisions, taking around ten million years before they eventually reach the surface and begin to travel freely. When we look to the Sun, the photons all emerged from a quite well-defined region, known as the photosphere, located about 700 000 kilometres from its centre. The visibility function for the Sun is therefore a function strongly peaked at that location.

An important use of the visibility function in *cosmology concerns the *cosmic microwave background, where it tells us the location from which the photons we detect emanate. According to the *standard cosmological model it is a double-peaked function, with approximately 10% of photons last scattering at *redshifts below about 10 from the free *electron population caused by *reionization, and the remaining majority component emanating at the epoch of *decoupling at a redshift of around 1 100.

It also receives a small contribution from *residual ionization. The width of the visibility function at redshift 1 100 indicates the thickness of the *last-scattering surface, which is important in calculating *cosmic microwave background anisotropies on small angular scales.

voids

The apparently empty regions in plots of the distribution of *galaxies (*see* CfA REDSHIFT SURVEY, Fig. 1) are known as voids. They appear to be roughly spherical in shape, are up to ~ 50 Mpc in diameter, and are delineated by the surrounding galaxies, which, in a relatively narrow slice of space, appear to lie along filaments or within flattened structures known as sheets. Voids are not so apparent when viewing projections of galaxies on the sky (*see* LICK GALAXY CATALOGUE, Fig. 1) since they are overlaid by superimposed foreground and background galaxies.

The discovery of extra-Galactic voids thus dates back to the first large *redshift surveys carried out in the 1970s. According to the review by Herbert Rood (see below), the word 'void' was first used in an extra-Galactic context in a paper by Guido Chincarini and Rood in 1980, although the phrase 'region devoid of galaxies' appears to date back to 1976.

The prominent voids visible in CfA Redshift Survey, Fig. 1, led some astronomers to question whether these bubble-like features could form via *gravitational instability in a *cold dark matter (CDM)-dominated Universe. Alternative models included *top-down structure formation, proposed earlier by Yakov *Zel'dovich, in which non-linear collapse of gaseous *density perturbations formed flattened structures known as 'pancakes', which later condensed into galaxies. Others suggested that cosmic explosions releasing ~ 10^{54} joules of energy were responsible for emptying the voids of gas from which galaxies could form, or that they were due to *phase transitions during *cosmological inflation. It has since become apparent from semi-analytic models of *biased galaxy formation, that *hierarchical structure formation in a CDM Universe is capable of producing the observed voids (*see* SEMI-ANALYTIC GALAXY FORMATION, Fig. 1), and these other models are now disfavoured.

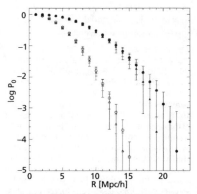

Fig. 1. The void probability function (VPF) measured from the two degree field galaxy redshift survey. Plotted is the log probability that a randomly placed sphere of radius R contains no galaxies brighter than $M_{b_J} = -19.3$ (open symbols) and $M_{b_J} = -20.2$ (filled symbols). Circles and triangles show the VPF in the Southern and Northern Galactic hemispheres respectively, showing that the VPF is consistent between the two.

The statistical properties of voids are most frequently studied with the **void probability function** (VPF). This is the probability that a randomly placed sphere of given radius contains no galaxies brighter than a given luminosity: see Fig. 1. Less luminous galaxies tend to be more uniformly distributed in space, and more numerous, than their more luminous cousins, and so their VPF drops more rapidly with radius. A related statistic is the number density of voids larger than a given radius.

Voids are not completely empty, but contain a much lower than average density of galaxies. In addition, the fact that a void is deficient in luminous galaxies (most large redshift surveys are only statistically complete in galaxies brighter than absolute *magnitude, $M \approx -16$), does not necessarily imply that it is deficient in *dwarf galaxies or in gas. However, most projects to detect dwarf galaxies and gas clouds in known voids have failed to find these objects in significant numbers, suggesting that voids are indeed underdense in most types of *baryonic matter.

Rood, H. J. *Annual Reviews of Astronomy and Astrophysics*, 26, 245, 1988.

wavenumber

In studying *density perturbations in the *Universe, it is common to describe them as a superposition of sinusoidal waves. It is conventional to use the wavenumber k rather than the wavelength λ to describe those waves, the relation being $k = 2\pi/\lambda$. In cases where the evolution with time is to be followed, it is always the *comoving wavenumber that is deployed, so that the expansion of the Universe is automatically incorporated.

The typical amplitude of density perturbations is given by their *power spectrum, which is usually expressed as a function of comoving wavenumber.

The condition for a wave to be of a scale corresponding to the *Hubble length at a given epoch is taken as $k = aH$ (also known rather inaccurately as the horizon-crossing condition). Waves usually evolve differently depending whether their wavenumber exceeds the Hubble length or not.

Wilkinson Microwave Anisotropy Probe (WMAP)

The Wilkinson Microwave Anisotropy Probe is a NASA space mission whose aim, as the name implies, was to measure *cosmic microwave background (CMB) anisotropies. It is led by Charles Bennett of the Goddard Space Flight Center, and named in honour of team member and microwave background pioneer David Wilkinson of Princeton University, who died shortly before the probe was launched. It made the first precision all-sky maps of the CMB, and is widely acknowledged to have ushered in the era of precision *cosmology.

The mission was conceived around 1996 following the success of the *COBE satellite, which made the first detection of the CMB anisotropies. WMAP (then named just MAP) was one of three proposals for such a mission and won the selection battle. For a major satellite mission, it was coordinated by an unusually small group of scientists, many of whom had been part of the COBE satellite team. A schematic of the satellite is shown in Fig. 1.

WMAP was launched in June 2001, and spent several months travelling to a location known as the Earth–Sun Lagrange 2 point—see Fig. 2. This corresponds to an orbit around the Sun, staying roughly one and a half million kilometres away from the Earth on a direct line joining the Earth and Sun. This placed it well outside the orbit of the Moon, meaning that the Earth, Sun, and Moon were all in roughly the same part of the sky so that the telescope could avoid pointing at any of them.

The satellite featured two separate mirrors of diameter roughly 1.5 metres, easily seen in Fig. 1. They receive light from regions of the sky separated by 140 degrees. In a similar strategy to that of COBE, the intensity of the light in the different directions was compared, and by building up the differences between vast numbers of pairs of directions it became possible to reconstruct all-sky maps of the microwave radiation intensity. WMAP operated with detectors at five different microwave frequencies, which enabled it to discriminate the CMB anisotropies, which have a *black-body spectrum, from emission from our own Galaxy and others. This procedure is shown in Fig. 3.

As well as the intensity, WMAP also measured the *polarization of the CMB radiation. The polarization had first been detected by the *Degree Angular Scale Interferometer (DASI) experiment, but WMAP was able to give higher precision and to measure polarization on much larger angular scales, which allows a probe of the *ionization history of the *Universe.

(For the full results of the WMAP satellite, *see* COSMIC MICROWAVE BACKGROUND ANISOTROPIES). Its first data release took place in February 2003 with the simultaneous submission of 14 science papers. In subsequent months cosmology research was dominated by researchers interpreting the WMAP data. A second data release, featuring data from the

Fig. 1. A computer-aided design rendering of the WMAP satellite.

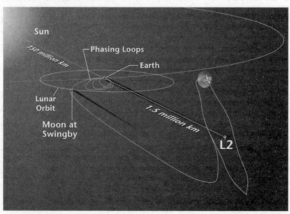

Fig. 2. A schematic of the WMAP satellite's approach and orbit at the Earth–Sun Lagrange 2 point.

Fig. 3. All-sky maps of the microwave radiation in five different frequency ranges, shown in Mollweide projection. Going clockwise from the leftmost panel, the frequencies are 23 GHz, 33 GHz, 41 GHz, 61 GHz, and 93 GHz. The horizontal feature is emission from the Milky Way Galaxy, and is more prominent at low frequencies. Information from the five separate frequencies can be combined to remove the effects of our galaxy and show the CMB anisotropies alone, in the centre panel.

first three years of observations, followed in March 2006. The full nominal mission is currently envisaged as eight years.

WMAP Home Page: http://map.gsfc.nasa.gov/

WIMPs (weakly interacting massive particles)

WIMPs is a generic name for a class of particles which are candidates to be the *cold dark matter in the *Universe. The name describes their key properties. They should be massive, so as to provide the gravitational attraction that is dark matter's role. And they should be weakly interacting, as otherwise they would have been detected by now and they haven't been. In this case, weakly interacting has the technical meaning that they interact only via the weak nuclear force.

Amongst the known particles in the *standard model of particle physics, the only particles that just exhibit weak interactions are the *neutrinos. While qualifying on that count, they fail to be WIMPs on the requirement of massiveness; upper limits on

neutrino masses from particle accelerators prevent them from fulfilling the cold dark matter role. They are still a candidate to be hot *dark matter, though they then cannot be all of the dark matter.

WIMP candidates must therefore be so-far-unknown particles motivated by particle theory, and by far the best candidates in this class come from *supersymmetry theory. The usual WIMP candidate in that context is the *neutralino. Supersymmetric WIMPs are widely regarded as the leading explanation for the dark matter, and considerable investment is being made worldwide in large underground detectors capable of directly detecting such particles.

It remains possible that the dark matter doesn't even interact via the weak force, but rather only gravitationally. Candidate particles of that type are not included under the title WIMPs.

WMAP satellite
See WILKINSON MICROWAVE ANISOTROPY PROBE (WMAP).

XMM–Newton satellite

The XMM–Newton satellite, XMM for short or X-ray Multi-Mirror satellite in full, is a satellite designed to measure *X-rays. As X-rays do not penetrate the Earth's atmosphere, satellite observations are the only way of exploring this part of the *electromagnetic spectrum. X-rays are towards the high-energy end of the spectrum, and primarily probe very energetic physical environments, including stellar and galactic *black holes, *supernovae, and the hot gas within galaxy *clusters. Like the *Hubble Space Telescope, it is an observatory mission with astronomers around the world able to apply to have their favourite objects studied.

XMM was launched into a highly elliptical Earth orbit by the European Space Agency in December 1999, and has a design lifetime of ten years. Originally known just as XMM, the name Newton was added after launch in honour of Sir Isaac Newton, for his discoveries of the theory of *gravity and of *spectroscopy. It is the largest satellite ever built within Europe, with a launch mass of nearly four tonnes

and a height of ten metres. Fig. 1 shows an artist's impression of what it might look like in orbit.

XMM was constructed with the aim of providing a very large collecting area to image faint sources, and with the ability to carry out X-ray spectroscopy by accurately measuring the energy of the incident X-ray *photons. Its abilities complement NASA's *Chandra X-ray satellite, which was designed to have exquisite imaging resolution but with less sensitivity and spectral resolution. It also possessed a small optical telescope for making simultaneous visible light observations of the X-ray targets.

A considerable challenge for X-ray telescopes is simply focusing the X-rays into an image. Normal lenses or mirrors do not work as the X-rays are powerful enough to penetrate the surface. XMM uses the technique of **grazing incidence**, whereby the X-rays hit the mirrors at an extremely shallow angle, less than one degree from parallel. In this configuration an incoming X-ray sees the row of atoms along the surface lined up in a

Fig. 1. An artist's impression of XMM–Newton in orbit. The light enters through the holes seen on the rear of the satellite.

way that obscures the gaps between them, rather in the manner of a person looking along a fence and being unable to see the spaces between the slats. This set-up is able to reflect the X-rays and bring them to a focus, though the long focal length of nearly ten metres demands a huge satellite. XMM uses three separate nested sets of 58 concentric mirrors to bring the X-rays to a focus, one of which is shown in Fig. 2, the successful launch being an astonishing technical achievement.

Despite being designed for a large field of view, XMM–Newton does not have enough viewing area to survey the entire sky, as would be desirable for many cosmological applications. On average, it takes about three images of the sky per day, and over a year the accumulated area amounts to about two hundred square degrees, which is about half a percent of the full sky. The vast majority of its observations are therefore pointed observations made towards specific objects, rather than surveying observations which seek new discoveries by imaging large areas of unknown content. Fig. 3 shows the *Coma cluster as imaged by XMM–Newton.

Surveying galaxy clusters is a particularly important cosmological application of XMM. While its limited observing area

Fig. 2. An image of the XMM focusing mirror array, consisting of 58 concentric mirrors. The outermost mirror has a diameter of 70 cm. The X-rays travel through this array being brought into focus by grazing incidence reflections along its length. Three such mirror arrays were deployed in XMM–Newton.

represents a problem for dedicated surveys, which have trouble covering enough sky area to see significant numbers of clusters, XMM is well suited to carry out serendipitous

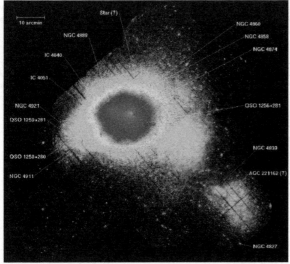

Fig. 3. A composite image of the Coma cluster by XMM–Newton, achieved by patching together several separate images across the cluster.

surveys, where objects are sought within the field of view of images taken for entirely different reasons. For example, in a targeted image of a supernova remnant within our Galaxy, the supernova itself might occupy only the very centre of the image, while in the background distant clusters of galaxies might be found. One such survey, the XMM–Newton Cluster Survey (XCS), aims to search the entire XMM data archive in order to find distant galaxy clusters observed by accident in this way.

XMM–Newton Home Page: http://xmm.esa.int

X-ray

The term X-ray refers to radiation with wavelengths shorter than 10^{-8}m (10 nm). X-ray radiation is often subdivided into 'soft' X-rays (around 10^{-8} to 10^{-10} m in wavelength), and 'hard' X-rays (wavelength shorter than 10^{-10} m). X-rays are used for medical imaging (flesh is transparent to them, revealing the bone structure underneath) and are produced by high-temperature gases, such as are found in *clusters of galaxies. Indeed, arguably the best way of finding galaxy clusters is to search in the X-ray part of the spectrum.

y-parameter

The *y*-parameter is a measure of the effect on a *black-body spectrum of radiation when it scatters from hot gas. It is the quantity used to describe the thermal *Sunyaev–Zel'dovich effect. There is no particular reason why it is called *y* rather than any other symbol; it just happens to be what someone chose when the effect was first written down and it stuck. Happily for us, it provides an entry for the letter *y*.

The radiation in question is the *cosmic microwave background radiation. If it encounters gas which is hot enough to be ionized (around 10 000 degrees Celsius or higher), there is a population of free *electrons that it can scatter from. Because the gas is much hotter than the radiation (whose present temperature is less than three kelvin), the net effect is to heat the radiation but without changing the total number of *photons. On average this shifts photons from the low-frequency part of the black-body spectrum to the high-frequency part in a characteristic way. The *y*-parameter describes this; $y = 0$ is the black-body case and increasing *y* indicates larger distortions.

Its full name is the Compton *y*-parameter. However this appears to be a misnomer: Compton scattering refers to photons scattering off electrons which have energies comparable to, or greater than, the electron rest-mass energy (more precisely, when the electron is relativistic in the centre-of-mass reference frame), whereas in fact the scatterings creating the *y*-distortion are almost always well described by Thomson scattering (the same process, but with non-relativistic electron motions).

One application is the all-sky average distortion, which has not been detected and is discussed in the cosmic microwave background topic. A second is to measure *y* as a function of position on the sky. It has been detected in the direction of *clusters of galaxies, whose environment contains copious hot gas, and is discussed in detail in the Sunyaev–Zel'dovich effect entry.

Zel'dovich, Yakov (1914–87)

Yakov Borisovich Zel'dovich is recognized as one of the true giants of *cosmology and relativistic astrophysics. Yet remarkably his work in cosmology amounts to only a small part of his contribution to science, with him entering the field around 1963 already aged nearly 50. The sheer volume and quality of his work led Stephen Hawking to remark, on meeting him, 'Before I met you here, I had thought that you were a collective of authors, as Bourbaki.' ('Bourbaki' was a pseudonym adopted by a group of mathematicians, mostly French, in the first half of the 20th century, in which name they published a series of rigorous textbooks on the foundations of mathematics. Occasionally newcomers mistake those as being the work of one prolific mathematician.) The basis of his success was an extraordinarily strong physical intuition, albeit backed up with a high level of mathematical skill.

Zel'dovich was born on 14 March 1914 in Minsk, now the capital of Belarus but at the time part of the Russian Empire. He was brought up in Petrograd/Leningrad (now St Petersburg), where he was largely self-taught. After high school, he was initially based at the rather literally named Institute for Mechanical Processing of Useful Materials, but in 1931 was transferred to the newly independent Institute of Chemical Physics in Leningrad; according to an autobiographical note, the rumour was that the director of his original institute traded him for a fuel pump! Through continuing self-education, he was able to receive a PhD in 1936 from the Institute of Chemical Physics, despite never taking an undergraduate degree. Later, he identified himself, as far as his theoretical work was concerned, as a protégé of the great Soviet physicist Lev Landau with whom he had a close relationship over many decades.

In 1943 the laboratory moved to Moscow (having been evacuated to Kazan early in the Second World War) and Zel'dovich with it. His work at this time focused on combustion and detonation, and led to involvement in the Soviet nuclear weapons programme from 1946 onwards. He was a leader of theoretical work on both fission and fusion bombs. The above work merits an extensive biography of its own, but here we focus on his achievements in cosmology, which began in 1963 when he left the nuclear weapons programme and became head of the relativistic astrophysics division at the Sternberg Astronomical Institute in Moscow.

Zel'dovich's main astrophysical contributions were in the physics of the hot *Universe, in theories of *structure formation, and in the physics of relativistic compact objects including *neutron stars and *black holes. His role in those subjects is discussed within many of the entries in this book. Particular highlights include his work with Rashid Sunyaev, a key member of his research group, on the *cosmic microwave background; they carried out research both on its thermal *spectrum and on its *anisotropies. They also discovered that hot gas, particularly in the environment of galaxy *clusters, can heat the cosmic microwave background as it passes through and give an observable effect; this *Sunyaev–Zel'dovich effect, now confirmed observationally, is set to be one of the most powerful tools for studying galaxy clusters, particularly at high redshift.

Another main area was his work on the initial stages of structure formation. He was a co-proposer in the early 1970s of a possible form for the initial conditions for structure formation, the *Harrison–Zel'dovich spectrum, which remains a viable description of observations to this day. (Although this terminology is universally used, Harrison's paper predated Zel'dovich's by two years. In fact, at a conference at Fermilab near Chicago attended by Zel'dovich in the mid 1980s, when someone wrote 'Harrison–Zel'dovich spectrum', Zel'dovich marched to the front, crossed out the word 'Zel'dovich', and returned to his seat.) He also developed key tools for following the evolution of structure once

Fig. 1. Yakov Zel'dovich.

non-linearities begin to develop. With Igor Kobzarev and Lev Okun, he was also the first to use cosmology to constrain physics at very high energies, by showing that a type of *topological defect known as a domain wall would give large distortions, much larger than observed, to the cosmic microwave background, and hence that the physics of the early Universe must ensure that such domain walls do not form.

He was a prolific author of textbooks as well as science articles. Within astrophysics his two-volume series on *Relativistic Astrophysics* with Igor Novikov remains in print more than thirty years after the first volume was produced. A collection of his most important works in astrophysical cosmology (part of a two-volume set) was published in 1993, including a fascinating autobiographical afterword in the second volume dating from 1984.

During his career he received a dazzling array of awards, including being made a Hero of Socialist Labour, the Soviet Union's highest civilian award, on three separate occasions. He was a member of the Soviet Academy of Sciences from 1958, and an honorary member of several overseas organizations including the US National Academy of Sciences and the UK Royal Society. However because of Cold War politics, he was not allowed to travel outside the Soviet Bloc until 1982, aged nearly 70, when he attended a conference in Greece.

Scientific terminology is littered with acknowledgements to his contributions.

Within cosmology alone, we have the **Harrison–Zel'dovich spectrum** of initial density perturbations, the **Sunyaev–Zel'dovich effect** by which galaxy clusters interact with the cosmic microwave background, the *Zel'dovich approximation for calculating the early stages of structure formation, and **Zel'dovich pancakes** which are the first structures to form. In other research fields, one finds the Zel'dovich–von Neumann–Dohring detonation theory, the Zel'dovich number in combustion theory, and the Zel'dovich mechanism for oxidation of nitrogen. COSPAR (the Committee of Space Research) awards the Zel'dovich medal to young scientists for 'excellence and achievements'. Rather incongruously, given that planetary science is one of the few areas of physics to which he did not contribute, asteroid 11438 Zeldovich is named after him.

Obituary (English translation) in *Soviet Astronomy Letters 14, 118*, 1988.

Ostriker, J. P. (Editor) *Selected Works of Yakov Borisovich Zel'dovich, volume II, Particles, Nuclei and the Universe*, Princeton University Press, 1993.

Zel'dovich approximation

The Zel'dovich approximation is a way of following the initial evolution of *cosmic structure when *density perturbations become too large for *cosmological perturbation theory to be valid. It was invented in the 1970s by Yakov *Zel'dovich, who used it to show that the first structures to form were large sheet-like structures which became known as Zel'dovich pancakes.

The approximation notes that while density perturbations are of small amplitude (the linear era), there is a particular relation between the *density and velocity of particles. Zel'dovich hypothesized that this relation might continue to hold even as structures began to form, leading to a simple description of how those first structures did form.

Nowadays, with the advent of large computer calculations, particularly *N-body simulations, the Zel'dovich approximation is not much used, except in order to specify accurate initial conditions for such calculations.

zone of avoidance

The zone of avoidance is that part of the night sky that is obscured by the plane of the *Milky Way. *Dust in the interstellar medium obstructs our view of around 20% of the sky in optical light, resulting in an incompleteness in all-sky optical *galaxy surveys.

One can overcome the obscuring effects of the zone of avoidance by observing at longer wavelengths. The plane of the Milky Way is only partially obscured in the near-*infrared and is effectively transparent at radio wavelengths. Infrared and radio surveys thus allow one to make galaxy maps over the whole sky (*see* TWO MICRON ALL SKY SURVEY, Fig. 1).

Zwicky, Fritz (1898–1974)

The maverick Swiss astronomer Fritz Zwicky made many significant contributions to *cosmology and astrophysics, amongst the most important of which were postulating the existence of neutron stars, the construction of a *galaxy catalogue, and the discovery of *dark matter.

Zwicky was born in Bulgaria to Swiss parents, and grew up in the Swiss canton of Glarus. He studied at the Swiss Federal Institute of Technology (ETH) in Zurich while Albert *Einstein was teaching there, and wrote his Diploma thesis under the supervision of the mathematician Herman Weyl. He moved to the California Institute of Technology (Caltech) in 1925 to work with Robert Millikan, who fifteen years earlier had made an accurate measurement of the charge carried by an *electron. Millikan had expected Zwicky to do theoretical work in *quantum mechanics, but Zwicky was more and more attracted to astrophysics, perhaps tempted by easy access to the Mount Wilson Observatory, home to the world's then largest telescope, a 2.5-metre (100-inch) reflector.

When the observational astronomer Walter Baade arrived at Caltech in 1931, he and Zwicky considered the problem of novae, stars which suddenly flare up to 10 000 times their original brightness, and then slowly fade. Some of these novae occurred in what were just then known to be other galaxies, much further away than novae in our own Galaxy. Zwicky and Baade, realizing that these extragalactic novae must be significantly more luminous than ordinary novae, coined the term *supernovae. To explain the origin of the enormous energy in a supernova explosion, they postulated in 1934 the existence of **neutron stars**, stars so dense that their constituent *protons and *electrons are forced together to form *neutrons. They also postulated that *cosmic rays would be emitted in the supernova explosion as an ordinary

Fig. 1. Fritz Zwicky.

star implodes to become a neutron star. The neutron star hypothesis only became widely accepted theoretically in 1939, and was not verified observationally until 1968 with the discovery of **pulsars**: rapidly rotating, magnetized neutron stars.

In 1935, Baade and Zwicky installed an 18-inch Schmidt telescope on Mount Palomar. Bernhard Schmidt's revolutionary new telescope made it possible to photograph large areas of the sky quickly, with little distortion. Zwicky used it to make the first rapid survey of the sky, mapping out hundreds of thousands of galaxies, and resulting in the Zwicky Galaxy Catalogue. From this catalogue, Zwicky was the first to notice that galaxies tend to *cluster, and he can thus be credited with inventing the study of *cosmic structure.

By analysing the velocities of galaxies within the *Coma cluster, Zwicky realized that there must be far more mass in the cluster than was apparent from its visible constituents. Zwicky was thus the first person to draw attention to the existence of what is now known as *dark matter.

In addition to his work in astrophysics, Zwicky was active in the aerospace industry, becoming director of research at Aerojet

Engineering 1943–9. He is credited by some as being the 'father' of the modern jet engine.

As well as his undoubtedly brilliant insights, Zwicky had numerous wacky ideas. One was that turbulent air (which degrades astronomical image quality) could be smoothed by firing a bullet through it. His 'tired light' theory of *redshift has since been ruled out (*see* ALTERNATIVE COSMOLOGIES).

Zwicky was infamous for his prickly nature, and anecdotes about his time at Caltech abound. On encountering students whose names Zwicky did not know, he would often greet them, 'Who the devil are you?' Other Caltech astronomers were known to Zwicky as 'spherical bastards', 'Because they were bastards, when looked at from any side.'

However, one wonders if this side of Zwicky's nature may have been self-exaggerated, as he was engaged in a number of charitable activities, including his work to help rebuild scientific libraries destroyed during the Second World War and participating in a programme to establish war orphan villages.

On being awarded the Gold Medal of the Royal Astronomical Society in 1973, Zwicky made the following response: 'I heard as a boy that there will always be an England, a place where debatable gentlemen will be recognized. I hope you have not made a mistake this time.'

Thorne, K. *Black Holes and Time Warps*, Norton, 1994.

Index

Picture Acknowledgements

The publishers are grateful to the following for their permission to reproduce illustration material. Although every effort has been made to contact copyright holders, it has not been possible to do so in every case and we apologise for any that have been omitted. For permission to reproduce figures not listed below, please apply to Oxford University Press.

Frontispiece: lithograph, Jean-Pierre Luminet (lithograph, 1992)

active galactic nuclei: AGN schematic, © Nick Strobel (www.astronomynotes.com)

ALMA: ALMA artists, European Southern Observatory

APM galaxy survey: APM survey, © Steve Maddox, Will Sutherland, George Efstathiou and Jon Loveday

biased galaxy formation: SDSS correlations, from I. Zehavi et al., 2005, ApJ, 630, 1, reproduced by permission of the AAS

Boomerang experiment: Boomerang launch photo, Boomerang collaboration; Boomerang orbit schematic, Boomerang collaboration; Boomerang CMB map, Boomerang collaboration

CfA redshift survey: CfA survey plot, Lars Christensen

Chandra satellite: Chandra satellite image, NASA/CXC/NGST; Chandra deep field, NASA/CXC/SAO

clusters of galaxies: coma cluster optical/X-ray, Digitized Sky Survey, ROSAT, and SkyView; Millennium cluster, V. Springel, Max Planck Institute for Astrophysics, Garching, Germany

coma cluster: coma image, Omar López-Cruz and Ian K. Shelton

correlation function: APM w(theta), from S. J. Maddox et al., 1990, MNRAS, 242, 43P

Cosmic Background Explorer satellite: COBE satellite schematic, NASA; COBE orbit schematic, NASA; COBE DMR photo, NASA; DIRBE image, NASA

cosmic dipole: COBE dipole map, NASA

cosmic infrared background: background radiation intensity, Ned Wright

cosmic microwave background: Penzias and Wilson photo, reprinted with permission of Alcatel-Lucent

cosmic microwave background anisotropies: COBE CMB map, NASA; WMAP CMB map, NASA/WMAP Science Team; Tegmark CMB map, Max Tegmark; COBE/WMAP comparison, NASA/WMAP Science Team; CMB observed spectra, NASA/WMAP Science Team; WMAP polarization map, NASA/WMAP Science Team

cosmic rays: Auger observatory map, Auger collaboration

cosmography: Milky Way map, *Norton Anthology of English Literature*

cosmological inflation: eternal inflation, Andrei Linde

dark energy: supernova cosmology project plot, S. Perlmutter et al., The Supernova Cosmology Project

dark matter: Millennium simulation plot, V. Springel, Max Planck Institute for Astrophysics, Garching, Germany

Degree Angular Scale Interferometer: DASI photo, John Yamasaki/DASI Collaboration

dust: dust distribution, David Schlegel, Douglas Finkbeiner and Marc Davis

Einstein, Albert: Einstein photos, © Bettmann/Corbis

electromagnetic spectrum: electromagnetic spectrum graphical, Advanced Light Source, Berkeley Lab

extra-galactic background light: EGB spectrum, from R. C. Henry, ApJ, 516, 49, reproduced by permission of the AAS

Faber-Jackson relation: Line-of-sight velocity plot, from S. M. Faber and R. E. Jackson, ApJ, 204, 668, reproduced by permission of the AAS

galaxy classification: tuning fork classification, NASA; elliptical galaxy NGC4636, Astrophysical Research Consortium (ARC) and the Sloan Digital Sky Survey (SDSS) Collaboration, http://www.sdss.org; morphology density, from A. Dresssler, 1980, ApJ, 236, 351, reproduced by permission of the AAS

galaxy groups: Hickson group, Astrophysical Research Consortium (ARC) and the Sloan Digital Sky Survey (SDSS) Collaboration, http://www.sdss.org

galaxy rotation curves: rotation curve, from L. Bergstrom, 2000, *Reports on Progress in Physics*, 63, 793, reproduced by permission of the author and IOP Publishing. This figure is a composite of a rotation curve from E. Corbelli and P. Salucci, 2000, MNRAS, 311, 441 and an optical image from the NASA/IPAC Extragalactic Database (NED) which is operated by the Jet Propulsion Laboratory, California Institute of Technology, under contract with the National Aeronautics and Space Administration

globular clusters: globular cluster image, Astrophysical Research Consortium (ARC) and the Sloan Digital Sky Survey (SDSS) Collaboration, http://www.sdss.org

gravitational lensing: Abell cluster lensing, NASA; Einstein cross, Geraint Lewis and Michael Irwin; reconstructed mass, from D. Clowe et al., 2006, ApJ, 648, L109, reproduced by permission of the AAS

gravitational waves: wave pattern schematic, Timothy Carnahan and Cherie Congedo/NASA-GSFC; LIGO observatory photo, LIGO Laboratory

halo occupation distribution: halo occupation distribution, from I. Zehavi et al., 2004, ApJ, 608, 16, reproduced by permission of the AAS

Hammer-Aitoff projection: Earth Hammer projection, NASA Earth observatory

hierarchical structure formation: merger tree schematic, from C. Lacey and S. Cole, (1993), MNRAS, 262, 627, reproduced by permission of Blackwell Publishing Ltd.

Hoyle, Fred: Hoyle photo, Chandra Wickramasinghe

Hubble Deep Field: Hubble Deep Field North, NASA; Hubble Ultra Deep Field, NASA

Hubble diagram: Hubble and Humason plot, from E. Hubble and M. L. Humason, 1931, ApJ, 74, 43, reproduced by permission of the AAS

Hubble Space Telescope: HST photo against Earth, NASA; HST photo with shuttle, NASA

inflationary perturbations: inflationary perturbations, NASA/WMAP Science Team

InfraRed Astronomical Satellite: IRAS all-sky map, NASA/JPL-Caltech

initial mass function: initial mass functions, Micol Bolzonella

intergalactic medium: hydrogen density, Renyue Cen

James Webb Space Telescope: JWST artists impression, NASA

K correction: four galaxy spectra, Michael R. Blanton

Keck Observatory: Keck observatory photo, © W. M. Keck Observatory

Lick galaxy catalogue: Lick catalogue map, from M. Seldner, B. Siebers, E. J. Groth, and P. J. E. Peebles, 1977, AJ, 82, 249, reproduced by permission of the AAS

Lyman alpha forest: quasar spectra, William C. Keel, University of Alabama

MACHOs: MACHO light curves, © MACHO Project

Milky Way: Milky Way schematic, NASA/CXC/M. Weiss

modified Newtonian dynamics: rotation curves, reprinted with permission from the Annual Review of Astronomy and Astrophysics, Volume 40 ©2002 by Annual Reviews www.annualreviews.org

Mollweide projection: Earth Mollweide, NASA Earth observatory

N-body simulations: Hubble Volume tie plot, August Evrard and the Virgo Consortium; Millennium simulation movie frames, V. Springel, Max-Planck-Institute for Astrophysics, Garching, Germany

neutrinos: superkamiokande photo, Kamioka Observatory, Institute for Cosmic Ray Research, University of Tokyo

nucleosynthesis: nucleosynthesis abundance predictions, David Schramm and Michael Turner

particle accelerators: CERN aerial photo, CERN; Z_0 decay, CERN

peculiar velocities: galaxy correlation function, © 2dF Galaxy Redshift Survey team, http://www2.aao.gov.au/2dFGRS

pixelization: HEALPix image, Krzysztof Gorski

Planck satellite: Planck satellite CAD image, ESA/Planck satellite collaboration

power spectrum: SDSS power spectrum, from M. Tegmark et al., 2004, ApJ, 606, 702, reproduced by permission of the AAS

quasar absorption systems: quasar spectrum graph, Wallace Sargent

quasars: composite quasar spectrum, from W. Zheng et al., 1997, ApJ, 475, 469, reproduced by permission of the AAS

radio galaxies: radio galaxy composite, © NRAO/AUI

redshift: redshift schematic, © Nick Strobel (www.astronomynotes.com)

redshift surveys: Picses–Perseus supercluster, © Riccardo Giovanelli; las campanas cone plot, from H. Lin et al., 1996, ApJ, 471, 617, reproduced by permission of the AAS

ROSAT: coma cluster image, NASA

satellite galaxies: Andromeda satellites, University of Basel, NOAO/AURA/NSF, W. M. Keck Observatory

semi-analytic galaxy formation: semi-analytic galaxy distribution, Andrew Benson

Sloan Digital Sky Survey: SDSS LRG correlation function, from D. J. Eisenstein et al., 2005, ApJ, 633, 560, reproduced by permission of the AAS

smoothed particle hydrodynamics: SPH simulation boxes, V. Springel, Max Planck Institute for Astrophysics, Garching, Germany

spectrum: SDSS galaxy spectrum, Astrophysical Research Consortium (ARC) and the Sloan Digital Sky Survey (SDSS) Collaboration, http://www.sdss.org

Spitzer Space Telescope: Spitzer composite, NASA

Square Kilometer Array: SKA artists impression, © ISPO and Xilo Studios

star-formation history: star formation rate, from A. J. Bunker et al. (2004), MNRAS, 355, 374, reproduced by permission of Blackwell Publishing Ltd.

starburst galaxy: M82 starburst, Mark Westmoquette, Jay Gallagher, Linda Smith, WIYN/NSF, NASA/ESA

Sunyaev-Zel'dovich effect: SZ cluster image, John Carlstrom; SZ simulation image, Antonio da Silva

superclusters: nearby galaxy distributions, from O. Lahav et al. (2000), MNRAS, 312, 166, reproduced by permission of Blackwell Publishing Ltd.

supermassive black holes: galaxy centre orbits, European Southern Observatory; M84 nucleus, NASA

supernova surveys: SN Hubble diagram, High-Z SN Search Team

supernovae: HST supernova image, High-Z SN Search Team; light curves, Perlmutter et al., The Supernova Cosmology Project

topology of the Universe: CMB topology, Alain Riazuelo and Jeff Weeks

Tully-Fisher relation: HI profile, from K. G. Begeman, 1989, Astronomy and Astrophysics, 223, 47

two degree field galaxy redshift survey: 2dfGRS cone plot, © 2dF Galaxy Redshift Survey team, http://www2.aao.gov.au/2dFGRS

two degree field quasar survey: 2dfQS cone plot, 2dF QSO Redshift Survey (2QZ); 2dfQS candidates, 2dF QSO Redshift Survey (2QZ)

Two Micron All Sky Survey (2MASS): 2mass all-sky map, NASA

Very Large Telescope: Paranal observatory, European Southern Observatory

voids: void probability function, from S. G. Patiri et al. (2006), MNRAS, 369, 335, reproduced by permission of Blackwell Publishing Ltd.

Wilkinson Microwave Anisotropy Probe: WMAP CAD image, NASA/WMAP Science Team; WMAP orbit schematic, NASA/WMAP Science Team; WMAP map combination, NASA/WMAP Science Team

XMM-Newton satellite: XMM artist's impression, ESA/D. Ducros; XMM mirror array, ESA; XMM coma image, U Briel/ESA

Zel'dovich, Yakov: Ze'ldovich photo, Russian Academy of Sciences

Zwicky, Fritz: Photograph of Fritz Zwicky, Fritz Zwicky Stiftung

Oxford Paperback Reference

A Dictionary of Chemistry

Over 4,200 entries covering all aspects of chemistry, including physical
chemistry and biochemistry.

'It should be in every classroom and library ... the reader is drawn
inevitably from one entry to the next merely to satisfy curiosity.'
School Science Review

A Dictionary of Physics

Ranging from crystal defects to the solar system, 3,500 clear and
concise entries cover all commonly encountered terms and concepts of
physics.

A Dictionary of Biology

The perfect guide for those studying biology – with over 4,700 entries
on key terms from biology, biochemistry, medicine, and palaeontology.

'lives up to its expectations; the entries are concise, but explanatory'
Biologist

'ideally suited to students of biology, at either secondary or university
level, or as a general reference source for anyone with an interest in the
life sciences'

Journal of Anatomy

Oxford Paperback Reference

A Dictionary of Psychology
Andrew M. Colman

Over 10,500 authoritative entries make up the most wide-ranging
dictionary of psychology available.

'impressive ... certainly to be recommended'
Times Higher Educational Supplement

'Comprehensive, sound, readable, and up-to-date, this is probably the
best single-volume dictionary of its kind.'
Library Journal

A Dictionary of Economics
John Black

Fully up-to-date and jargon-free coverage of economics. Over 2,500
terms on all aspects of economic theory and practice.

A Dictionary of Law

An ideal source of legal terminology for systems based on English law.
Over 4,000 clear and concise entries.

'The entries are clearly drafted and succinctly written ... Precision for the
professional is combined with a layman's enlightenment.'
Times Literary Supplement

Oxford Paperback Reference

The Concise Oxford Dictionary of Art & Artists
Ian Chilvers

Based on the highly praised *Oxford Dictionary of Art*, over 2,500 up-to-date entries on painting, sculpture, and the graphic arts.

'the best and most inclusive single volume available, immensely useful and very well written'

Marina Vaizey, *Sunday Times*

The Concise Oxford Dictionary of Art Terms
Michael Clarke

Written by the Director of the National Gallery of Scotland, over 1,800 entries cover periods, styles, materials, techniques, and foreign terms.

A Dictionary of Architecture
James Stevens Curl

Over 5,000 entries and 250 illustrations cover all periods of Western architectural history.

'splendid ... you can't have a more concise, entertaining, and informative guide to the words of architecture'

Architectural Review

'excellent, and amazing value for money ... by far the best thing of its kind'

Professor David Walker

OXFORD

Oxford Companions

'Opening such books is like sitting down with a knowledgeable friend. Not a bore or a know-all, but a genuinely well-informed chum ... So far so splendid.'

Sunday Times [of *The Oxford Companion to Shakespeare*]

For well over 60 years Oxford University Press has been publishing Companions that are of lasting value and interest, each one not only a comprehensive source of reference, but also a stimulating guide, mentor, and friend. There are between 40 and 60 Oxford Companions available at any one time, ranging from music, art, and literature to history, warfare, religion, and wine.

Titles include:

The Oxford Companion to English Literature
Edited by Margaret Drabble
'No guide could come more classic.'

Malcolm Bradbury, *The Times*

The Oxford Companion to Music
Edited by Alison Latham
'probably the best one-volume music reference book going'

Times Educational Supplement

The Oxford Companion to Western Art
Edited by Hugh Brigstocke
'more than meets the high standard set by the growing number of Oxford Companions'

Contemporary Review

The Oxford Companion to Food
Alan Davidson
'the best food reference work ever to appear in the English language'

New Statesman

The Oxford Companion to Wine
Edited by Jancis Robinson
'the greatest wine book ever published'

Washington Post

OXFORD

Great value ebooks from Oxford!

An ever-increasing number of Oxford subject reference dictionaries, English and bilingual dictionaries, and English language reference titles are available as ebooks.

All Oxford ebooks are available in the award-winning Mobipocket Reader format, compatible with most current handheld systems, including Palm, Pocket PC/Windows CE, Psion, Nokia, SymbianOS, Franklin eBookMan, and Windows. Some are also available in MS Reader and Palm Reader formats.

Priced on a par with the print editions, Oxford ebooks offer dictionary-specific search options making information retrieval quick and easy.

For further information and a full list of Oxford ebooks please visit: www.askoxford.com/shoponline/ebooks/

Oxford Paperback Reference

The Kings of Queens of Britain
John Cannon and Anne Hargreaves

A detailed, fully-illustrated history ranging from mythical and pre-conquest rulers to the present House of Windsor, featuring regional maps and genealogies.

A Dictionary of Dates
Cyril Leslie Beeching

Births and deaths of the famous, significant and unusual dates in history – this is an entertaining guide to each day of the year.

'a dipper's blissful paradise ... Every single day of the year, plus an index of birthdays and chronologies of scientific developments and world events.'

Observer

A Dictionary of British History
Edited by John Cannon

An invaluable source of information covering the history of Britain over the past two millennia. Over 3,600 entries written by more than 100 specialist contributors.

Review of the parent volume
'the range is impressive ... truly (almost) all of human life is here'
Kenneth Morgan, *Observer*

OXFORD

Oxford Paperback Reference

The Concise Oxford Dictionary of English Etymology
T. F. Hoad

A wealth of information about our language and its history, this
reference source provides over 17,000 entries on word origins.

'A model of its kind'

Daily Telegraph

A Dictionary of Euphemisms
R. W. Holder

This hugely entertaining collection draws together euphemisms from all
aspects of life: work, sexuality, age, money, and politics.

Review of the previous edition
'This ingenious collection is not only very funny but extremely
instructive too'

Iris Murdoch

The Oxford Dictionary of Slang
John Ayto

Containing over 10,000 words and phrases, this is the ideal reference for
those interested in the more quirky and unofficial words used in the
English language.

'hours of happy browsing for language lovers'

Observer

OXFORD

Oxford Paperback Reference

Concise Medical Dictionary

Over 10,000 clear entries covering all the major medical and surgical specialities make this one of our best-selling dictionaries.

'"No home should be without one" certainly applies to this splendid medical dictionary'

Journal of the Institute of Health Education

'An extraordinary bargain'

New Scientist

'Excellent layout and jargon-free style'

Nursing Times

A Dictionary of Nursing

Comprehensive coverage of the ever-expanding vocabulary of the nursing professions. Features over 10,000 entries written by medical and nursing specialists.

An A-Z of Medicinal Drugs

Over 4,000 entries cover the full range of over-the-counter and prescription medicines available today. An ideal reference source for both the patient and the medical professional.

Oxford Paperback Reference

The Concise Oxford Companion to English Literature
Margaret Drabble and Jenny Stringer

Based on the best-selling *Oxford Companion to English Literature*, this is an indispensable guide to all aspects of English literature.

Review of the parent volume
'a magisterial and monumental achievement'

Literary Review

The Concise Oxford Companion to Irish Literature
Robert Welch

From the ogam alphabet developed in the 4th century to Roddy Doyle, this is a comprehensive guide to writers, works, topics, folklore, and historical and cultural events.

Review of the parent volume
'Heroic volume ... It surpasses previous exercises of similar nature in the richness of its detail and the ecumenism of its approach.'

Times Literary Supplement

A Dictionary of Shakespeare
Stanley Wells

Compiled by one of the best-known international authorities on the playwright's works, this dictionary offers up-to-date information on all aspects of Shakespeare, both in his own time and in later ages.